The Engines of Life

Structure and Function of Proteins at Work

Bernardo Pérez-Ramírez

Adjunct Professor, Department of Biomedical Engineering,
Tufts University, Medford, Massachusetts. Honorary Professor,
Department of Biosciences, University of Nottingham, United Kingdom.

Novo Codex

Novo Codex Editions 2024

THE ENGINES OF LIFE:
Structure and Function of Proteins at Work

Published by Novo Codex Editions
6 Liberty Square, Suite 2405,
Boston MA 02109

Library of Congress Control Number: 2023922789

ISBN (hardcover): 979-8-9889711-0-8

Proteins are like people; they have
a similar skeleton but a different personality.
Our bodies are part of nature,
run by molecules, and we are not a separate creation
from plants and animals.
We are all members of the same family,
a historical accident of evolution.

CONTENTS

PREFACE

I have been involved in the understanding of biological molecules, in particular proteins, for more than forty years, with emphasis on protein's structure function relationships. The more I study proteins, the more complex they become. I have participated in developing proteins as therapeutic agents (medicines) to treat bone fractures and spinal fusion, and others for the treatment of rare, life-threatening diseases as well as immune-mediated diseases and oncology. Throughout the years, I became convinced that proteins are like people; they have common features, but what sets them apart is their "personality," expressed by their different solution behavior, which is hard to predict. Proteins are complex molecules, with various structural and functional properties. They have regions with different electrical potentials, which are determined by the distribution of charged amino acids (positively charged, negatively charged, or uncharged) within their structure. These electrical potentials play a crucial role in the function of proteins, such as enzyme catalysis, binding to ligands, and protein-protein interactions.

The idea for this book was initiated around 2015 while teaching a graduate class on drug product development for therapeutic proteins at Tufts University. However, several years passed before I started a more systematic writing of the book. This was prompted by an invitation to present a lecture for nonscientist diplomats at the European School of Administration in Brussels in December 2020 on the topic of the development of medicines. The idea was to simplify the complexity of protein behavior and pharmaceutical development to provide a general overview for nonexperts. After the presentation in Brussels, I thought that the information put together could be beneficial as a general overview to serve as a complement to my students at Tufts University and the series of lectures that I give every year as an honorary professor at the University of Nottingham in the United Kingdom. The more I thought about the book, the more an idea of an overview expanded with more details. Being at home during the COVID-19 pandemic from 2020 to 2023 helped me focus on writing the book.

I did not attempt to be comprehensive in the coverage of proteins, and I decided not to include detailed medical and pharmaceutical aspects, though they are certainly described where relevant. Comprehensive experimental details have been skipped so as not to cloud the general principles. There are many analytical assays that can be used to characterize proteins. It is beyond the book's scope to discuss the principles of the many assays employed to characterize proteins. Nevertheless, an overview of some common assays based on first principles is discussed in relation to the physical-chemical profile of the protein under study. Detailed information can be found in the references provided.

Chapter 1 deals with the evolution of macromolecules and how building blocks could have been assembled into polymeric biological macromolecules on early Earth to give rise to the multienzyme complexes and origination of primitive organisms. Chapter 2 is focused on the pioneering research that led to the discovery of proteins and their amino acid components, the scientific challenges, and the personality of the scientists involved. In Chapter 3, the fundamental principles of protein structure are discussed. Protein behavior in solution is governed by physical-chemical and thermodynamic laws that are discussed in Chapter 4. All proteins interact with other proteins and ligands that could be stabilizers or destabilizers, impacting the biological activity. Those principles are discussed in Chapter 5.

Chapter 6 discusses the energetics of protein solutions that are affected by conformational stability and solution variables. The dogma in the protein world is that it has been the tight relation between structure and function. However, in the last twenty years, a series of proteins have been discovered that only obtain their final conformation upon binding to the target. We can refer to those proteins as intrinsically disordered proteins or "dancing proteins," discussed in Chapter 7. Chapter 8 shows a summary of the variety of key proteins that play important functions in vivo. Chapter 9 presents a general overview of the development of proteins as therapeutic medicines and a comparison with small chemical molecules.

The book is directed at advanced undergraduates or beginning graduate students with general knowledge of chemistry and physics and can serve as an introduction to the world of protein structure and function with applications to the biopharmaceutical development of therapeutic proteins, an area that continues to grow. The book emphasizes general principles, avoiding a detailed mathematical treatment of the subjects. In my experience, I have seen many students who have worked to understand the mathematical concepts and have not obtained a good grasp of the fundamental concepts, and others do not have the deep mathematical skills required for more complex situations. In Chapter 5, where the multicomponent solution thermodynamic is described, the mathematical treatment cannot be avoided, but it has been simplified to help in the understanding of the subject matter.

The biopharmaceutical development of proteins was initiated three decades ago when the first monoclonal antibody was approved for therapeutic uses. This unleashed the age of biological-based drug discovery and development for biopharmaceutical companies. These therapies will continue to provide benefits to patients in the future. The book represents a comprehensive overview of proteins as the engines of life. The information provided in the book is the cornerstone to expanding and applying that knowledge to develop therapeutic proteins.

Drug substance development, including formulation and process development, is a key area that could benefit from scientists and engineers who have a strong background in protein biochemistry and biophysics. Students using this book may have a biology, chemistry, or physics background, but I have tried to write a text accessible to all. I have provided references useful to readers to understand a particular topic further. A deliberate attempt was made to provide, in some instances, old articles that I deemed classics of scientific literature and that every scientist should be familiar with. Also, to help understand many of the concepts used, a glossary is provided at the end of the text.

I want to thank everybody who helped me along the way to make this book possible. Special thanks to Dr. Juan Carlos Slebe from the Universidad Austral de Chile for his love of enzymes and all the support he provided me during the beginning of my career as a protein biochemist. I am also grateful to my undergraduate mentors, the late Dr. JC Vera and Dr. Luis Burzio, for their teaching on experimental biochemistry and the power of keeping up with the scientific literature. My PhD advisor, Dr. Marino Martinez-Carrion, for his insight into enzymology and protein structure function. Thanks to my postdoctoral mentor, Professor Serge Timasheff, for the wonderful years that I spent in his laboratory learning the fundamentals of physical biochemistry of proteins and linkages to solution variables; and many colleagues and students who have worked with me over the years in the quest to understand the behavior of proteins. Finally, I want to thank my daughter Ariana, my wife Mariana, and my son Alex for being there, while I was immersed in the world of proteins.

Brookline, MA, August 7, 2023

PROLOGUE

During my four decades of teaching undergraduates and postgraduate students in biochemistry or biomolecular science and technology, there has been an absence of a suitable introductory text on the structure and function of proteins pitched at the final year undergraduate or master's/PhD level, and in particular, students wishing for careers in biopharma. Two texts have come close: Klaus Dieter Schwenke's *Eiweissquellen der Zukunft* and Olav Smidsrød and Inger-Lill Andresen's *Biopolymerkemie.* But these are now rather old and not available in English, which makes the readership limited. It was therefore a considerable delight when my good friend and colleague Bernardo Pérez-Ramírez presented me with a draft of his *The Engines of Life: Structure and Function of Proteins at Work,* and of course, it was a great honor to accept his invitation to write a prologue.

Over the last hundred years, our knowledge of the structure and function of proteins has increased in an exponential way. We did not even know whether proteins were discrete molecular entities with a clearly defined molecular weight, an uncertainty that was soon to be resolved by the osmotic pressure measurements of G.S. Adair and the analytical ultracentrifuge measurements of T. Svedberg and coworkers. The explosion of knowledge that followed is captured wonderfully in Professor Pérez-Ramírez's book and perfectly balances a consideration of structure and properties with the techniques that have been developed to understand these. Often, Nobel Prizes in Chemistry and Medicine have been awarded as much to the development of new techniques and methodologies as to their application to the molecules themselves. Further, without scientists, there would be no science, and so the key players behind these discoveries have been highlighted: this should be a great encouragement to students, as all the eminent scientists who made these discoveries were once students themselves.

In his book, Professor Pérez-Ramírez takes us in a clear way from first principles—the natural synthesis of the first amino acids and the origins of life—through the key discoveries and the unique physical, chemical, and thermodynamic/energetic properties of proteins to how they interact with each other and the other molecules of life. In this way we can understand how these molecular engines of life work, what happens when they don't, and how they can form the basis of and be developed into biopharmaceutical products in the treatment of disease, overcoming problems such as toxicity and aggregation. The book wonderfully introduces us to these issues and the complex issues involved in the development of monoclonal antibodies and drug products.

I can therefore thoroughly recommend *The Engines of Life: Structure and Function of Proteins* to any student embarking on a postgraduate course in protein science and biotechnology. The book will also be a most useful text for academics in universities and researchers in the biopharma industry.

Stephen E Harding
Nottingham, December 2023

INTRODUCTION

Chemistry and physics were the disciplines that brought tremendous technological advances in the biological sciences during the twentieth century. They created the industry and information technology that underlay the economic progress of the world and health care. Most of the biological medicines that are on the market today are proteins. Proteins carry out almost all the catalytic functions in the cell, as well as directing and forming the structural framework, so we can call proteins the engines of life.

Alamy image ID: F4MYDT. Standard license, Royalty-free – JPEG and vector. All rights, all media, worldwide in perpetuity. IY03688895 August 7, 2023.

We will provide, for now, a simple and basic definition of proteins that we will refine as we advance in the book. Proteins are large molecules composed of one or more long chains of amino acids and are an essential part of all living organisms. Because of their large size, we can call them macromolecules. When we talk about large molecules, we refer to those as having molecular masses[1] higher than 10,000 g/mol, compared, for example, with acetylsalicylic acid, also known as aspirin (a small molecule), whose molecular mass is 180 g/mol. Monoclonal antibodies are a type of therapeutic protein with a mass of around 150,000 g/mol. This means that monoclonal antibodies are more than eight hundred-fold bigger than a molecule of aspirin. For each protein, there exists a unique amino acid sequence, and that sequence is codified by the genes. If a wrong amino acid is inserted into the sequence, the result could have serious consequences for health.

We are made of proteins, like plants and animals. Foods with high-protein content, such as milk, beans, meat, fish, cheese, eggs, and nuts, give us energy and building blocks (amino acids) to grow and maintain a healthy body. Some amino acids are synthesized by the body and others are obtained from the diet. One serving of beans provides the same amount of protein as 30 g of meat. The proteins we eat are broken into their individual amino acids building blocks. We reuse these amino acids to build new proteins. Collagen is the most abundant protein in mammals—which holds our skin, tendons, muscles, and bones together—making up from 25 percent to 35 percent of the whole-body protein content. Proteins have been around for millions of years. The oldest protein known is preserved in a terrestrial vertebrate found inside the vascular canals of a rib of a one-hundred-and-ninety-five-million-year-old sauropodomorph dinosaur, where blood vessels and nerves would normally have been present in the living organism (Lee et al. 2017).

Different arrangements of amino acids (the sequence) can make extremely strong proteins, as in silk fibers, flexible and elastic, the elastin in our skin, or others with a highly loose conformation, like “dancing proteins” (Chapter 7). Like pieces of a jigsaw puzzle, proteins can “socialize” and interlock with other molecules. For example, each type of antibody in our blood has a unique arrangement of amino acids at its tips that can attach to a specific pathogen, marking it for destruction by the immune system. Proteins need to physically interact with each other and “socialize” with other molecules to do their work. These interactions might activate an enzyme, turn on a gene, or communicate a message from one cell to another. Interactions between proteins depend not just on their shape and three-dimensional structure but also on their chemical properties: positively and negatively charged amino acids in proteins attract each other; hydrophobic residues cluster together, away from water. These physical properties allow proteins to interact in specific ways, as we will see in Chapter 4.

The complexities of even the simplest organisms can be analyzed using the physical laws of nature. Thus, equilibrium thermodynamics is relevant to the study of proteins. It provides the fundamental variables if a process will occur, like the conformation of a macromolecule—like a protein—the catalysis of a chemical reaction, or the association of a protein into supercomplexes with unique biological functions. Thermodynamics is the science of work, heat, temperature, and the subtle relationships between these magnitudes. The laws of thermodynamics are established as balance equations between these observables. However, because of these laws, certain properties of the system appear on the scene whose variation in a process only depends on the initial and final conditions. Any science that prides itself on being well constructed must be based on the smallest number of basic concepts and above all on the correct choice of them. Such concepts must be based on the fundamental laws or principles of physical theory. The result of a correct choice is to obtain the highest levels of generalization and applicability of the concrete results derived from these laws.

Proteins play a critical role in all life processes, with remarkable specificity. They also serve as key structural elements in the cells and tissues, where they can control the movement and structure of cells such as myosin, actin, kinesin, and collagen. Proteins as enzymes catalyze the formation of new macromolecules as well as the digestion of food. Antibodies constitute the defensive protein system of the body and protect us against diseases. Proteins are membrane receptors that allow the cell to communicate with the exterior and regulate ion transport, hormone activation, and intercellular recognition. Repressors regulate gene expression. DNA polymerases replicate and repair genes. RNA polymerases catalyze the formation of the messenger RNA, and histones help in the packing of the DNA into the chromosomes. Thus, proteins are involved in every aspect of life. In vivo proteins work in clusters; their interactions with one another and other molecules mediate metabolic and signaling pathways, cellular processes, and organismal systems.

Due to their central role in biological function, protein interactions also control the mechanisms leading to healthy and diseased states in organisms. Diseases are often caused by mutations of proteins or, in some cases, by the lack of synthesis or low-level production of critical enzymes affecting the binding interface or leading to biochemically dysfunctional changes in proteins. Many neurodegenerative diseases in humans result from protein misfolding and aggregation. Protein misfolding is believed to be the primary cause of Alzheimer's disease (β-amyloid protein and tau protein), Parkinson's disease (α-synuclein accumulation in the brain), Huntington's disease (misfolded huntingtin protein forms amyloid aggregates), Creutzfeldt-Jakob disease (an abnormal isoform of a cellular glycoprotein known as the prion protein), amyotrophic lateral sclerosis (ALS), cystic fibrosis, Gaucher's disease, and many other degenerative and neurodegenerative disorders. Still, we need to continue learning how those proteins form and become toxic to humans to create effective therapies.

Infectious diseases have followed humans since the dawn of time. At the beginning of the twentieth century, infectious diseases were a major preoccupation; we have only to remember the pandemic of 1918 and its devastating consequences on the population. It was not until the COVID-19 pandemic became a world problem in 2020 (Pérez de Buerres Ramírez, 2020) that the scientific community realized we need to be continually vigilant of new infecting viruses arriving when we least expect them and have the right tools to combat them. Today's miracle vaccines result from an entirely new approach called RNA vaccines, which are highly successful in preventing death due to COVID-19.

The advances in immunology are built on hundreds of years of scientific work—Edward Jenner's discovery of the smallpox vaccine in 1796; Louis Pasteur's discovery in 1870 of immunization for animal diseases, including cholera, anthrax, and others; and Jonas Salk's polio vaccine in 1950. As we are in the twenty-first century, the relationships between cancer and immunology are starting to be understood. Programmed cell death protein 1, known as PD1, and its ligand have become a new target for cancer therapy. PD1 is found on the surface of T and B cells that have a role in regulating the immune system's response to the cells of the human body by downregulating the immune system and promoting self-tolerance by suppressing T cell inflammatory activity.

Today, as information age technology arrives at an intersection with bio-pharmaceutical technologies and the development of better protein-based medicines, the benefits can be reached by millions across the global village. Proteins are complex macromolecules that work together as the *Engines of Life.* Understanding the functions of proteins could give us hints to answer the question, “what is life?” and how to cure devastating diseases.

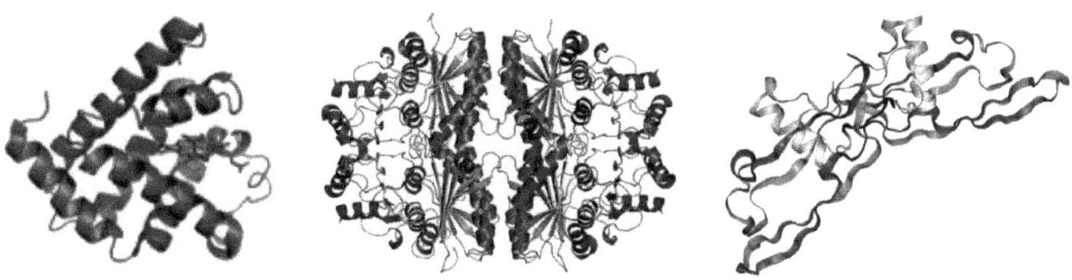

Helices, turns, loops, and β-strands motifs illustrate the complex three-dimensional structure of proteins.

CHAPTER 1

LIFE AND EVOLUTION

Shortly after Earth's formation, about 4.5 billion years ago, the planet's surface was very different from the one we know today. Large asteroids hit the earth's crust. Temperatures were much higher, and plate tectonics—which governs continental movements—was starting to evolve. Understanding how, in this hostile context, the processes that gave rise to the first organisms could appear is a mystery that science is still trying to solve. The first lifeforms that emerged billions of years ago have little in common with today's organisms.

Aleksandr Oparin and J. B. S. Haldane (Tirard 2017) both separately proposed in 1920 (Oparin 1957) that life on Earth could have arisen step-by-step from nonliving matter through a process of "gradual chemical evolution" (Oparin-Haldane hypothesis). The hypothesis assumes that the early earth had a reducing atmosphere, meaning an oxygen-poor atmosphere in which molecules tend to donate electrons. Hence, simple inorganic molecules could have reacted (with energy from lightning or the sun) to form building blocks like amino acids and nucleotides, which could have accumulated in the oceans, making a "primordial soup." Those building blocks combined, forming more complex structures like proteins and replicating structures such as DNA.

Thirty-three years later, Stanley Miller and Harold Urey tested the Oparin-Haldane hypothesis of the so-called primordial pool (Miller and Urey 1959). They built a closed system containing a heated pool of water and a mixture of gases that were thought to be abundant in the atmosphere of early Earth, such as methane, water, hydrogen, and ammonia. To simulate the lightning that might have provided energy for chemical reactions in Earth's early atmosphere, Miller and Urey sent sparks of electricity through their experimental system (Miller 1953,1955; Miller and Bada 1988; Miller and Lazcano 1995). The experiment then supported Oparin and Haldane's hypothesis that putative reducing conditions on the primitive earth favored chemical reactions that synthesized more complex organic compounds from simpler inorganic precursors. As a result of the experiment, various types of amino acids, sugars, lipids, and other organic molecules formed. Large, complex molecules like DNA and protein were missing, but the Miller-Urey experiment showed that at least some of the building blocks for these molecules could form spontaneously from simple compounds.

How Could Building Blocks Have Been Assembled into Polymeric Biological Macromolecules on Early Earth?

It is difficult to imagine how the transition from amino acids and nucleotides to polypeptides, proteins, DNA, and RNA took place in the primordial soup. As we know today, the formation of polymers as messenger RNA serves as a shuttle that carry the genetic information to be translated into proteins in the ribosomes. Building blocks like amino acids may have been able to spontaneously form polymeric structures under the conditions found on early Earth.

Polymerization reactions take place through dehydration reactions. A molecule of water is lost in forming a peptide from two amino acids joining together (Chapter 3). Dehydrating agents are used to initiate polymerization. In the 1950s, biochemist Sidney Fox and his colleagues found that if amino acids were heated without water, they could link together to form protein-like molecules—"proteinoids." (Fox and Harada 1958; Fox 1964; Fox and Windsor 1970). The polyamino acids they formed are not random molecules unrelated to life. They have distinct catalytic activities. Fox suggested that, on early Earth, ocean water carrying amino acids could have splashed onto a hot surface like a lava flow, boiling away the water and leaving behind a protein (Fox 1964). However, as we know today, proteins are necessary for replicating nucleic acid molecules, and nucleic acids are needed to make proteins in the cell. The question has been which came first, proteins or nucleic acids? There is no easy answer to this question.

Before 1970, the dogma was that one gene could only originate one protein, and all the enzymes are proteins. With the discovery of the ability of the messenger RNA to have enzymatic activities to catalyze its own splicing, the old dogma of molecular biology was rewritten. Now, one gene can give origin to many proteins, and RNA can be enzymes (ribozymes)[2]. If RNA can store the information necessary to codify a protein and be an enzyme, based on that, it has been postulated that at the beginning, there was an "RNA world" populated by RNA molecules that could replicate themselves and carry enzymatic reactions. As part of evolution and natural selection, the RNA molecules in the primordial soup could have been replaced by a more stable molecule like DNA to store genetic information. By the same principles of evolution, proteins are central in catalyzing enzymatic reactions since they are better suited than RNA and can adopt a broad range of conformations, creating conditions for the specific and proper interactions between substrate and enzyme.

Apart from the primordial soup as the origin of life, other theories have been postulated after the discovery of thermophilic organisms in association with deep-sea hydrothermal systems in late 1970 (Corliss et al.1979). This led to the idea that life might have originated in hydrothermal systems on primitive Earth (Takai et al, 2006; Martin et al. 2008; Sousa et al. 2013). The perceived benefits afforded to early life in these environments included protection from intense asteroid bombardment and UV radiation, and a source of thermal and chemical energy, along with potential catalytic minerals. In support of this hypothesis, biologists have demonstrated that the thermophilic microbial inhabitants of the active seafloor populate the deepest branches of the universal phylogenetic tree (Reysenbach and Shock 2002). Thus, deep-sea hydrothermal systems favor the abiotic synthesis of biochemically significant organic molecules. The question is—are these thermophiles the universal ancestors or survivors of the high-temperature regimen?

Another source of organic compounds that could have generated the building blocks of life may have them delivered by extraterrestrial objects (meteorites, interplanetary dust particles). It could also be possible that life originated elsewhere in the universe, where conditions were favorable, and transferred to the earth. The apparently short time span between the cooling of the earth and the appearance of modern microbes 4.2 billion years ago has been taken as evidence supporting this idea (Nicholson 2009). However, it is unclear how amino acids accumulated and formed the macromolecular structures we know today as proteins.

So far, there is no overwhelming possible scenario that can explain all the stages of the origin of life. All living organisms, plants, and animals have the same molecular basis. They all use the same building blocks in their protein, consistent with twenty amino acids, always the L isomer. However, the genetic code would permit more than the twenty universal amino acids it contains. Numerous lines of evidence suggest that ancient proteins relied on a limited alphabet of ten "early" amino acids, and that the ten "late" amino acids were products of biosynthetic pathways (Makarov et al. 2023).

The individual amino acids of proteins are encoded at the genetic level by a sequence of three so-called nucleotide bases (a base triplet and a codon) each. The alphabet of the genetic substance DNA contains four different bases, and the number of possible triplets from these is sixty-four (four to the power of three). With the genetic code, up to sixty-four different amino acids could be described. Instead, nature "only" programs a limited set of twenty amino acids and a redundant code: typically, individual amino acids are each described by several triplets (codons). One reason is probably that this redundancy cushions the impact of mutations in the DNA, which was evidently an advantage during evolution. With the repertoire of twenty universal amino acids, evolution achieved a momentary optimum (Mueller et al. 2013).

Apart from the twenty universal amino acids, a more complex incorporation mechanism allows the use of a twenty-first amino acid called selenocysteine. It is found in many living organisms, including humans. In certain microorganisms, the archaea, there is even a twenty-second amino acid, pyrrolysine, which is a lysine in which a pyrroline ring is linked to the end of the lysine side chain (Dinman 2012). This leads scientists to conclude that the genetic code has continued to evolve over the last billion years. Thus, evolution is not over, and additional amino acids could be incorporated into the genetic code in some distant future.

The role of DNA and RNA in protein synthesis is remarkably similar in all organisms. Therefore, it can be postulated that all living organisms are descendants of a common ancestor, and further differentiation is a consequence of natural selection due to environmental pressure as was first proposed by Charles Darwin.

> *One may say there is a force like a hundred thousand wedges trying [to] force every kind of adapted structure into the gaps in the economy of Nature, or rather forming gaps by thrusting out weaker ones.*
>
> —CHARLES DARWIN
>
> Darwin's manuscripts: Notebook D 135e, 28 September 1838. Cambridge University Library.

The fundamental biochemical processes have been conserved because their alteration could have become lethal. Thus, evolutionary variations of a protein give us information about the protein itself. Some proteins are very similar in structure and perform different functions. Conversely, there are proteins with different structures that can have identical functions. It is known that function and three-dimensional structure are intimately related, but there may be a limited number of ways that a protein can fold due to the physical-chemical constraints; we will discuss this later in the book. Hence, families of different evolutionary enzymes with unrelated functions may share a common structure. Thus, structural similarity does not necessarily imply an evolutionary relationship.

Most folds have a related homologous family (proteins that have evolved from a common ancestor) and can be essentially considered functional folding. Proteins and genes could be homologous or not, depending on if they descend or not from a common ancestor. Hence, two proteins with 80 percent identical sequences might not be 80 percent homologous. In this case, we are in the presence of *convergence,* where under the evolutive pressure to preserve biological function, two or more unrelated sequences become similar. Thus, in comparing related proteins, it is important while comparing the sequence to analyze the chemical nature of the amino acids. In related proteins, usually the change through mutations will incorporate chemically similar amino acids. Hence, the acidic aspartic and glutamic will replace each other, and the aromatic amino acids tyrosine, tryptophan, and phenylalanine will be interchangeable.

Chymotrypsin, subtilisin, and serine carboxypeptidase II are considered examples of convergent evolution because they independently evolved to do similar functions. These enzymes are both serine proteases, meaning they use an active site serine residue to help catalyze the nucleophilic breakdown of peptide bonds. They share a similar structural motif in the absence of sequence homology. In subtilisin and serine carboxypeptidase II, a catalytic triad of serine-aspartic-histidine is observed that may imply a serine protease, but the arrangement of residues within their amino acid sequences (primary structure) is different to chymotrypsin, and these proteins differ in their overall structure. It is very unlikely that these three proteins have arisen from a common ancestor of chymotrypsin or a related serine protease, since the order of residues in the catalytic triad differs as well as other structural features present in each of the enzymes.

With the help of X-ray crystallography, it has been possible to show that in proteins that fold to form domains, those domains are conserved in different environments and proteins. The extraordinary similitude among proteins that catalyze different reactions strongly indicates that they originated from a common gene. It is possible that this occurred without major changes in the three-dimensional structure. On the other hand, convergent evolution can lead to different enzymes sharing a mechanism of action that could be due to different factors. In general, most of the mutations in proteins generate nonfunctional proteins. However, on certain occasions, a mutation may create a protein with a new function. Comparing proteins within the same species, one can see that there are families of molecules closely related. It is not surprising that those enzymes that catalyze similar reactions have homologous sequences.

In 1974, Michael Rossmann (Rao and Rossmann 1973; Hanukoglu 2015) discovered that certain enzymes that have different three-dimensional structures all contain a domain of seventy amino acids that have the same folding. The enzymes had different functions, but all had in common the binding of certain coenzymes, such as the dinucleotide of nicotinamide adenine (NAD), the mononucleotide of flavine (FMN), and the monophosphate of adenosine. The three molecules had a mononucleotide binding motif in their structure known as the Rossmann fold. Rossmann thought this motif was a ghost of an ancient protein whose capacity to bind nucleotides was critical and, therefore, conserved.

There are other proteins that we can trace back potentially to the creation of the first polymer structures. One of those proteins is ferredoxin, which binds iron and sulfur compounds, which were abundant at that early stage on Earth. In addition to amino acids, compounds of carbon, nitrogen, oxygen, sulfur, and hydrogen early proteins probably used metals, mainly iron. Iron can be used to move electrons, which must have been crucial to electron transfer processes in early metabolisms; indeed, ferredoxins transport electrons around cells to drive metabolic processes. Electrons flow through solids, liquids, gases, and power living systems. It has been speculated that the same electrical charge must be present in any other planetary system that can potentially support life. Together with Rossmann's fold, these two processes would have laid the foundation for the first building blocks with which to create life.

Mutations in Proteins Are not Random and Respond to Evolutionary Pressure

Most mutations in proteins are the result of selective pressure. The preference in the mutations is that the new amino acid substitution must function similarly to the original amino acid or better; otherwise, this could lead to a nonfunctional protein. Thus, similar physical and chemical amino acid mutations are likely neutral at the protein level. Analysis of large protein families has shown that serine, asparagine, and alanine are more mutable, whereas tyrosine, tryptophan, and cysteine are less mutable. Of particular interest is cysteine, which plays a unique role in stabilizing the three-dimensional structure of a protein and joining different subunits together by forming disulfide bonds that no other amino acid can, like we will see in Chapter 3. The low mutability of tryptophan and tyrosine has been attributed to their unique role in protein folding.

An international team, including University of Oxford professor Justin Benesch and DPhil student Shane Chandler from the Department of Chemistry, has revealed that complexity can evolve through surprisingly simple mechanisms. They identified the evolutionary "missing link" through which hemoglobin—the protein complex that transports oxygen in our blood—evolved from simple precursors (Pillai et al. 2020). They found that the emergence of modern hemoglobin's structure and function was triggered by just two mutations more than four hundred million years ago (figure 1).

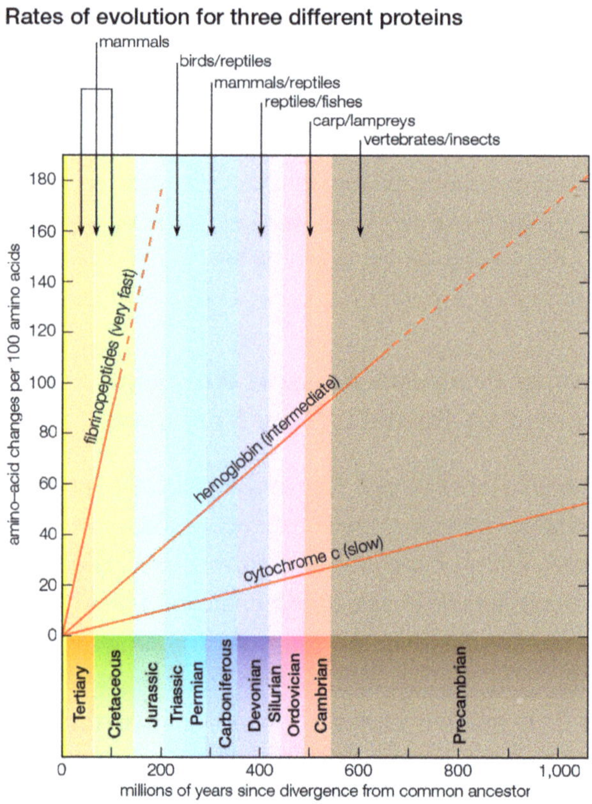

Figure 1.0. **Rates of evolution for fibrinopeptides, hemoglobin, and cytochrome c.** *Image ID: BNB5FX. Universal Images Group North America LLC / Alamy Stock Photo. Order reference: OY72287243, 27 August, 2022.*

Chance and Necessity

The universe is not pregnant with life, nor is the biosphere with man ... Man at last knows that he is alone in the universe's unfeeling immensity, out of which he emerged only by chance.

—JACQUES MONOD,
Chance and Necessity (1970)

As scientists, we have taken a reductionist approach, looking at the most minor pieces of biological organisms to try to understand life, and the progress has been evident. There are universal physical laws that govern the behavior of biological systems. If we were to become as small as a water molecule and enter the cell (the fundamental unit of life), we would witness internal chaos in a viscous environment like the cell's cytosol. In that aqueous environment, molecules are tumbling, dancing, and crushing against each other as a molecular storm. If the molecules were frozen in their positions, no reaction would take place, and life would not be possible. Hence, a delicate interplay between the molecular storm and the physical laws led to the formation of sophisticated structures that dissipate energy. Thus, through the process of jigsaw puzzle pieces first, organisms could appear in this hostile context as a pure *chance* for the origin of life, and evolution is a *necessity,* constantly acting over the current living forms.

Life can be recognized as a thermodynamic process that could only arise and proliferate by dissipating energy as suggested by a most recent theory (Michaelian 2011). Briefly, all irreversible processes arise, persist, and even evolve in response to a well-characterized thermodynamic imperative of generating entropy through the dissipation of the conservative thermodynamic quantities, mass-energy, momentum, charge, etc. The production of entropy is not incidental but the sole origin, persistence, and evolution of an irreversible process (Michaelian 2011). Thus, the dissipating theory of the origin of life suggests that the twists and turns in every new direction in evolution taken by life on Earth or any planet in the part of the cosmos, be it the formation of RNA and DNA, encapsulation in lipid vesicles, formation of superstructures, multicellular plants, and animals, all happen for one and only one fundamental reason, which is thermodynamics. In other words, to increase the global entropy production. If life is based on universal laws like thermodynamics, then it is possible that life similar or not too like our own could exist in the universe where the organic elements are present, along with a source of energy and a dissipative solvent medium like water.

CHAPTER 2

THE BEGINNING OF THE STUDY OF PROTEINS

The word protein derives from the Greek *proteios*, meaning fundamental or first principle, and refers to the god Proteus, son of Poseidon. It was the proper name for a substance that was found in all living things and was speculated to be the fundamental primordium for life. Prophetic abilities were attributed to the god Proteus, and he used to appear to mortals in different forms. Thus, the name seemed appropriate for substances believed to be of unique composition and could be found in nature under various guises. The name was proposed by the Swedish chemist Jöns Jacob Berzelius (1779–1848) and used for the first time by his colleague, Gerardus Johannes Mulder (1802–1880), who was a Dutch organic and analytical chemist. Mulder employed the term "protein" in a publication entitled: "On the composition of some animal substances," originally in French 1838, but translated in 1839 to German (Mulder 1838; Harold 1951).

Organic Substances in all Animals and Plants

The organic substances which are present in all constituents of the animal body, also as we shall soon see, in the plant kingdom, could be named protein from πρωτειος [proteios], primaries, which have the composition $C_{400}H_{620}N_{100}O_{120}$....." (G.J. Mulder, 1802–1880).

While at Utrecht University, Mulder described the chemical composition of proteins in the following terms: "I have claimed that albuminous substances such as in eggs and blood are made up of a common radical, protein, and that protein had the same empirical formula except for some variation in amounts of sulfur and phosphorus." Mulder's idea of proteins was developed long before the polymer nature of proteins was recognized after work by Staudinger and Carothers. Studies on proteins led Mulder to his protein theory in 1838. He proposed that all albuminous substances consist of a radical compound (protein) of carbon, hydrogen, nitrogen, and oxygen, in combination with varying amounts of sulfur and phosphorus. The differences among proteins resulted from the multiplication of the protein units in conjunction with the two other elements.

Around 1860, a fundamental role was assigned to albumin and proteins in general as the physical basis of life, but there needed to be a clear concept of whether albuminoid structures had a stable structure, and they were called *colloids* due to the charge they possessed. The idea that proteins are composed of amino acids was yet to be accepted, even though by that time many of the amino acids were already known, and they appeared as a product of the degradation of proteins. The first amino acid discovered was asparagine, isolated by the French chemist-pharmacist Louis Nicolas Vauquelin in 1806 from asparagus juice concentrate (Vauquelin 1806). During the nineteenth century and the beginning of the twentieth century (ca 1938), most of the amino acids were identified.

Figure 2.0. **Emil Fischer (1852--1919). Circa 1895.**
The photo was in the general public domain in its source country on January 1, 1996, and in the United States.

In 1902, Emil Fischer was awarded the Nobel Prize in Chemistry for his work on sugar and purine syntheses. In the field of enzymology, Fischer developed "the lock and key" model as a mechanism of substrate binding, and he was one of the first, together with Franz Hofmeister, to postulate that proteins are built of amino acids joined by the peptide bond.

Amino Acids as the Building Blocks of Proteins

It was in the first half of the twentieth century that amino acids became accepted as the building blocks of proteins. Franz Hofmeister and Emil Fischer first proposed in 1902 that proteins are made up of linear chains of amino acids joined by what is called a peptide bond. The understanding of proteins was done from a strictly chemical point of view, since advances in organic chemistry led many notable chemists to venture into the study of proteins, formulating chemical hypotheses. The birth of a specialized area of chemistry, called protein chemistry, was possible with the studies of Emil Fischer and Hofmeister.

Before the advent of protein chromatography, fractional precipitation (salting out) was the major method used to purify proteins. The separation of proteins by fractional precipitation using ammonium sulfate led Hofmeister to propose the famous Hofmeister series in 1888 (Hofmeister 1888a, b). Due to their complex structures, proteins have a more complex solubility dependence on salts than simple organic molecules. At very low salt concentrations, proteins are poorly soluble in water and tend to agglomerate. Small increases in salinity improve solubility, and proteins are "salted in." It is assumed today that agglomeration occurs because charged amino acid residues on the surface of the proteins are poorly shielded in pure water, resulting in strong intermolecular attractions that eventually cause the aggregates to precipitate. With the increasing concentration of ions in the medium, electrostatic attractions are shielded, and the protein becomes more soluble.

As he experimented with different salts, Hofmeister observed that some were more efficient at salting out (i.e., they could cause proteins to precipitate at lower concentrations) than others. He empirically found that common anions could be ordered from highest to lowest precipitating capacity:

$$CO_3^{2-} > SO_4^{2-} > S_2O_2^{2-} > H_2PO_4^- > F^- > Cl^- > Br^- \approx NO_3^- > I^- > ClO_4^- > SCN^-$$

Similarly, he ranked common cations from highest to lowest protein precipitation capacity:

$$(CH_3)_4N^+ > Cs^+ > Rb^+ > NH_4^+ > K^+ > Na^+ > Li^+ > Mg^{2+} > Ca^{2+}$$

Franz Hofmeister, from the German University in Prague, published his results in 1888 in a seminal paper entitled "Concerning regularities in the protein-precipitating effects of salts and the relationship of these effects to the physiological behavior of salts" (Hofmeister 1888a, b). The article presented a set of experiments ordering various salts according to their ability to precipitate egg white proteins from aqueous solutions. By using salts with a common cation or anion, Hofmeister separated cationic and anionic effects, thereby establishing what later became known as the lyotropic or Hofmeister series for ions. Depending on the anion, different concentrations were required to precipitate a prescribed concentration of the proteins. The salts could be ordered in a sequence of effectiveness.

Hofmeister extended his studies to additional proteins and colloidal particles and tried to connect the observed ordering of ions with their strength of hydration, denoted in his time as the water-absorbing effects of salts (Hofmeister 1888b). However, the explanation leaves the chemical details of the surfaces of proteins or other hydrated solutes out of the picture. Without considering the nature of the protein itself, it is impossible to properly rationalize not only the regular ordering of ions, but also the well-documented and long-known exceptions to Hofmeister's behavior with several proteins (Traube 2002).

The relative effectiveness of different anions in crystallizing proteins follows a reversed Hofmeister sequence for pH < pI and a direct Hofmeister sequence for pH > pI. What emerges is that for pH < pI (where the anions are counterions) the repulsive double layer forces increase in the order NaSCN < NaI < NaCl, while at higher pH > pI (where anions are co-ions) the forces increase in the order NaCl < NaI <NaSCN (Boström et al. 2005). Hofmeister's explanation for

the ionic ordering was eventually framed into the 1930s–1950s theory of structure-making and structure-breaking ions (Cox and Wolfenden 1934; Gurney 1953). Within this model, ions are divided into structure makers (kosmotropes) and structure breakers (chaotropes), with the former having and the latter lacking the ability to order water molecules beyond their immediate solvation shells (Marcus 2009).

$$SO_4^{2-} \sim HPO_4^{2-} > F^- > \vdots\ CL^- > Br^- > I^- \sim ClO_4^- > SCN^-$$

(stabilizing) (destabilizing)

Figure 2.1. **Protein stabilizing and destabilizing ions.**

According to this picture, strongly hydrated kosmotropes effectively "steal" water from the protein, leading to a salting-out effect, whereas weakly hydrated chaotropes do not possess this ability (Hribar et al. 2002). Hofmeister tried interpreting his observations based on the theory of electrolytic dissociation developed previously by Arrhenius and Ostwald during the 1880s. Briefly, when an electrolyte is dissolved in water, it is separated into two types of charged particles—one with a positive charge and the other with a negative charge. When an electrolyte is dissolved in a solvent, these forces are weakened, and then the electrolyte undergoes dissociation in ions and the ions are dissolved.

The mechanism of interaction between Hofmeister ions and the peptide group is not well understood. Hofmeister ions are generally thought to exert their effects indirectly by changing the hydrogen-bonding properties of water (Baldwin 1996). A nonspecific salting-in interaction is known to occur between simple ions and dipolar molecules; it depends on ionic strength, not on position in the Hofmeister series. A theory by Kirkwood predicts the strength of this interaction and indicates that it depends on the first power of the ionic strength as reviewed by Robert Baldwin (Baldwin 1996).

Baldwin (1927–2021) was another giant biochemist. Baldwin studied biochemistry at the University of Oxford as a Rhodes scholar, receiving his D.Phil.

degree (the British equivalent of a PhD) in 1954. Baldwin was interested in the problem of how proteins fold. He developed tools and designed rigorous experiments to show that instead of simply switching in binary fashion from their initial linear conformations into correctly folded final shapes, as had been postulated, proteins transition into their final shapes through short-lived intermediate structures. This demonstration and Baldwin's follow-up findings simplified the analysis of proteins, accelerating the pace of discovery in medicinal chemistry and drug development.

Hofmeister also did studies on the well-known biuret reaction discovered in 1833 and produced by the interaction of copper sulfate with proteins. A biuret reaction is a chemical reaction that creates a purple color when most protein or biuret is revealed to copper sulfate in an alkaline solution (a biuret reagent is a solution consisting of potassium hydroxide, hydrated copper II sulfate, and potassium sodium tartrate). We know today that it is a direct reaction with the peptide bonds in proteins that produces the purple color. Studies of the interaction of proteins with the biuret reagent led Hofmeister to postulate that proteins were made up of sequences of amino acids since the biuret reagent does not test positive for protein degradation products or free amino acids.

Meanwhile, between 1899 and 1908, Emil Fischer made his great contributions to the knowledge of proteins, including enzymes; at the time, he was a well-known chemist for his work on the structure of sugars, which transformed the knowledge of these compounds. Proteins were known to be composed of amino acids, as proposed by Hofmeister, but Fischer postulated that these amino acids in proteins are linked together by amide bonds, which he called "peptide bonds." He proceeded to use chemical reactions to put together long chains of amino acids held together by such bonds to create artificial protein-like substances. He sought by analysis effective methods of separating and identifying the individual amino acids, discovering a new type of them, the cyclic amino acids: proline and oxoproline. He also studied the synthesis of proteins by obtaining the various amino acids in an optically active form in order to unite them. In 1901, he discovered, in collaboration with Fourneau, the synthesis of the dipeptide glycyl-glycine, and in that year, he also published his work on the hydrolysis of casein. Amino acids occurring in nature were prepared in his laboratory. His synthesis of the oligopeptides culminated in an octo decapeptide, which had many characteristics of natural proteins. This and his subsequent work led to a better understanding of the proteins and laid the foundations for later studies of them. When Fischer died in 1919, the Emil Fischer Memorial Medal was instituted by the German Chemical Society.

Crystalloids and Colloids

During the second half of the nineteenth century, the Scottish chemist Thomas Graham worked on the diffusion of one liquid into another. He divided particles into two classes—crystalloids, such as common salt, having high diffusion, and colloids, such as gum arabic, having low diffusion. He devised dialysis, a method for separating colloids from crystalloids, and proved that the process of liquid diffusion causes partial decomposition of certain chemical compounds. He invented many of the terms used in colloid chemistry, and he proposed the first theory to explain the structure of proteins as colloids. Graham used the term colloid to refer to a substance between solid and liquid, large enough not to pass through the pores of the dialysis membrane, but not crystallizing.

However, the controversy about the nature and structure of proteins extended until the first half of the twentieth century when Sanger published the sequence of insulin in 1952. For this purpose he used the "Sanger reagent" fluorodinitrobenzene (FDNB) to react with the exposed amino groups in the protein and with the N-terminal amino group at one end of the polypeptide chain. He then partially hydrolyzed the insulin into short peptides, either with hydrochloric acid or using an enzyme such as trypsin. The mixture of peptides was fractionated in two dimensions on a sheet of filter paper, first by electrophoresis in one dimension and then, perpendicular to that, by chromatography in the other. The different peptide fragments of insulin, detected with ninhydrin, moved to different positions on the paper, creating a distinct pattern that Sanger called "fingerprints" (Sanger 1945; Sanger and Thompson 1953a, b).

Chemists working on proteins thought that proteins should be molecules with molecular sizes no greater than one thousand, and in no case did they think they could be macromolecules with molecular sizes greater than ten thousand since they would not be stable.

Figure 2.2. **Frederick Sanger—seen with an atomic model structure of insulin—in the Cambridge laboratory.**
(Credit Image: Keystone Pictures USA/ZUMA PRESS) Photo Alamy (order reference: OY72841269, September 10, 2022.). Image ID: E0R9ED.

As we indicated before, the colloidal theory of proteins proposed by Graham continued to be the main influence in explaining the structure of proteins in solution since the beginning of the twentieth century. Emil Fischer, the great chemist of the time, also believed that proteins could not have a large polymeric structure due to the perception that these chains would be very difficult to synthesize by living organisms. At that time, the evidence that proteins constituted large molecular masses was evident from the studies of Mulder and others. In 1871, chemists Heinrich Hlasiwetz and Josef Habermann used a pioneering technique such as sequential hydrolysis to show that the casein protein could be subjected to several consecutive hydrolytic attacks, which pointed to a protein of considerable molecular size (Hlasiwetz and Habermann 1873).

Caseins have been shown to have molecular masses of twenty-five thousand to thirty-five thousand daltons[3] (Da). Frederick Sanger perfected the technique, which allowed him to sequence the hormone insulin before developments in the separation and quantitation of amino acids and when the nature of the peptide bond was unknown. This milestone served to end the resistance of the followers of the colloidal theory. Later in the book, we will return to the concept of colloidal stability, which is deeply rooted in biopharmaceutical development. However, based on what we have discussed related to the controversy of whether proteins are colloids, we shall use the *energetics of the solution* rather than *colloidal stability* as a better descriptor (Pérez-Ramírez 2020).

In 1922, the chemist Hermann Staudinger (1881–1965) developed the macromolecular theory for the case of some organic molecules, which he characterized as polymers and was another support for the possibility that sequences of amino acids could form proteins. Like the colloidal theory, the macromolecular theory was developed for the case of organic polymers. The reigning theory was that all chemical compounds were low molecular weight entities that sometimes "aggregated" to display characteristics like those found in rubber and polystyrene polymers. In the early 1920s, Staudinger published a series of papers on his new theory, presenting chemical evidence that such compounds were covalently bonded molecules of high molecular weight formed by the linking of molecules. Polymers were, as he termed them, "macromolecules" (Staudinger 1920). He was awarded the Nobel Prize in Chemistry in 1953 for his discoveries in the field of macromolecular chemistry. Further advances in analytical development were required to accept the macromolecular theory of proteins.

Ultracentrifugation, Electrophoresis, and Chromatography Facilitated the Knowledge of Proteins

At the University of Upsala, Théodor "Thé" Svedberg (1884–1971), the Swedish chemist, was doing research on colloids and proteins using the analytical ultracentrifuge that he developed to distinguish pure proteins from others. In 1923, Svedberg traveled to the University of Wisconsin-Madison, US, to lecture on colloid science. On the journey across the Atlantic, he began sketching designs for what he called an ultracentrifuge. This technique measures the physical properties of sedimented particles and how they behave in the solution state. The seminal role played by the analytical ultracentrifuge in the development of protein biochemistry has been summarized by Steve Harding (Harding 2017). Svedberg invented the analytical version of ultracentrifugation and won the Nobel Prize in Chemistry for the research he did with it.

Developing the analytical ultracentrifuge was a long process (Svedberg and Pedersen 1940; Schachman 1957). First, it needed to have a rotor to spin at a very high angular velocity and keep the system under refrigerated conditions to prevent the degradation of the biological material. In addition, the analytical ultracentrifuge has a detection system that permits monitoring of the sedimentation of the sample as a function of radial distance. There is the famous quote when Svedberg's associates were running an overnight experiment: His assistant, Fahraeus, was on the late shift and when a Schlieren[4] peak appeared. He phoned through the news to Svedberg—apparently in bed—"Thé, I see a new dawn." A postage stamp was produced to commemorate the occasion (Steve Harding, personal communication to the author).

In modern versions of the analytical ultracentrifuge, samples are centrifuged in a cell, and an optical system projects an image of the cell onto a computer in real time. Most common types of experiments based on analysis are sedimentation velocity and equilibrium (Pérez-Ramírez and Stecker 2005), which provide information on the shapes, sizes, and conformational changes of macromolecules (Laue 1995; Schuck 1998; Rivas et al.1999; Demeler and van Holde 2004; Harding and Rowe 2010; Kingsbury and Laue 2011). In addition, the analytical ultracentrifuge is one of four main techniques that can provide the absolute mass of a macromolecule, the others being membrane osmometry, mass spectrometry, and light scattering.

In an ultracentrifuge, particles/macromolecules are made to move by centrifugal force, and the distribution in concentration of the particles/macromolecules along the length of the ultracentrifuge tube is determined at one or more times. There are two types of experiments that can be performed in the analytical ultracentrifuge, namely *sedimentation velocity* and *sedimentation equilibrium.*

In a *sedimentation velocity* experiment, a relatively high speed is employed to sediment the particles/macromolecules from the center of rotation to be monitored as the rate of movement of a sedimenting boundary (Schuck and Rossmanith 2000; Pérez-Ramírez and Stecker 2006). A measure made while the molecules are moving along the centrifugal axis is called a *sedimentation velocity determination.* The result obtained is the *sedimentation coefficient*—a value that gives information about the molecular weight and shape of the particle/macromolecule.

In a *sedimentation equilibrium* experiment, the concentration distribution is measured under conditions that the distribution no longer changes with time; the particles/macromolecules are said to have reached *sedimentation equilibrium.* Hence, the experiments are performed at a lower speed so that the sedimentation and back-diffusion forces are of comparable magnitude and therefore give rise to an equilibrium distribution of solute concentrations (Van Holde and Baldwin 1958; Yphantis 1964; Wills et al. 2000). This type of measurement yields data about molecular weight, density, and composition. In addtion, sedimentation equilibrium permits the study of protein systems with weak to moderate association constants.

Sedimentation velocity is preferred over sedimentation equilibrium for: i) analyzing unstable proteins, ii) identifying protein intermediates along assembly pathways, iii) analyzing proteins that self-associate very slowly, iv) analyzing protein or protein complexes that undergo conformational transitions without self-association.

Figure 2.3. Théodor Svedberg (1884–1971).

Photo from 1926. This work is in the public domain. Published anywhere (or registered with the US Copyright Office) before 1928 and public domain in the US.

Svedberg tested several proteins in his analytical ultracentrifuge, and all of them had large molecular masses, incompatible with the colloidal theory. The analysis of hemoglobin in the analytical ultracentrifuge showed molecular masses of 67,000 Da; four times larger than the value obtained with other techniques, suggesting that hemoglobin was composed of four subunits. Despite Svedberg being a devoted follower of colloidal theory, believing that protein could not be larger than 500 Da, his results were another piece of evidence to prove the macromolecular nature of proteins. Svedberg's experiments with the ultracentrifuge also contributed to the acceptance that proteins in solution could be understood as particles of a spherical nature, giving rise to the concept of globular proteins. We know that proteins are electrically charged macromolecules having electrostatic interactions, but to begin understanding the electrical charge in proteins, we needed to wait until the development of the Debye-Hückel theory.

Figure 2.4. Model E analytical ultracentrifuge.
(Photo courtesy of Dr. Steve Harding, University of Nottingham, United Kingdom).

In 1923, the Debye-Hückel theory was proposed by Peter Debye and Erich Hückel as a technical deviation of the ideality of electrolytes and plasma solutions (Debye and Hückel 1923). In the chemistry of electrolyte solutions, an ideal solution is a solution whose colligative properties are proportional to the concentration of the solute. Real solutions show deviations from this type of ideality, except for the lowest concentrations. The theory describes how ions in solution screen or attenuate the electrostatic interactions between other ions. This screening effect is particularly important in explaining the deviation of real electrolyte solutions from ideal behavior. To accommodate these effects in the thermodynamics of solutions, the concept of activity was introduced: the properties are proportional to the activities of the ions. In an ideal mixture, the microscopic interactions between each pair of chemical species are the same (or macroscopically equivalent, the enthalpy change of solution and volume variation in mixing is zero) and, as a result, properties of the mixtures can be expressed directly in terms of simple concentrations or partial pressures of the substances present e.g. Raoult's law. Activity is the effective or thermodynamic concentration of a particular ion or ionic molecule in the solution. The activity, *a*, is proportional to the concentration, *c*. The constant of proportionality is known as an activity coefficient, γ (Wright 2007; Robinson and Stokes 2002).

$$a = \gamma c \tag{2.0}$$

The theory of Peter Debye and Erich Hückel provides theoretical expressions for single-ion activity coefficients and mean ionic activity coefficients in electrolyte solutions. The Debye-Hückel theory is essential for understanding the behavior of electrolyte solutions, particularly in the context of ionic strength, ion activities, and the impact of charged species on solution properties. The expressions in one form or another are useful for extrapolating quantities, including mean ionic activity coefficients, to low solute molality or infinite dilution. The only interactions the theory considers are the electrostatic interactions between ions. These interactions are much stronger than those between uncharged molecules, and they die off more slowly with distance. If the positions of ions in an electrolyte solution were completely random, the net effect of electrostatic ion-ion interactions would be zero because each cation-cation or anion-anion repulsion would be balanced by a cation-anion attraction.

Based on the Debye-Hückel theory, the Danish scientist Kaj Ulrik Linderstrøm-Lang (1896–1959) postulated that proteins would be electrically charged spheric ions, with multi charges. With the knowledge that proteins behave like charged particles, another breakthrough was the development of electrophoresis by Arne Tiselius, a student of Svedberg.

Electrophoresis is a laboratory technique that separates DNA, RNA, or protein molecules based on their size and electrical charge. An electric current moves molecules through a gel or other matrix. The pores of the gel or matrix act like a sieve, allowing smaller molecules to move faster than larger molecules. To determine the size of molecules in a sample, size standards are used as we know them today. However, it took decades to perfect the technique for the identification of a proper sieving system to help separate the molecules, while avoiding their diffusion or heating of the sample.

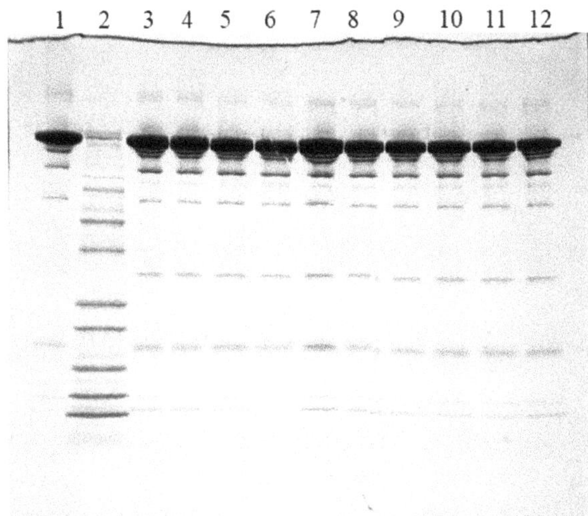

Figure 2.5. Polyacrylamide gel electrophoresis separating an enzyme under different experimental conditions. Line 1 is the enzyme as control. Line 2 are molecular weight markers, and lines 3–12 represent the enzyme subjected to different solution conditions.

One of the most transformative experimental techniques in the rise of modern molecular biology and protein biochemistry was the development in 1970 of high-resolution sodium dodecyl sulfate (SDS) polyacrylamide gel electrophoresis (PAGE), which allowed the separation of proteins—including structural proteins—in complex mixtures according to their molecular weights (see King 2022). SDS is an anionic detergent; it imparts a negative charge to all the proteins in the sample. SDS swamps the inherent charge of the proteins and gives every protein the same charge-to-mass ratio. SDS binds to most proteins at a constant ratio of 1.4 g SDS/g protein (Reynolds and Tanford 1970), except for positively charged proteins such as histones and highly glycosylated proteins.

Thus, if protein binds a constant ratio of SDS to protein, the separating gels in electrophoresis have sieving properties, and mobility becomes a function of molecular mass. Its development was intimately tied to investigations of the control of virus assembly within phage-infected cells. The method was developed by Ulrich K. Laemmli, working in the virus structural group led by Aaron Klug at the famed Medical Research Council Laboratory for Molecular Biology (LMB) in Cambridge, UK, while Laemmli was tackling T4 head assembly.

Laemmli's original paper (Laemmli 1970) has been cited almost three hundred thousand times (King 2022). The other important development made by Laemmli was the inclusion of another polyacrylamide gel on top of the separating gel, called the stacking gel, allowing for the focusing of the samples and getting better resolutions.

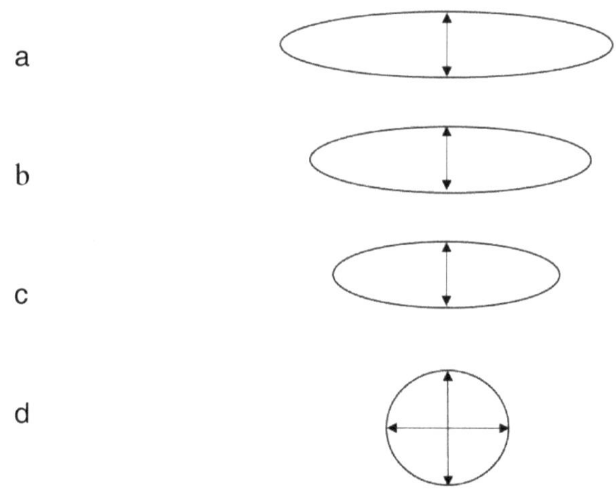

Figure 2.6. Schematic representation of ellipsoids formed by the interaction of SDS with proteins. When the protein is <16,000 Da, as in (d), the binding of SDS to protein resembles a sphere, and therefore the migration of the protein in the gel is affected.

At the same time, when improvements in electrophoresis were obtained, allowing better resolution and characterization of proteins, chromatographic techniques were also being developed. Chromatography is a laboratory technique for separating a mixture into its components. Chromatography gets its name from a technique first used in the late nineteenth century to separate pigments in complex mixtures. The mixture is dissolved in a fluid solvent (gas or liquid) called the mobile phase, which carries it through a system (a column, capillary tube, plate, or sheet) on which a material called the stationary phase is fixed.

By altering the mobile phase, the stationary phase, and/or the factor determining the speed of travel, a wide variety of chromatographic methods have been created, each serving a different purpose and ideal for different mixtures: i) in *thin-layer chromatography* (TLC), the stationary phase is a thin layer of solid material, usually silica-based, and the mobile phase is a liquid in which the mixture of interest is dissolved, ii) in *liquid,* the mixture of interest is dissolved in a liquid and passed through a solid stationary phase, which is often made of a silica material.

Several varieties of liquid chromatography exist, depending on the relative polarities of the mobile and stationary phases (normal phase versus reverse phase) and whether the mobile phase is pressurized (high performance). In the case of the normal phase, the most used is gel chromatography or *molecular sieve chromatography* (size-exclusion chromatography, abbreviated as SEC). It is a special case of partition chromatography in which separation is based on molecular size. In this case, a column of tiny particles of an inert substance (silica) containing small pores is prepared, containing molecules of various dimensions. The sample to separate is passed through the column; molecules larger than the pores move only in the space between the particles and therefore are not retarded by the column material. However, molecules smaller than the pores diffuse in and out of particles, with the probability of increasing with decreasing molecular size; in this way, they are slowed in their movement down the column.

Assuming that the protein to be separated is not adsorbed to the particles in the column, the probability of penetration in the particles is the main factor determining the rate of movement through the column and separation. The charge of the protein could also affect the rate of movement and interaction with the particles in the column. To avoid that problem, the concentration of salt in the buffer used should be optimized, i.e., the ionic strength (see Chapter 3). *Ion exchange chromatography* separates the components of a mixture based on their charge, in addition to or instead of their size. In essence, positively (cations) or negatively (anions) charged ions are separated using different stationary phases and different pH mobile phases.

An *ion exchanger* is a solid that has chemically bound groups to which ions are electrostatically bound. It can exchange these ions for ions in an aqueous solution. Ion exchangers can be used in column chromatography to separate molecules according to charge. The principle of ion exchange chromatography is that charged molecules *adsorb* to ion exchangers reversibly so that molecules can be bound or eluted by changing the ionic environment. This can be accomplished by changing the acidity of the medium (pH) or the ionic strength (concentration of salt). An ion exchanger contains covalently linked charged groups. If a group is negatively charged, it will exchange positive ions and is a *cation exchanger (e.g.,* SO_3^-, carboxyl groups, phenyl). If the charge group is positive, it is an *anion exchanger.* For example, a quaternary amine group would be a strongly basic *anion exchanger.*

The most common anion exchangers are aliphatic or aromatic amino groups. The resolving power of a column increases with the length of the column and the number of theoretical plates per unit length. The columns with a high number of theoretical plates are considered more efficient. Theoretical plates are calculated per meter length of the column and are usually called N_m.

The following formula is used to calculate the theoretical plates:

$$N = 16 (V_e/W_b)^2$$

Where,
N = Theoretical plates
V_e = Retention time
W_b = Peak width

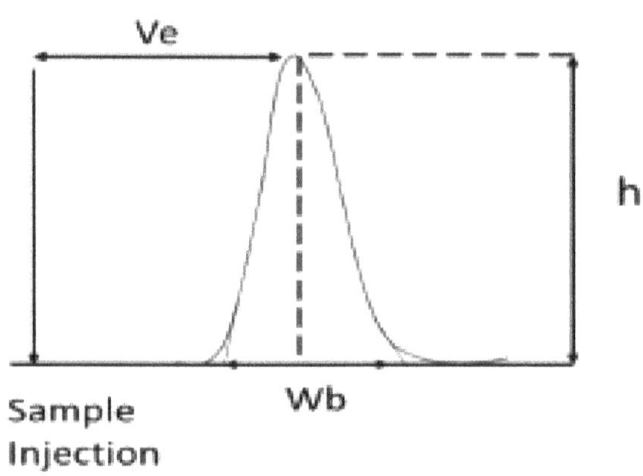

Figure 2.7. Schematic representation of a column chromatography.

As the matrix particles become smaller, the theoretical plates increase. However, the flow rate of a column drops as the particle size decreases, and this allows time for significant diffusional spreading, leading to a loss in resolution. The diffusional spreading can be minimized by establishing a pressure difference across the column's top and bottom to force the liquid through the bed. Thus, using small particles (10 μm) in diameter and high pressure to maintain an adequate flow rate is called high-performance or high-pressure liquid chromatography, or HPLC.

With the advances in chromatography came the ability to identify amino acids in each protein. From 1948, the biochemists W.W. Stein and Stanford Moore, together with D.H. Spackman, engaged in research on amino acid analysis by chromatography using starch particles and ion-exchange resins, and in 1958 they developed an automatic amino acid analyzer at the Rockefeller Institute for Medical Research in New York (Moore et al.1958; Spackman et al.1958).

The amino acid analyzer opened an entirely new field for research, permitting the elucidation of the composition of proteins. The rapid expansion of biochemistry would have been impossible without it. The Rockefeller Institute team realized this and turned to the Spinco Division of Beckman, with which they already had a good relationship, to transfer the design to production. The amino acid analyzer designed by Moore and colleagues, using ion-exchange chromatography and automated by Spackman, used the reaction with ninhydrin as a detection method. The automated amino acid analyzer allowed the rapid quantitation of the composition of amino acids in a protein hydrolyzed. In addition, the amino acid analyzer allowed Stein and Moore to sequence the first enzyme, pancreatic ribonuclease A.

In 1951, the famous Lowry protein quantitation method was developed by Oliver Howe Lowry (1910–1996) and colleagues (Lowry et al. 1951). As of January 2004, the paper has been cited two hundred and seventy-five, six hundred and sixty-nine times. Briefly, the assay uses copper that reacts with the peptide bond under alkaline conditions, forming a monovalent copper ion that reacts with the Folin reagent (phosphomolybdic-tungstic acid), facilitating the reduction of the Folin reagent, which gives rise to several reduced species with a characteristic blue color that can be quantitated using a spectrophotometer. Thus, the intensity of the blue color is proportional to the concentration of protein. A standard of bovine serum albumin (BSA) is usually employed.

X-rays and Crystal Structures: the Path to Protein Conformation

Protein crystallization was discovered by chance nearly two hundred years ago and was developed in the late nineteenth century as a powerful isolation tool and for demonstration of chemical purity. The crystal structure describes the ordered arrangement of atoms, ions, or molecules in a crystalline material. Ordered structures occur from the intrinsic nature of the constituent particles to form symmetric patterns that repeat in matter along the principal directions of three-dimensional space.

The first protein to be crystallized was the enzyme urease, done by James B. Sumner in 1926 (Sumner 1926). In the following years, other enzymes were crystallized, such as peptidase and ribonuclease A. These studies led to the acceptance that enzymes were indeed proteins, and that in the crystalline state, the enzyme can maintain an active stay if kept in the same solution (liquid) where the protein was crystallized.

German physicist Wilhelm Conrad Röntgen discovered X-rays in 1895 while experimenting with electron transfer when two plates were subjected to an electric current in a vessel with a certain vacuum to investigate the violet fluorescence produced by cathode rays. Under those conditions, he identified a new radiation that he called X-rays. X-rays are ionizing electromagnetic radiation of the same nature as radio waves, microwaves, infrared rays, visible light, ultraviolet rays, and gamma rays (figure 2.8).

Electromagnetic radiation consists of particles called photons that travel in waves, and it includes a vast array of different types of radiation, each with different properties and characteristics. These radiations differ in terms of wavelength and frequency, and they encompass a wide range of energies and applications. Figure 2.8 shows the main regions of the electromagnetic spectrum from the shortest wavelengths (gamma rays) to the longest (radio waves).

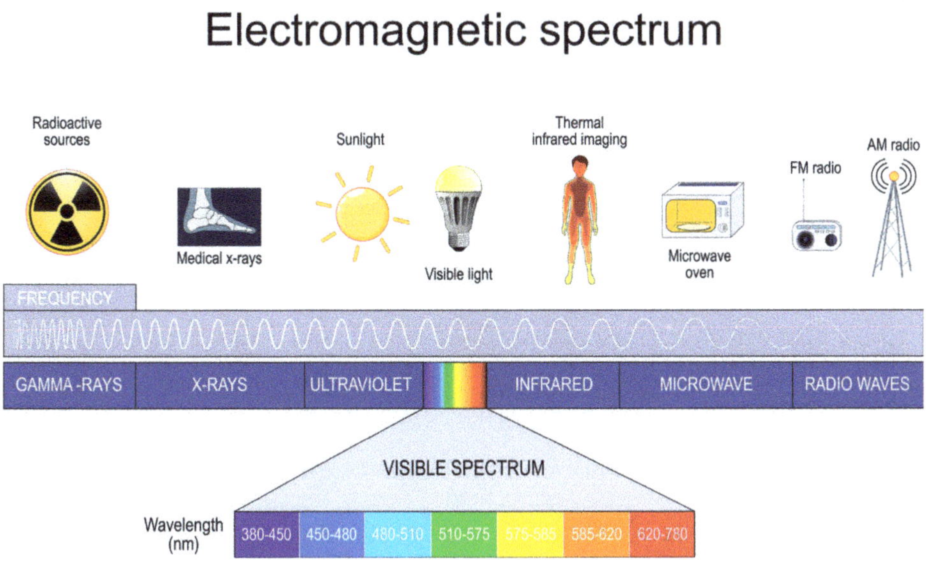

Figure 2.8. Electromagnetic spectrum from short to long waves.

Source: Alamy, standard license, royalty-free. Image ID: 2R1RYM6. All rights, all media, worldwide in perpetuity.

The energy (*E*) associated with radiation is defined by Planck's law.

$$E = h\upsilon \quad \text{where} \quad \upsilon = c/\lambda \qquad (2.1)$$

C is the velocity of light, *h* is Planck's constant, and υ is the frequency of the radiation.

The study of X-rays as waves and a new radiation source led to the development of a very powerful technique, X-ray diffraction. Max von Lue reasoned that if X-rays were high-energy waves of a very small wavelength, they could be diffracted into atoms that are at a distance like the wavelength of X-rays. The interpretation would be easier if those atoms were arranged in a crystal with a repeated structure. Diffraction information is collected on a photographic plate, appearing in the form of dots called order reflections that depend on the type of crystal formed (figure 2.9).

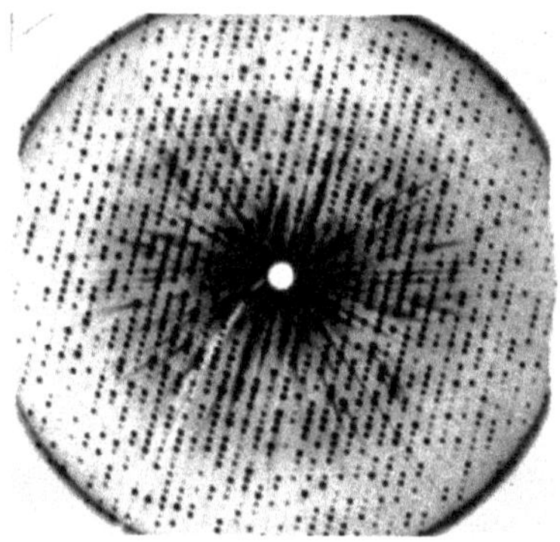

Figure 2.9. X-ray photograph of a myoglobin crystal (the late 1950s).
Photograph by John Kendrew. Courtesy of Medical Research Council Laboratory of Molecular Biology. Reproduced with permission of the copyright holder John Wiley and Sons License Number 5398511106226, September 29, 2022.

The great advances in X-ray crystallography were due to the British scientist William Bragg and his son Lawrence, who realized that X-rays could be used to study the nature of crystals, and they are considered the founders of the English school of molecular biology. In that group was also Max Perutz, about whom we will talk later in the chapter. Bragg suggested that the diffraction of X-rays by crystalline molecules could be used to calculate the distance between the atoms of these molecules and therefore elucidate their three-dimensional structure.

X-ray (neutron) crystallography is a high-resolution method (up to 3Å) that allows solving very small molecules such as NaCl to very large ones, such as ribosomes, providing a static image of the structure. It is based on the phenomenon of diffraction of light from crystal structures. Before the structure of a macromolecule can be determined, it must be obtained in large quantities in order to crystallize it (which is not always possible). In addition, the sample must be of great purity and contain only the molecules that we want to crystallize, and we must avoid that, even if it is the same molecule in different conformations (cis or trans).

Protein crystallography is a form of very high-resolution microscopy, which allows scientists to “see” at atomic resolution. A molecule’s atoms are surrounded by a cloud of electrons, which effectively define its shape. In X-ray crystallography, X-rays are diffracted by electrons, so it is this electron cloud that we see, and it is known as an *electron density map*, the extent to which the diffract crystal determines what kind of resolution can be obtained in the image. With very good crystals, you can see individual atoms.

Today, pharmaceutical and biotechnological researchers need to understand how the disease is controlled at the molecular level within related proteins and macromolecules in order to target specific entities based on this knowledge. A prerequisite for this approach is the ability to obtain a high-resolution three-dimensional molecular structure from the macromolecule under investigation. Hence, X-ray crystallography is a powerful tool, aiding in the development of therapeutic proteins.

In 1958, an article in *Nature* presented the first low-resolution model of a globular protein derived by direct structure determination (Kendrew et al. 1958). The work was part of an ongoing effort by John Kendrew (1917–1997) and his colleagues at the Medical Research Council Unit for Molecular Biology in Cambridge (UK) that, two years later, led to the publication of the first atomic structure of the same protein, myoglobin (Kendrew et al. 1960).

Kendrew came to Cambridge in late 1945 and joined the Cavendish laboratory, where there was a small group, mainly consisting of Max Perutz, working under the tutelage of Lawrence H. Bragg, the director of the Cavendish Laboratory, that was engaged in applying X-ray crystallography to determine the structure of proteins. Protein crystallography was just one of many approaches employed to study proteins. It was carried by the idea that knowing the structure was a way to understand function. The question was how to interpret the complex diffraction images. In the mid-1940s, the hope to make some headway was very much based on the expectation that there was a general plan of protein structure, and that the solution of one protein would give clues to the structure of proteins more generally (De Chadarevian 2018).

In the mid-1950s, as early as Fred Sanger published the first complete sequence of insulin, Kendrew started searching intensively for someone who would undertake a sequence analysis of myoglobin. Advances in protein sequencing threatened to render protein crystallography obsolete, as it was believed at the time that knowledge of the sequence alone would make it possible to predict the three-dimensional structure of a protein (Kendrew 1964). Protein crystallographers became the biggest clients for protein sequences, and the early atomic structure analyses all relied on sequence information to help with the interpretation of the electron density maps (Kendrew 1961). The other great man in the field of crystallography was the Austrian-born Max Perutz (1914–2002). A chemist by training, Perutz went to learn crystallography in J. D. Bernal's lab in Cambridge in 1936 and began work on hemoglobin in 1937. The scientific work of Perutz on the structure of hemoglobin started as a result of a conversation with F. Haurowitz (1896–1997) in Prague in September 1937.

G.S. Adair (1876–1979), who was a protein chemist who used osmotic pressure to establish that hemoglobin was a tetramer, made the first crystals for Perutz from horse hemoglobin, and Bernal and Fankuchen showed him how to take X-ray pictures and how to interpret them. Early in 1938, Bernal, Fankuchen, and Perutz published a joint paper on X-ray diffraction from crystals of hemoglobin and chymotrypsin to a 2 Å resolution (Bernal et al. 1938). "In science, truth always wins" (Ferry 2007). This was a favorite aphorism of Max Perutz, according to Georgina Ferry, Perutz's biographer. Ferry invokes it to good effect as an example of Perutz's endearing combination of optimism and naïveté. An earlier interest of Perutz was the structure and mechanical properties of ice. He was an enthusiastic mountaineer and skier, at home with both granite and glaciers (Dickerson 2008).

Nobel Prizes in chemistry and medicine were awarded for work on crystallography in 1962: in chemistry to Max Perutz and John Kendrew for their work on the protein hemoglobin and myoglobin, and in medicine to Francis Crick, James Watson, and Maurice Wilkins for discovering the DNA structure.

Linus Pauling and the Nature of the Chemical Bond

At the same time that hemoglobin and myoglobin were analyzed by X-ray crystallography, one of the greatest American scientists of the time, Linus Pauling (1901–1994), was using concepts from physics and chemistry to apply to the study of biological macromolecules. For an interview with Pauling later in his life, see Kauffman and Kauffman, 1996. Pauling studied chemistry at the University of Oregon and the California Institute of Technology (Caltech), where he learned about X-ray diffraction. He decided to become more familiar with the new technique of X-ray diffraction. He traveled to Europe, where he interacted with some of the greatest scientists of the time, such as William Bragg, Erwin Schrödinger, and Niels Böhr, who allowed him to develop a new perspective to view biological macromolecules.

Figure 2.10. Linus Pauling lecturing on the structure of proteins, 1957. Linus Pauling (1901–1994), American chemist who won the 1954 and 1962 Nobel Prize in Chemistry.
Source: Alamy Stock Photo Image ID: EX6PPB.World History Archive / (order reference: OY72342866, August 29, 2022).

Pauling became interested in the forces that maintain the structure of biological molecules and in 1939 published the book *The Nature of the Chemical Bond*, where he describes the quantum mechanical nature of the chemical bond (Pauling 1939; Pauling and Niemann 1939). By 1947, the exact nature of a protein structure had yet to be determined, but there was clear evidence that amino acids combine in proteins, leading to long chains. Pauling suggested that two glycine molecules could be condensed, removing a water molecule and forming a glycine-glycine; the bond thus formed is called a peptide bond. Pauling also realized that molecules could be stabilized by a series of weak bonds or interactions, distinguishing them from the covalent bonds. The concept of weak bonds in proteins allowed him to describe, in 1936, the phenomenon of protein denaturation (Edison 2001).

While a visiting lecturer at Oxford University in 1948, Pauling had a sudden insight regarding the fundamental structure of proteins, an insight that had eluded him for over a decade. Working with a sheet of paper that he folded over, where he knew from theoretical considerations that the chain could bend, he found that the polypeptide chain formed from sequences of amino acids would coil into a particular helical structure, which he named the alpha helix. He based this theoretical configuration on chemical-bonding considerations plus X-ray diffraction evidence from certain fibrous proteins.

This proposal, as well as a companion concept of a related "β-pleated sheet" structure, proved correct (Pauling and Corey 1951; Pauling et al. 1951; Pauling 2015a, b). Essential in the model for his α-helix and β-sheet was the planarity of the peptide bond, the result of ~40 percent N–C′ double bond character arising from two dominant resonance structures, allowing for a great simplification in the understanding of protein structures.

Of the three repeating protein backbone dihedral angles, only φ and ψ need to be considered when assigning secondary structure, because the peptide bond dihedral angle ω is generally considered fixed at one hundred and eighty degrees (see Chapter 3). Linus Pauling's prediction of the α-helix, one of the greatest achievements in structural biology, was made by assuming (i) that the peptide bond is planar, (ii) all amino acid residues are equivalent with respect to backbone conformations, and (iii) each amide proton is hydrogen bonded to an oxygen atom of another residue with an N–O distance of 2.72 Å (Pauling and Corey 1951). Similar criteria allowed Pauling to predict several other protein secondary structural elements and parallel and antiparallel β-sheets (Edison 2021). Pauling and colleagues deduced these fundamental building blocks from the properties of small molecules, known from crystal structures and Pauling's resonance theory of chemical bonding, which predicted planar peptide groups. Earlier attempts by others to build models for protein helices failed by including nonplanar peptides and by insisting on helices with an integral number of units per turn.

The Pauling-Corey-Branson models were astoundingly correct, including bond lengths that were not surpassed in accuracy for more than forty years. However, they did not consider the hand of the helix or the possibility of bend sheets. They also proposed structures and functions that have yet to be found, including the gamma helix. Subsequent X-ray diffraction studies have found that the alpha helix is a major component of both globular and fibrous proteins and extensively controls their structure and function—now known to form the backbones of tens of thousands of proteins.

Perutz did not realize that the chemical groups on either side of a peptide bond were likely coplanar. This cost Perutz and colleagues the opportunity to arrive at the now well-known nonintegral α-helical structure Linus Pauling published in May 1951. However, based on Pauling's publications, Perutz deduced that the α-helix, with its regular repeat of residues with a 1.5 Å periodicity, should give rise to axial X-ray reflections at that spacing, providing a powerful diagnostic test. Such reflections would not have been visible in the usual X-ray diagrams for technical reasons. But on the same day that Perutz read Pauling's paper, he took an X-ray photograph, with appropriately altered camera configurations, and recorded a clear reflection, never seen before, at 1.5 Å, confirming Pauling's structure.

Pauling became the only person to have won two unshared Nobel Prizes. His first prize (1954) was awarded for research into the nature of the chemical bond and its use in elucidating molecular structure; the second (1962) recognized his efforts to ban the testing of nuclear weapons.

Modification of Amino Acids in Proteins

Amino acids in proteins react with various chemical reagents to form covalent bonds. Proteins can increase their diversity in function by modifying their amino acids differently.

Acetylation: Many proteins are acetylated. Acetylation occurs as a co-translational and posttranslational modification of proteins. The histone proteins that wrap DNA are acetylated, regulating gene expression (Eberharter and Becker 2002; Gräff and Tsai 2013).

ADP-ribosylation (ADPr): A reversible posttranslational modification of proteins, which controls several biological processes, including DNA damage repair, cell proliferation and differentiation, metabolism, stress, and immune responses. Histones can undergo ADPr (Burzio et al. 1979; Riquelme et al. 1979).

Hydroxylation: Adding hydroxyl groups to proline residues in collagen increases the stability of the triple helix and influences collagen fibril formation (Yamauchi and Sricholpech 2012 Rappu et al. 2019). The scurvy disease is characterized by the inability to produce the hydroxylation of proline due to deficiency in vitamin C (ascorbic acid) that in turn destabilizes the structure of collagen.

Carboxylation: Many proteins get modified by carboxylation, like the addition of carboxyl groups in prothrombin, which is part of the blood coagulation cascade (Stenflo et al. 1974; Hao et al. 2020). Prothrombin is a vitamin K–dependent blood coagulation protein that undergoes posttranslational gamma-carboxylation. If the glutamate residues in prothrombin are not modified, bleeding cannot be stopped, resulting in hemorrhage.

Glycosylation: Carbohydrates are added to increase polarity and hydrophilic nature to interact better with other proteins and ligands. The target of glycosylation is usually asparagine residues or serine and threonine (Spiro 2002; Lowe and Marth 2003; Stanley 2016; Eichler 2019; Schjoldager et al. 2020).

Phosphorylation: An amino acid modification key for turning on or off many processes in the cell (Tan et al. 2010; Bilbrough et al. 2022; Adnan et al. 2023). For example, the insulin receptor is phosphorylated upon binding to insulin to regulate glucose in the blood (Rosen et al. 1983; Herrera et al. 1986;).

In some cases, when peptides are hydrolyzed (cleavage), they can become active or inactive. This is the case of digestive enzymes like chymotrypsinogen, which is activated and converted into its active form by another enzyme called trypsin (Bode et al. 1978; Mátrai et al. 2004) or the blood clotting factor fibrin (Carlin et al. 1980; Weisel et al. 2017).

Chemical Probes of Structure and Function

Before the birth of molecular biology, a specialized approach using covalent chemical modification of proteins was developed to understand the fundamental principles of the structure and function of enzymes and other proteins and in studies related to industrial needs, foods, and nutrition. This general approach to understanding the mechanism of enzymes, critical catalytic residues, and structural contributions of crucial amino acid residues was prevalent until 1990. In the last ten years of the twentieth century, advances in molecular biology led to the mutation of critical residues by genetic engineering, an improved approach to understanding their role in the structure and function of proteins.

For several years, the covalent modification of the α-amino groups of hemoglobin with cyanate was studied intensively as a possible therapy for the sickle cell trait. Interest was based on the finding that the modification displaces the oxygen dissociation curve of hemoglobin to the left, reducing the molecule's tendency to give up oxygen and be converted to its deoxy form (in which it is liable to the gelation that results in the sickling of the erythrocyte). Unfortunately, orally administered cyanate had forbidding side effects in volunteer patients and did not appear to reduce the incidence of the disorder's "painful crisis" (Harkness 1976).

The common denominator is the effects of chemical modification of the side chains of the amino acids on the properties of the proteins. The reagents must be specific and result in the quantitative covalent derivatization of the functional group (i.e., primary amine, imidazole ring, sulfhydryl group of a single unique amino acid in a protein) without any effect on any other functional groups or the conformation of the molecule. However, this was an ideal and, in some cases, difficult to obtain. It is unlikely that chemical modification of a single amino acid residue can be accomplished without any concomitant conformational change in an enzyme.

Furthermore, demonstrating a specific modification of a particular residue was challenging. It must be established that the modification of one residue mole per mole of protein has occurred without modification of another residue (e.g., modification has occurred only with lysine and not a sulfhydryl group). The approach of Ray and Kosland (1961) established a relationship between the rate of loss of biological activity (catalytic activity) and the rate of the modification of amino acid residues. Another approach will be the cleavage of the protein, separating the peptides by chromatography and isolating the peptide containing the modified amino acid (Pérez-Ramírez et al. 1994) based on the spectroscopic properties of the probe, radioactivity tracing, fluorescence emission, etc. Furthermore, demonstrating that the effects of specific chemical modification are reversible lends strong support against the arguments that such effects result from nonspecific conformational change.

The development of suicide inhibitors for enzymes started with the pioneering work of Professor Robert Abeles (1926–2000) from Brandeis University. Despite his debilitating Parkinson's disease, when I met him while I was a postdoctoral fellow in Timasheff's laboratory at Brandeis University, he continued to be active in his laboratory. The suicide inhibitors get activated by contact with the active site of an enzyme. Suicide inactivators are relatively unreactive molecules that bind to the active site where, as a result of the catalytic capabilities of the enzyme, they are converted to active species, which irreversibly modify the active site.

The work of Abeles in transition-state or reaction-intermediate analogs, most notably trifluoromethyl ketones, has substantially impacted the design of novel pharmaceutical compounds. This structural similarity of suicide inhibitors with the usual substrates enables the molecule to become bound to the active site of the enzyme, and the labile functional groups allow it to react with any suitable nucleophile group, such as carbonyl, phosphoryl, and sulfhydryl groups. The resulting covalent adduct is catalytically inactive (Hedstrom et al.1984). The best-known example is TPCK, an inhibitor of chymotrypsin. Suicide inhibitors are of interest to many biopharmaceutical companies as a potential use for therapeutic purposes.

An essential characteristic of suicide inhibitors is that the inactivating species are generated at the active site and are not molecules that react from the solution. The design of these inhibitors is based upon the attachment of a chemically reacted group, generally an electrophile (electrophiles "love electrons" and are positively charged or electron-deficient groups), to a molecule with a structural feature resembling that of the enzyme's standard substrate (Abeles 1978, 1981). This structural similarity allows the molecule to become bound to the enzyme's active site, and the labile functional group allows it to react with any suitably located nucleophile of the enzyme's active site. Nucleophile means "loving the nucleus." The atomic nucleus is positively charged, and nucleophiles are generally charged or electron-rich groups that can interact with positively charged electrophiles.

The principle of suicide inhibition is illustrated by the affinity labeling of chymotrypsin with *tos*-L-phenylalanine chloromethyl ketone (TPCK). TPCK resembles substrates like *tos*-L-phenylalanine methyl ester, but the chloromethyl ketone group of TPCK is an alkylating reagent (Schoellmann and Shaw 1962,1963). Enzymes containing the coenzyme pyridoxal 5-phosphate (PLP), such as aspartate aminotransferase (Martinez-Carrion 1975), are targets for suicide inhibition (Silverman and Abeles 1976). The aldehyde group of PLP reacts with amine groups of amino acids to form a Schiff's base as shown in figure 2.11 (Edger and Rifkin 1977; Martinez-Carrion 1975; Dwyer 1988; Pérez-Ramírez and Martinez-Carrion 1989). The pyridine ring of PLP acts as a versatile electron sink that facilitates the stabilization of negative charges like in carbanions (carbanion is an anion in which carbon is trivalent and bears a formal negative charge).

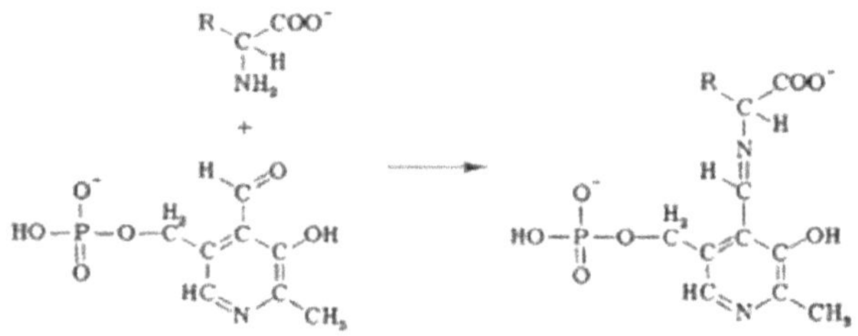

Figure 2.11. PLP forms a Schiff's base with amino acids.

Pyridoxal phosphate has also been employed as an anionic probe for protein amino groups exposed on the outer and inner surfaces of intact human red blood cells (Cabantchik et al. 1975) and the regions of the acetylcholine receptor exposed to the outer (Dwyer 1988) and inner surfaces (Pérez-Ramírez and Martinez-Carrion1987,1989). The popularity of using PLP as a probe of structure and function is because it reacts with free amine groups with a high degree of specificity, forming a reversible Schiff's base that can be fixed as an irreversible covalent bond upon reduction. Formation of PLP proteins derivatives can be monitored by ^{3}H, ^{14}C, ^{32}P-radiolabeled PLP (Edger and Rifkin 1977; Lui et al. 1986), electronic absorption or fluorescence spectroscopy (Means and Feeney 1978; Farach et al. 1983; Roberts et al. 1988); or ^{31}P nuclear magnetic resonance (Martinez-Carrion1975). These advantages make PLP a unique probe for investigating structural and functional aspects of macromolecules.

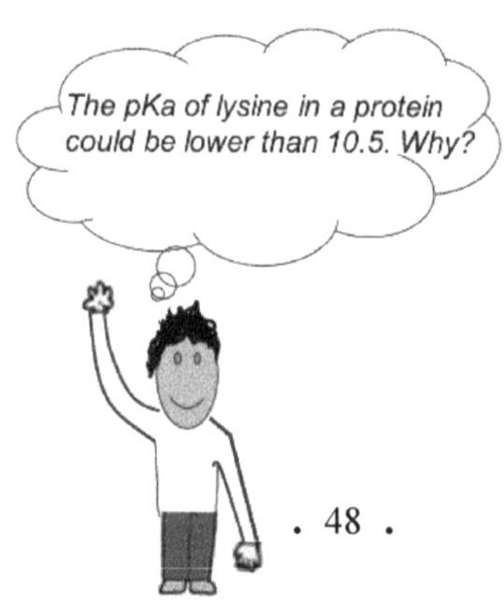

The PLP-containing enzyme aspartate aminotransferase has a reactive lysyl group in the active site, forming a Schiff's base with PLP. Lysyl residues in peptides show averages of pKa ~ 10. Several proteins have shown anomalous pK values for lysyl residues, which can be as low as 5.9 for oxalate decarboxylase (Schmidt and Westheimer 1971) and 7.7 in glutamate dehydrogenase (Veronese et al. 1972). This anomalous pKa for lysyl groups explains the high reactivity with PLP, requiring a nonprotonated ε-amine group of lysine.

Unprotonated lysyl groups at physiological pH can also be found in enzymes where the lysyl group is surrounded by a positively charged pocket or in hydrophobic regions. Abnormal lysyl residues have been detected in aspartate aminotransferase by following the reaction rate of the amine groups with potassium cyanate that produces carbamylation of the reactive lysyl 258 in the active site groups. Carbamylation at different pH values permitted the identification that lysine 258 in the active site of aspartate aminotransferase has a pK value of 7.98 (Slebe and Martinez-Carrion 1976).

Reductive methylation of protein amino groups with formaldehyde and sodium cyanoborohydride can be chosen to modify selective lysyl groups in proteins (Jentoft and Dearborn 1979). The active site lysyl residue of aspartate aminotransferase can also be selectively modified by reductive methylation to get further insight into the role of lysine 258 in catalysis (Roberts et al. 1988). Methylation does not significantly alter the net charge of proteins; there is only a small change in pK observed. α-amino groups typically exhibit pK values of six to eight (Tanford et al. 1955), while monomethylated derivatives have pK values of seven to nine, and dimethylated derivatives have pK values of seven to eight (Jentoft and Dearborn 1983). ε-Amino groups with a normal pK of 10.2 (Tanford et al. 1955) have a pK of 11.0 when monomethylated and a pK of 10.2–10.4 when dimethylated (Jentoft and Dearborn 1983).

At the active site of aspartate transaminase, Lys-258 is preferentially modified, and the main reactive product is the dimethyl derivative making the enzyme inactive (Roberts et al. 1988). Among other properties of the modified enzyme is its ability to bind tightly to the cofactor PLP, but it cannot form Schiff base adducts with the aldehyde group of the cofactor. Since, in the presence of glutamate, there is a conversion from the spectroscopic form characteristic of active site-bound PLP to that of active site-bound pyridoxamine phosphate (PMP), it is unlikely that major steric constraints are created through methylation, preventing active site-substrate interaction (Roberts et al. 1988).

Reductive methylation has also been employed to identify amino groups in α-bungarotoxin interacting with nicotinic acetylcholine receptors (Garcia Borron et al. 1987). The approach has been to compare the chemical reactivities of amino groups in free and acetylcholine receptor–bound α-bungarotoxin (snake toxin) to identify the regions in the α-bungarotoxin molecule that become masked upon binding to the receptor (Garcia Borron et al. 1987). The methylated toxin is cleavage with proteases. The resulting peptides are resolved by reversed-phase, high-performance liquid chromatography, and the modified residues identified by amino acid analysis.

Selective chemical modifications have been utilized extensively to elucidate the interrelationship between the structure and function of the acetylcholine receptor. Thiol group modifying agents elicit marked alterations in several of the properties associated with the receptor. Labeling of functionally sensitive sulfhydryl-containing domains of acetylcholine receptor from Torpedo californica membranes have been done using N-(1-Pyrene) maleimide, a fluorescent, lipophilic, alkylating agent (Clarke and Martinez-Carrion 1986). Fluorescent probes are molecules that absorb light at a specific wavelength and emit it at a specific longer wavelength that can be detected (see Chapter 3: Fluorescence Spectroscopy). Preincubation with N-(1-pyrene) maleimide under nonreducing conditions inhibits agonist-induced cation permeability of the acetylcholine receptor-enriched membranes.

Amino acid analysis and sequencing data of the N-(1-Pyrene) maleimide-labeled peptides identified cysteine-222 as the only residue labeled by the probe on the α-subunit primary structure (Marquez et al. 1989). Cysteine-222 is in the middle of a hydrophobic domain designated MI, which contains a homologous class of cysteines (Cys-241 in the aligned sequences) conserved in the four subunits of the acetylcholine receptor. For further examples of reagents for modification of specific residues in proteins such as histidine, arginine, tryptophan, carboxyl group, or cross-linking, see Lundblad and Noyes, 1984.

Around 1987, I was doing my PhD in protein chemistry using chemical probes to study the topology of the acetylcholine receptor. It was the time when everybody was working with DNA (deoxyribonucleic acid), and many students were enrolled in PhD programs in the field of molecular biology. They used to say I should be cloning something; otherwise, I would never get a job. Frankly, I did not have much interest in DNA and molecular biology. I found proteins fascinating since the moment I started working as an undergraduate with the enzyme fructose 1.6 bisphosphatase in the laboratory of Professor Juan Carlos Slebe at the Universidad Austral in Valdivia, Chile.

This sudden popularity of the molecular biology field was due to the many advances in the area and the belief that understanding genes through molecular biology will give us a better understanding of life's processes. This revolution of genomics, as we now know it, began in the 1970s with the technical advances done by Frederick Sanger. He developed in 1977 a technique for DNA to sequence the first complete genome, called phiX174 virus, which opened the doorway to the possibility in the field of genomics.

Almost Every Cell in a Human's Body Contains a Complete Copy of the Genome

Genomics is the study of the complete set of DNA (including all its genes) in a human or other organism. It all started with the discovery by Friedrich Miescher in 1869 of a substance called "nuclein" (now known as DNA), an associated protein present in the cell's nucleus. In 1910, Albrecht Kossel was awarded the first Nobel Prize in Physiology or Medicine for his discovery of the five nucleotide bases, adenine (A), cytosine (C), guanine (G), thymine (T), and uracil (U). Erwin Chargaff in 1950 found that concentrations of thymine and adenine, and cytosine and guanine, are always found in equal amounts in samples of DNA. This suggested that A always pairs with T, and C with G.

In 1953, the young American James Watson and the English physicist Francis Crick (1916–2004) at Cambridge University reported in *Nature* a model for the structure of DNA (the famous double helix) based on an X-ray photograph taken by Rosalind Franklin (Watson and Crick 1953). Watson, Crick, and Maurice Wilkins were awarded the 1962 Nobel Prize in Physiology or Medicine "for their discoveries" concerning the molecular structure of DNA. However, Watson and Crick failed to credit Rosalind Franklin (1920–1958) for her contributions to the discovery of DNA.

In 1961, Marshall Nirenberg (1927–2010), Har Gobind Khorana (1922–2011), and colleagues discovered the "code for life." They identified how the letters in DNA are read in blocks of three, called a "codon." Each codon specifies an amino acid that is added to the protein during synthesis. They shared the 1968 Nobel Prize for Physiology and Medicine with Robert Holley (1922–1993), who first sequenced the alanine tRNA.

Figure 2.12. Rosalind Franklin (1920–1958) was an English chemist and DNA pioneer.

Source: Alamy Stock Photo. Contributor: Pictorial Press Ltd / Order reference: OY80623969. Invoice number: IY03463332 April 25, 2023.

New sequencing procedures brought advances to molecular biology. In 1980, Fred Sanger shared the Nobel Prize for Chemistry with Wally Gilbert and Paul Berg for pioneering DNA sequencing methods. In 1983, the polymerase chain reaction (PCR) was developed—a technique used for amplifying DNA—by Kary Mullis (1944–2019) at the Cetus Corporation in California, USA. Michael Smith, who had developed other essential ways of manipulating DNA, shared the Nobel Prize in Chemistry in 1993. However, seventeen years before the formal invention of the PCR, Khorana's laboratory had developed a process that they called "repair replication" (Kleppe et al. 1971) with similar principles as the current PCR.

The Human Genome Project was launched in 1990 with the aim to sequence all three billion letters of a human genome. Chromosome 22 was the first chromosome to be sequenced as a part of this project in 1999. The project was completed in 2003 and confirmed that humans have twenty thousand to twenty-five thousand genes. In 1996, an international team completed sequencing the genome of yeast, Saccharomyces cerevisiae, and the first cloned animal, Dolly the sheep, was born at the Roslin Institute, part of the University of Edinburgh in Scotland. In the year 2000, the full genome sequence of the model organism Drosophila melanogaster (fruit fly) was completed.

The mouse was the first mammal to have its full genome sequence completed in the year 2002. The project was carried out by the International Mouse Genome Sequencing Consortium. The mouse genome is 14 percent smaller than the human genome, but over 95 percent of the mouse genome is like ours. The chimpanzee genome was completed in 2005. The Neanderthal genome was published in *Nature* in 2006 (Green et al. 2006), showing that Neanderthal DNA is 99.7 percent identical to present-day human DNA and 98.8 percent identical to chimpanzee DNA. Present-day human DNA is also 98.8 percent identical to chimpanzee DNA. In August 2023, the first complete sequence of a human Y chromosome was finalized—the final human chromosome to be fully sequenced (Rhie et al. 2023).

Genome databases and informatics were growing fast due to computer and software development and the internet; databases could be easily accessed and used to mine information. Despite all the progress obtained by genomics, there were limitations. The molecular biology methods had restrictions when it came to the analysis of molecular functions of the products of the genes. It was thought that DNA was the blueprint that would allow us to understand most of the secrets of life. However, DNA is just "the highway needed for the cars to travel." The cars transport the individuals to their destination to play a function. The information in our DNA only encodes the amino acid sequence; it does not encode the final three-dimensional shape of the protein. The three-dimensional shape is the result of the energy landscape (see Chapter 3, 4), which is determined by physical forces (hydrophobic and electrostatic forces, binding energies, etc.) acting on sequences of amino acids and affected by the environment of the protein (pH, ionic strength, excipients, etc.).

Therefore, a large amount of information to get the final structure of a protein in solution is not contained in the DNA information. This information is contained in the physical laws governing thermodynamics, kinetics, charges, and in some degree of randomness for the amino acids chain to find the optimal shape to give rise to the final functional protein structure. Thus, the interest in proteins, the products of the genes, had a rebirth toward the end of the twentieth century, and proteomics was born.

Proteomics

Proteomics is the measurement of the spatial and temporal distribution of protein expression patterns and interactions. It is the complete evaluation of the function and structure of proteins to understand an organism's nature. Proteomics is a multistep technique in which every step should be very well controlled to avoid nonbiological factors interfering with protein expression and interaction. An investigation of large-expression patterns of proteins, proteomics requires methods that are high throughput, as well as accuracy and precision, using mass spectrometry (MS) and bioinformatics to analyze and assemble the MS data.

Two methods can be used in proteomics: top-down and bottom-up workflows. The bottom-up method is sometimes called peptide-based proteomics. Here, the protein is digested by trypsin and separated by a specific column, followed by an analysis of the peptides by MS (Pandey and Mann 2000).

The bottom-up approach can be classified into two groups according to the fractionation step. The first approach uses two-dimensional electrophoresis (2-DE) to isolate the proteins from the gel. Then the proteins are digested into peptides that MS can identify. The second approach is called "shotgun" proteomics. Here, the digestion of protein occurs without fractionation, and liquid chromatography is used to separate the peptides identified by MS (Ortea et al. 2016).

In top-down proteomics, whole proteins or polypeptides are immediately assessed by MS. The molecular mass of proteins is sometimes calculated by using electrospray ionization (ESI), followed by matrix-assisted laser desorption/ionization (MALDI) MS (Chait and Kent 1992). Top-down proteomics can identify proteins with a molecular mass of > 200 kDa (Han et al. 2006).

MS has evolved to serve a critical role in proteomics research. MS is a physical technique that characterizes molecules by measuring the masses of their ions. The technique allows the rapid identification of components of a complex mixture of proteins, sequencing of proteins and nucleic acids, and analysis of posttranslational modifications or substitutions relative to an expected sequence. MS can also measure the extent of hydrogen-deuterium exchange to reveal the solvent exposure of individual sites. This provides information about conformation and interactions.

In MS, protein molecules are ionized, and their mass is calculated according to mass-to-charge ratios. The mass spectrometer has three main components: an analyzer, an ion source, and a detector. The methods used for ionization are ESI (electrospray ionization using an electrospray in which a high voltage is applied to a liquid to create an aerosol) and MALDI (Krishnankutty et al. 2016). In MALDI, a chemical matrix is mixed with the peptides and dotted onto a metal multiwell microtiter plate to make a crystal lattice. The matrix chemicals pass the energy to the samples after absorbing it. Then peptide ions are detected by a mass analyzer. MALDI creates mostly singly charged ions that help to determine the m/z value (Smith et al. 2006).

In ESI, the power is activated in the protein sample to create charged droplets that increase gaseous ion production, which then are analyzed with a mass analyzer (Lahm and Langen 2000). The advantages of ESI are its high reproducibility and high elasticity to combine many categories of MS. Furthermore, ESI can be fixed to time-of-flight (TOF) MS, quadrupole ion traps, and Fourier transform ion cyclotron resonance. On the other hand, the disadvantages of ESI are that it cannot be applied for molecular imaging, it requires a large quantity of samples, and multiple peaks are produced due to the many charged ions that result in the complexity of MS/MS spectra (Chiou and Wu 2011).

Mass spectrometry is an essential tool that is used for profiling proteins in the cell, and the coupling of liquid chromatography procedures to mass spectrometry has become very common. It has been important to have a high throughput method of analysis, and ESI proved to be able to analyze peptides in a high throughput fashion when coupled to reverse-phase HPLC. Algorithms like MASCOT and SEQUEST were developed that allowed peptide detection to be linked to gene-based databases. Mass spectrometers measure mass better than any other technique and can give information about chemical structures to identify, verify, and quantitate metabolites, recombinant proteins, proteins isolated from natural sources, oligonucleotides, drug candidates, peptides, synthetic organic chemicals, and polymers.

Mass is defined by using carbon-12 (^{12}C) as a reference point. Then, one unit of mass is defined as a dalton (Da). One dalton is defined as one-twelfth the mass of a single carbon-12 atom. Thus, one ^{12}C atom has a mass of 12.0000 Da. We also must take in consideration the isotopes. Most elements have more than one stable isotope. For example, most carbon atoms have a mass of 12 Da, but in nature, 1.1 percent of C atoms have an extra neutron, making their mass 13 Da.

Table 2.0: Stable isotopes of the most abundant elements of peptides.

Element	Mass (Daltons)	Abundance (percent)
Hydrogen	1.0078 2.0141	99.985 0.015
Carbon	12.0000 13.0034	98.89 1.11
Nitrogen	14.0031 15.0001	99.64 0.36
Oxygen	15.9949 16.9991 17.9992	99.76 0.04 0.20

Adapted from Steve Carr Fundamentals of Biological Mass Spectrometry and Proteomics. Broad Institute of MIT and Harvard. https://www.broadinstitute.org/files/shared/proteomics/Fundamentals_of_Biological_MS_and_Proteomics_Carr_5_15.pdf

There are many challenges in proteomics. The major challenge is the broad change in protein expression with the environment and cell type (Bichsel et al. 2001). Also, there is no comparable proteomics method, unlike genomics, which uses polymerase chain reaction (Blackstock and Weir 1999). Moreover, protein activities are highly regulated post-transnationally, which adds difficulty in proteomics (Srinivas et al. 2001). Proteomics provides a better understanding of the structure and function of the organism than genomics. This led to significant advances, with strong applications in the biopharmaceutical industry, where experts in protein characterization, enzymologists, and protein chemists were in high demand to develop protein-based drug products: in particular, monoclonal antibodies as targeted therapeutic agents.

CHAPTER 3

PRINCIPLES OF PROTEIN STRUCTURE

The skeleton of proteins is built from amino acids. Amino acids are organic acids where one carbon has been substituted by an amine group. Thus, they have amine (-NH_2) and carboxyl (-COOH) groups together with a side chain specific to each amino acid. There are twenty common amino acids that are the constituent of proteins in all cells. The amino acids are synthesized by living organisms (nonessential amino acids) from simple compounds or are adsorbed as nutrients, the so-called essential amino acids. Mammals can only synthesize about half of the required amino acids. The amino acids constitute the building blocks of proteins.

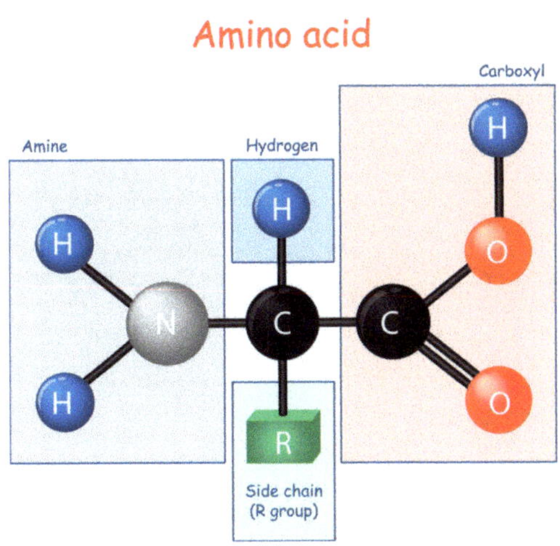

Figure 3.0. Amino acid structure.

Source: Alamy, ID: 2DJ64WW, Tetiana Zhabska / Alamy Stock Vector. order reference: OY72206947, August 27, 2022.

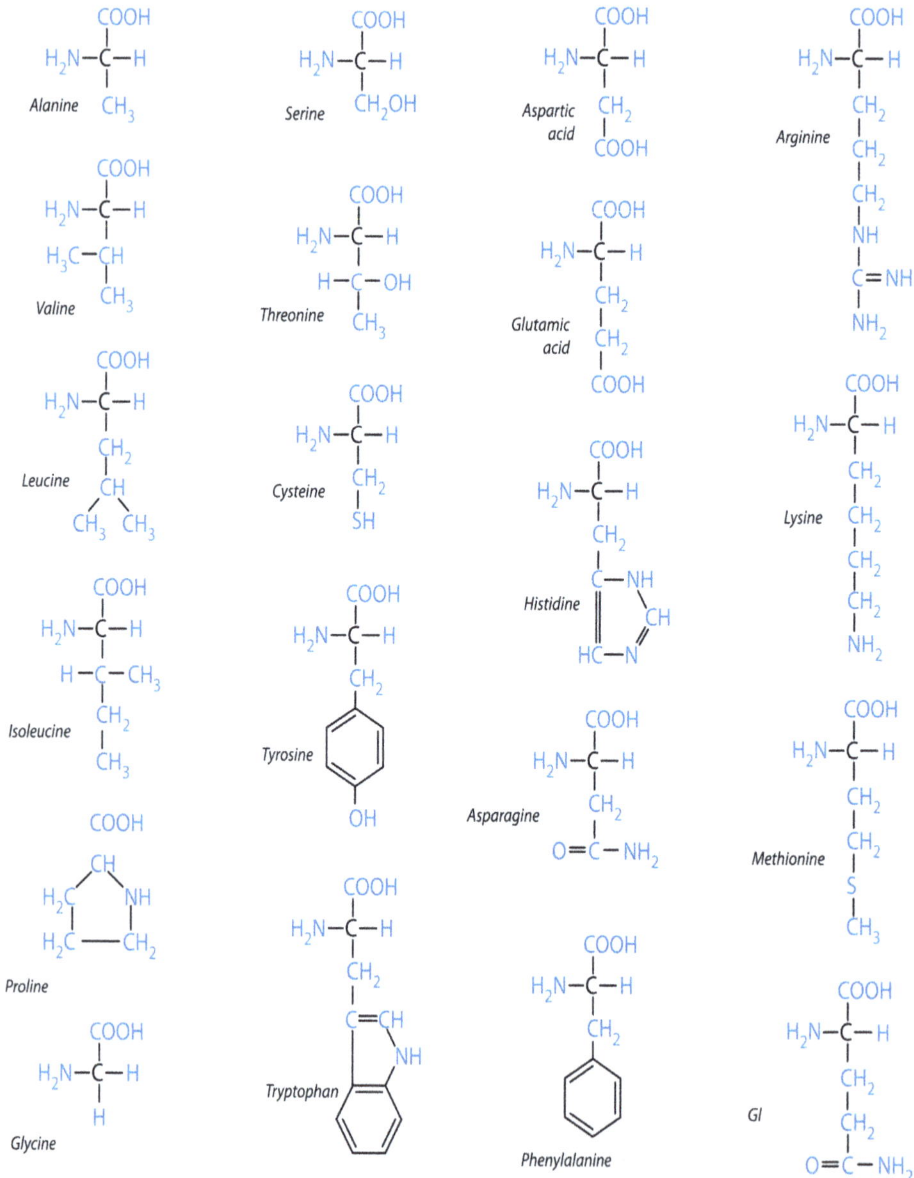

Figure 3.1. The twenty common amino acids that make proteins.

The chemical structures of the twenty amino acids that make up proteins.

Credit: Technology Networks. https://www.technologynetworks.com/applied-sciences/articles/essential-amino-acids-chart-abbreviations-and-structure-324357

The skeleton of proteins is built head to tail from amino acids. Thus, the carboxyl group of one amino acid unit forms a covalent bond with the amine group of the following unit. The bond is formed by eliminating one molecule of water and forming the peptide bond CO-NH-, creating the polypeptide chain defined as the *primary structure* of the protein (figure 3.2). The peptide bond formed between the carboxyl and amino groups of two amino acids is a bond with little intrinsic mobility because of the partial double bond character. The peptide bond length is shorter than expected for a single C-N bond. On average, a peptide bond length is 1.32 Å compared to 1.45 Å for an ordinary C-N bond. The average bond length associated with a C=N double bond is 1.25 Å.

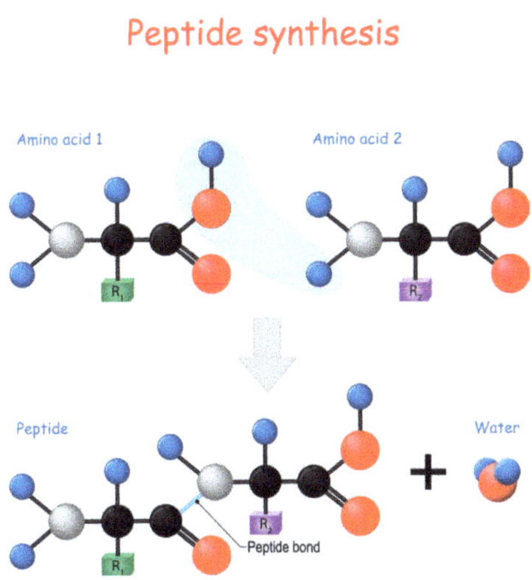

Figure 3.2. Peptide synthesis. Two amino acids are combined into a peptide to form a water molecule and a peptide bond.

Source: Alamy Stock Vector. Image ID: 2DJ64WF. order reference: OY72287324. Tetiana Zhabska

The peptide bond has a certain rigidity affecting the three-dimensional structure of the protein, as was predicted by Pauling in 1951 (see Chapter 2). The lateral chains of the amino acids can adopt different orientations (by interacting with the solvent) or other lateral changes affecting the three-dimensional structure of the protein. If there is a polar lateral chain, it will mostly interact with a polar environment, namely water. If the lateral changes are hydrophobic, they will exclude the water and will tend to be in the interior of the protein. One exception to this rule is the proteins inserted in the nonpolar lipid cell membranes. The exterior of the protein is composed of hydrophobic or nonpolar residues that establish interactions with the lipidic chains and hydrophobic interactions.

Protein Charges

At pH 7, the amino and carboxyl groups are charged, but over a pH range of 1 to 14, these groups show a series of equilibria involving binding and dissociation of a proton, a manifestation of these groups as weak acid or bases. The amino group characterized by a basic pKa value of approximately nine is a weak base. While the amino group ionizes around pH 9.0, the carboxyl group remains charged until a pH ~2.0 is reached. At this pH, a proton binds, neutralizing the charge of the carboxyl group (see Appendix for pK of amino acids). The acid-base behavior of amino acids is important since it influences the properties of proteins.

Georg Bredig (1868–1944) was a German scientist known as the father of catalytic chemistry. Bredig introduced the idea of amino acids as Zwitterion, a dipolar ion with at least one positive and one negative functional group and a zero net charge (Tanford and Reynolds 2001). The electric charge of the proteins is given by the different lateral chains of the amino acids (see amino acids in Figure 3.1).

William Hardy (1864–1934) was the first to demonstrate that proteins have an electric charge. He studied the colloidal properties of proteins, beginning with his discovery (in 1899) of the amphoteric behavior of albumin particles in an electric field. Hardy concluded that the mobility of proteins could change as a function of the degree of acidity of the medium. He coined the term *isoelectric point* (pI) for the proteins that did not move depending on the solution's acidity.

It is important to indicate that there are several methods for *in-silico* determination of the pI. However, the preferred approach for measuring the pI of a protein is experimental, since the *in-silico* determinations usually do not account for posttranslational modifications and could give errors up to four pH units.

The degree of acidity of solutions was difficult to interpret until the Danish chemist Soren Sørensen (1868–1939) developed the pH scale in 1909, which is a measure of the concentration of protons, namely H^+, or strictly speaking, the concentration of hydronium ions (H^+_3O) since in an aqueous solution the protons are hydrated. Sørensen developed the pH scale while running experiments with beer at the world-renowned Carlsberg research laboratory in Copenhagen. He was one of the first to measure the charge of proteins at different pH values, demonstrating that proteins could act as acids or bases and have multiple charges, a property common to amino acids. The belief was that proteins were spherical and had one major charge, but not multiple charges of different signs. Scientists began to wonder how the structure of proteins was obtained and what type of forces were keeping them together.

pH Is the Fundamental Variable in Protein Structure Function

Before we continue our journey of understanding protein's solution behavior, we shall review some important concepts. An understanding of acid-base chemistry is essential to appreciate the properties of proteins and biological systems in general. Many metabolites in living organisms are acids and bases, and thus have the potential to ionize. The electrical charge in proteins affects their solution behavior. Hence, one definition of acid and bases useful in biochemistry is the Brønsted concept of conjugate-acid-conjugate base pairs, where an acid is defined as a substance that donates protons (hydrogen ions). A base is a substance that accepts protons. When a Brønsted acid loses a proton, a Brønsted base is produced. The original acid and resulting base are referred to as a conjugate acid-conjugate base pair. In every ionization of an acid or base, two conjugate acid-conjugate base pairs are involved.

$$\underset{[\text{conjugate acid}]_1}{HA} + \underset{[\text{conjugate base}]_2}{B^-} \rightleftharpoons \underset{[\text{conjugate base}]_1}{A^-} + \underset{[\text{conjugate acid}]_2}{HB} \qquad (3.0)$$

A strong acid is a substance that ionizes almost 100 percent in aqueous solution. For example, HCL in solution is essentially 100 percent ionized to $H^+{}_30$ and Cl^-.

$$HCL + H_2O \longrightarrow H^+{}_3O + Cl^- \quad (3.1)$$

H_3O^+ (the hydronium ion, or conjugate acid of water) is the actual form of the hydrogen ion (proton) in solution. For all practical purposes, $H^+{}_30$ and H^+ mean the same thing.

The ionization of water can be considered in two ways: simple dissociation to yield H^+ and OH^- ions in terms of the Brønsted conjugate-acid-conjugate base pairs. Water is amphoteric: it yields both H^+ and OH^-, and it can donate and accept protons. The ionization of water can be described by a "dissociation" constant, Kd, an "ionization" constant, Ki, and a specific constant for water, Kw.

Simple dissociation *Conjugated acid-Conjugate base*

$$HOH \rightleftarrows H^+ + OH^- \qquad HOH + HOH \rightleftarrows H^+{}_3O + OH^- \quad (3.2)$$

$$Kd = \frac{[H^+][OH^-]}{[HOH]} \ (eq.\ 3.3) \qquad Ki = \frac{[H^+{}_30]\,[OH]}{[HOH]^2} \ (eq.3.4)$$

Note that water produces two conjugate acid-conjugate base pairs:

$$HOH/OH^- \text{ and } H^+{}_30/HOH \quad (3.5)$$

For every mole of H^+, 1 mole of OH- is produced. In pure water, $[H^+] = 10^{-7}$. Therefore, the concentration of $[OH^-]$ is 10^{-7}. The molarity (*M*) of water can be calculated as follows:

$$M = \frac{\text{Number of moles of } H_20}{\text{Liter}} \quad (3.6)$$

Number of moles = wt_g/MW

A liter of water weighs 1000 g. The MW of H_2O is 18.

$$M = \frac{1000/18}{1\ liter} = 55.6 \tag{3.7}$$

The molarity of water is 55.6 M (original concentration) minus 10^{-7} M (the amount that ionized). The amount is so close to 55.6 M that we can neglect the 10^{-7} M. We can now substitute the above values into the Kd and Ki expressions (eq, 3.3 and 3.4)

$$Kd = \frac{(10^{-7})(10^{-7})}{55.6} = \frac{10^{-14}}{55.6} = 1.8 \times 10^{-16} \tag{3.8}$$

$$Ki = \frac{(10^{-7})(10^{-7})}{(55.6)^2} = \frac{10^{-14}}{3.09 \times 103} = 3.24 \times 10^{-18} \tag{3.9}$$

We can define a new constant for the dissociation or ionization of water, Kw, which combines the two constants (Kd and $[H_2O]$ or Ki and $[H_2O]^2$).

$$\begin{aligned} Kw &= Kd \times [H_2O] \\ Kw &= (1.8 \times 10^{-16})(55.6) \\ Kw &= 1 \times 10^{-14} = [H^+][OH^-] \end{aligned} \tag{3.10}$$

If we examine *Kw* as a function of Ki:

$$\begin{aligned} Kw &= Ki \times [H_2O]^2 \\ Kw &= (3.24 \times 10^{-18})(55.6)^2 \\ Kw &= (3.24 \times 10^{-18}(3.09 \times 10^3) \\ Kw &= 1 \times 10^{-14} = [H^+_3O][OH^-] \end{aligned} \tag{3.11}$$

In all aqueous solutions, the equilibrium for the ionization of water must be satisfied:

$$[H^+][OH^-] = Kw = 10^{-14} \tag{3.12}$$

Taking logarithms:

$$\log[H^+] + \log[OH^-] = \log Kw$$

$$-\log[H^+] - \log[OH^-] = -\log Kw \tag{3.13}$$

$$-\mathrm{Log}[H^+] = pH; \quad -\log[OH^-] = pOH; \quad -\log Kw = pKw \tag{3.14}$$

We have then that $pH + pOH = 14$ (3.15)

pH is the abbreviated form of designating the hydrogen ion activity of a solution. We have defined pH as the negative logarithm of the hydrogen ion activity. Similarly, pOH is the negative of the hydroxyl ion activity. It is acceptable to use concentration for diluted solutions, but when dealing with more concentrated solutions, the difference in the observed concentration and the calculated concentration in equilibrium increases.

$$pH = -\log a_H+ = \log 1/a_H+ \tag{3.16}$$

$$pOH = -\log a_{OH}- = \log 1/a_{OH}- \tag{3.17}$$

The H^+ concentration is describing properties of water solutions affecting the solution behavior of proteins.

Now, let us look at the ionization of weak acids. In an aqueous solution, a weak acid ionizes to a limited extent as follows:

$$\underset{\text{[Conjugate acid]1}}{HA} + \underset{\text{[Conjugate base]2}}{H_2O} \rightleftharpoons \underset{\text{[Conjugate acid]2}}{H^+_3O} + \underset{\text{[Conjugate base]1}}{A^-} \tag{3.18}$$

The reversible ionization reaction can be described by an equilibrium constant, $K_{i.}$

$$Ki = \frac{[H_3O][A^-]}{[HA][H_2O]} \tag{3.19}$$

Because $[H_20]$ is itself a constant, we can define a new constant, K_a, which combines K_i and $[H_2O]$.

$$K_i\,[H_2O] = K_a = \frac{[H_3O][A^-]}{[HA]} \tag{3.20}$$

Because $[H^+{}_3O]$ is the same as the hydrogen ion concentration $[H^+]$, then Ka is usually written:

$$K_a = \frac{[H^+][A^-]}{[HA]} \tag{3.21}$$

And:

$$HA \rightleftharpoons H^+ + A^-$$

Solving for H^+ in (3.21):

$$[H^+] = Ka\frac{[HA]}{[A\ ]} \tag{3.22}$$

$$log[H^+] = log\ Ka + log\ \frac{[HA]}{[A\ ]} \tag{3.23}$$

Multiplying both sides by -1:

$$-log\,[H^+] = -log\,Ka - log\frac{[HA]}{[A^-]} \qquad (3.24)$$

$$pH = pKa - log\frac{[HA]}{[A^-]}$$

$$\boxed{pH = pKa + log\frac{[A^-]}{[HA]} \text{ (Henderson-Hasselbalch equation)}} \qquad (3.25)$$

The Henderson-Hasselbalch equation is used to calculate the amount of acid and conjugate base to be combined for the preparation of a buffer solution with a particular pH. In chemical terms, a buffer is a substance or mixture of substances that permits solutions to resist large changes in pH upon adding small amounts of H^+ or OH^- ions. A buffer helps maintain a near-constant pH. In general, the ability of a buffer to maintain a near-constant pH depends on:

i) The concentration of the buffer. It is not always possible to use a high-concentration buffer.
ii) The pH of concentrated stock solutions may change on dilution.
iii) The pKa value of a buffer should be influenced as little as possible by the buffer concentration, the temperature, and the ion composition of the medium. Choosing the right buffer is based on the pKa (see figure 3.3).
iv) The impact of temperature on ionization. Tris is an example of a buffer for its large dissociation constant shift with temperature, expressed as a change in pKa per degree centigrade.
v) Complex formation. When a buffer forms complexes with metal ions, protons are released, which causes the pH value to decrease. The buffer should therefore be inert.
vi) Around three pH units' variation in phosphate buffer under cold conditions.
vii) Buffers should be as easy as possible to manufacture and purify. Purity is extremely important since contamination (e.g., heavy metals) can easily interfere with sensitive biological systems.

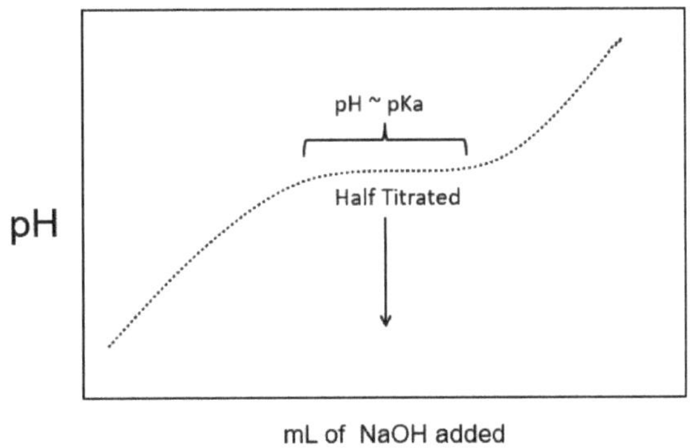

Figure 3.3. Buffer behavior. Maximal buffering capacity is when pH = pKa.

If a strong acid is added to the equilibrium mixture of a conjugate acid-base:

$$HA \rightleftarrows H^{+} + A^{-}$$

The equilibrium is shifted to the left in accordance with Le Chatelier's principle. Consequently, the amount of hydrogen ions increases less than expected.

Most biological happenings occur in the pH range of 5.2 to 7.5. Although the pH is a critical variable, the different buffer species could have an impact on the protein of interest at the same pH. There is a tendency to believe that 10 mM sodium phosphate, 150 mM NaCl buffer at pH 7.4 (PBS) is the perfect buffer for storing a protein given that it has a "physiological" pH and an osmolality (300 mOs). However, the sodium salt of the buffer tends to precipitate in the frozen state, with the consequence of a decrease in pH of 4 units.

When protein conformation is protonation dependent—in other words, the enthalpy of denaturation (see Chapter 4 for thermodynamic terms) or the association constant (for binding between protein and ligand) varies with pH—the observed binding enthalpy or denaturation varies with the kind of buffer employed (Bradshaw and Waksman 1998). This variation is given by the fact that the experimentally measured enthalpy (ΔH_{obs}) at a given pH is determined by: i) the ionization enthalpy of the buffer used (ΔH_{ion}) and ii) the enthalpy of binding or denaturation of the protein corrected for buffer effects (ΔH_{buffer}).

$$\Delta H_{obs} = \Delta H_{buffer} + (n)\, \Delta H_{ion} \qquad (3.26)$$

n is the number of protons released (positive sign) or taken up (negative sign) by the buffer during the binding or denaturation.

We do not know how many experiments have failed due to the wrong preparation of the buffer or its instability. Also, we do not know exactly the physiological environment of a protein in vivo.

Measuring pH

The pH meter is the typical instrument found in a biochemical laboratory and one of the most mistreated instruments. There are two parts to a pH meter: the electrode and the voltmeter. The glass electrode is connected to a voltmeter (or the voltmeter could be integrated with the electrode) to measure the voltage and convert the voltage output into pH units. The glass electrode is a thin and fragile glass bulb that contains a dilute acid solution and a connecting wire. When the glass electrode is immersed in the solution to be tested, protons from the solution stick to the surface of the glass electrode and generate a voltage across the glass membrane. The voltage generated is measured in reference to the second electrode, the Ag/AgCl electrode or the calomel electrode also placed in the solution to be tested.

The second electrode is usually combined with the glass electrode surrounded by a KCL solution. Most glass electrodes have a top-up hole where, periodically, saturated KCL should be added (Figure 3.4). If the electrode shows slow response or no response to different pH solutions, it should be replaced. Also, if the pH is drifting when measuring the standards or the solution, it is a signal that the electrode needs to be replaced.

Any substance that can be adsorbed by the H^+ permeable glass of the glass electrode can affect the pH reading by affecting the permeability to H^+ ions. Protein solutions can form a film on the glass electrode, affecting the pH reading. Treatment with detergents or acids can remove such a film. Solutions containing Na^+ could give anomalous pH values since glass electrodes can be permeable to Na^+ ions. Thus, the pH could decrease if the Na^+ is increased. Special Na^+ impermeable electrodes are commercially available.

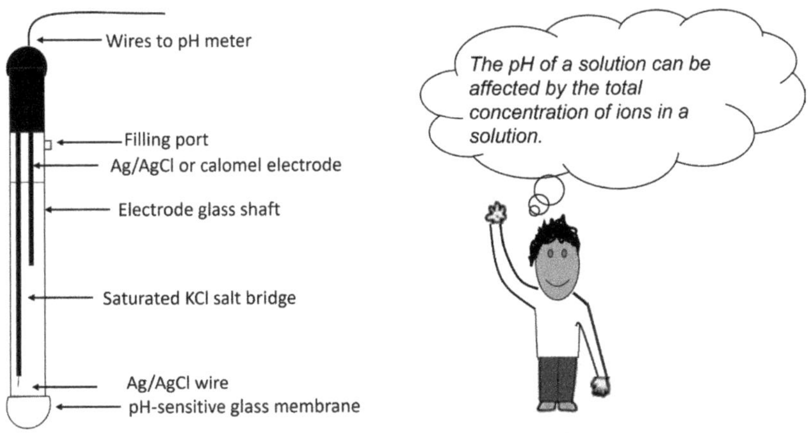

Figure 3.4. Glass electrode in a pH meter.

Protein Conformation

Depending on the sequence of amino acids (primary structure), the polypeptide chain can fold, forming a helix as was demonstrated by Linus Pauling and colleagues in a breakthrough paper published in the *Proceedings of the National Academy of Science* in 1951 (Pauling et al.1951; Pauling and Corey 1951). The α-helix has 3.6 amino acid residues per turn and is stabilized by hydrogen bonds (figure 3.5), forming a right-handed helix (figure 3.6).

Another configuration adopted by proteins is the β-sheet structure. The β-sheet structure is formed by stretches of amino acids known as β-strands that are embedded within a single polypeptide chain. These are side-by-side polypeptide chains that could be parallel (N to C) or antiparallel (C to N). The strands in the β-sheets assume an extended zigzag conformation in which carbonyls and amides of successive residues point in opposite directions (figure 3.7). Here the hydrogen bonds form between the backbone atoms of separate β-strands.

The α-helix and β-structure are the most predominant *secondary* structures in proteins. The secondary structure results from hydrogen bonding between peptide bonds. The hydrogen in a peptide bond has more charge than the hydrogen and oxygen in a water molecule. The consequence is that the interactions between two peptide bonds are stronger than the interaction of a peptide bond with water. In α-helical regions of secondary structure, the polypeptide chain is organized in a right-handed corkscrew with all the carbonyl oxygen atoms along the helix pointing toward the C-terminus, and all the amide hydrogens pointing toward the N-terminus.

Measuring from the carbonyl carbon at one end of the turn of the helix (360°) is 3.6 residues in length. The rise of the α-helix per amino acid is 1.5 Å. Thus, a thirty amino acid–long helix would be approximately 45 Å. Proteins could be mixtures of secondary structures, namely α-helix and β-sheets, or could contain only one type of secondary structure. Secondary structure types can be connected by loops formed mainly by proline and glycine.

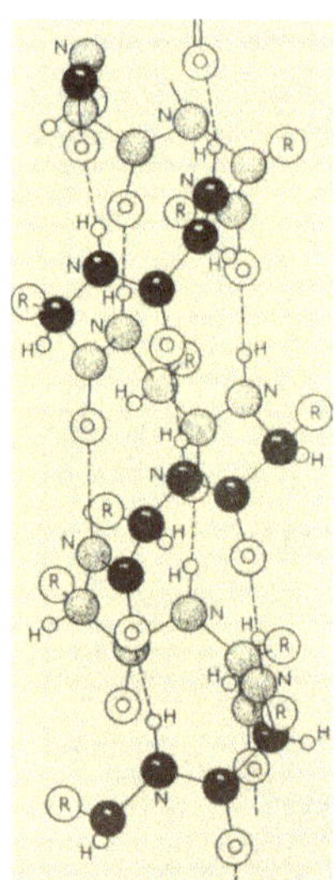

Figure 3.5. Right-handed helix. Left-handed α-helices are also possible. Pauling and Corey demonstrated that a helix with 3.7 amino acid residues per turn, a diameter of 6.8 A, and turns spaced 5.4 A apart would be permitted by the observed bond angles.

Source: Library book collection /Alamy Stock photo.

License reference OY72342866, August, 29, 2022. Image ID RHJYBA.

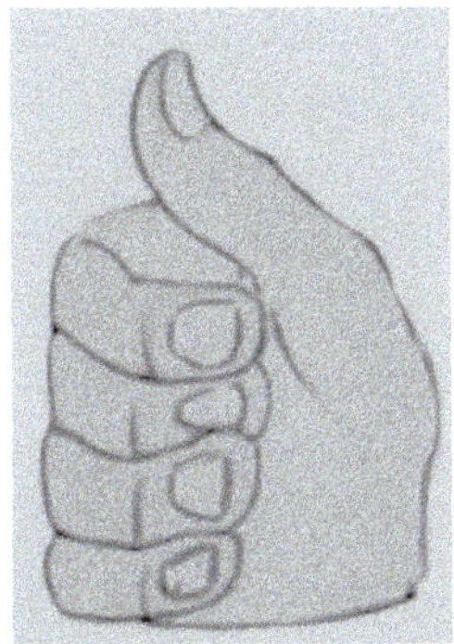

Figure 3.6. In a right-handed coordinate system, when the thumb of the right hand points up, the curled fingers point from the left to right, so it is a right-handed helix. A left-handed helix bends in the opposite direction—it is the mirror image of a right-handed helix, so a helix is chiral.

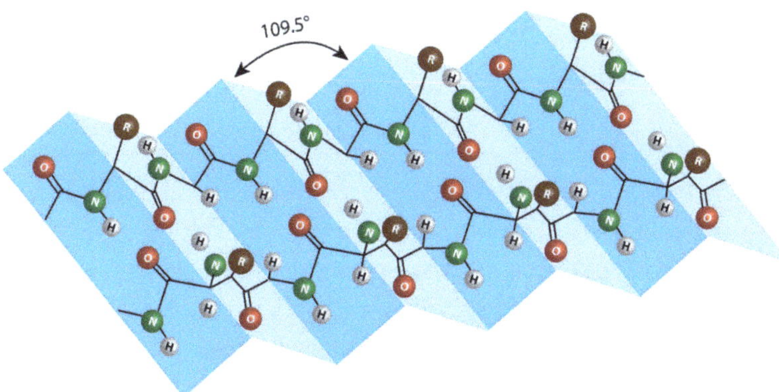

Figure 3.7. β-structure model of a protein.

Source: elementix / Alamy Stock Photo. Image ID: GNE34A, Order reference: OY73667666.

The three-dimensional structure becomes more complex as the secondary structure folds into compact structures. Thus, the spatial arrangement of amino acid residues far apart in the sequence comes in close contact. These superstructures result from interactions between the variable (R) groups of the polypeptide chain and constitute the so-called *tertiary structure* of the protein. The tertiary structure of proteins could be divided into structural and functional units. Thus, certain combinations of secondary structures or *motifs* are common to many proteins with a common function. Intra and intermolecular interactions help to stabilize the tertiary structures and quaternary structure.

Several proteins have an additional level of structure beyond the *tertiary structure* formed by several polypeptide chains joined by weak interactions (ionic bonds, hydrogen bonds, and van der Waals interactions, which we will review later in the chapter); in other cases, those polypeptide chains are joined by covalent bonds such as disulfide bridges between cysteine residues. Each polypeptide chain in such a protein is called a subunit. The rearrangement of the subunits is referred to as a *quaternary structure.* Hemoglobin, the oxygen-carrying protein in the blood, consists of two subunits of one type (named α) and two subunits of another type (designated β). The hydrophobic effect is the principal force stabilizing tertiary and quaternary structure.

Measuring the Secondary and Tertiary Structure of Proteins by Circular Dichroism

A property of many biological molecules is molecular asymmetry or chirality. That is, molecules that are not identical to their mirror image. This arises from the presence of asymmetric carbons. A carbon that is tetrahedrally bonded to four different atoms or atomic groups can exist in two different structures that are mirror images of each other, as shown in figure 3.8. These molecules are generally named as left-handed and right-handed forms (review organic chemistry textbooks). Molecules that have asymmetry in their structure also exhibit chirality in their optical properties. Optical activity was first observed in quartz crystals in 1811 by the French physicist François Arago. Another French physicist, Jean-Baptiste Biot, found in 1815 that liquid solutions of tartaric acid or sugar are optically active, as are liquid or vaporous turpentine (Britannica 2018).

1. *Optical rotation*: It depends on the difference in the refractive indices of a substance for left or right circularly polarized light. Polarized light is light that has one direction of propagation (figure 3.9), unlike natural light, which propagates in different directions. If plane-polarized light propagates through a trans-

parent chiral sample, the emerging light is also plane-polarized, but its plane of polarization (even including the direction of propagation) will have rotated by an angle ϕ with respect to the direction of the polarization of the incident ray. If the rotation is clockwise, as seen from the observer, ϕ is positive. In counterclockwise rotation, the angle ϕ is assigned a negative value. The electrical vectors of the two components of circularly polarized light propagate through a chiral sample with different speeds. Consequently, one is ahead of phase with respect to the other. Hence, a substance that is optically active transmits right circularly polarized light with a different velocity from left circularly polarized light. Thus, optical rotatory dispersion (ORD) is the wavelength dependence of the optical rotation.

2. *Circular dichroism* (CD) spectroscopy: A form of light absorption spectroscopy that measures the difference in absorbance of right and left circularly polarized light. Circular dichroism causes incident linearly polarized light to become elliptically polarized. Optical rotatory dispersion and CD are related, as are ordinary absorption and dispersion. In many biological molecules, asymmetry is caused by the asymmetric environment, and CD provides information on these structures. The CD bands' position and intensity depend on their conformation. Above 250 nm, only the aromatic residues and cysteine absorb light. This is the part of the spectrum that gives information on the tertiary structure.

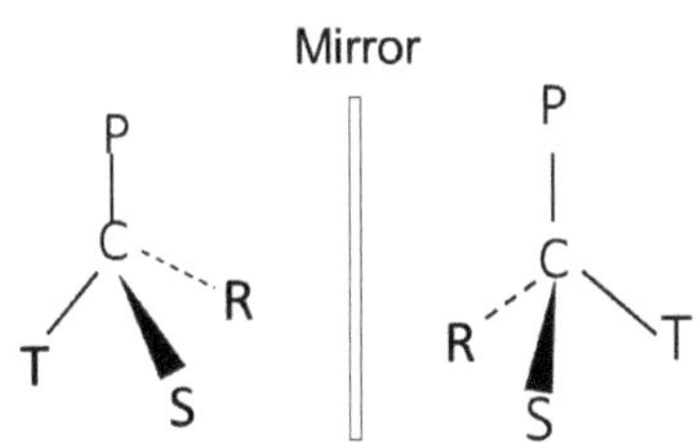

Figure 3.8. Arrangement of groups around a tetrahedral carbon atom in asymmetric molecules.

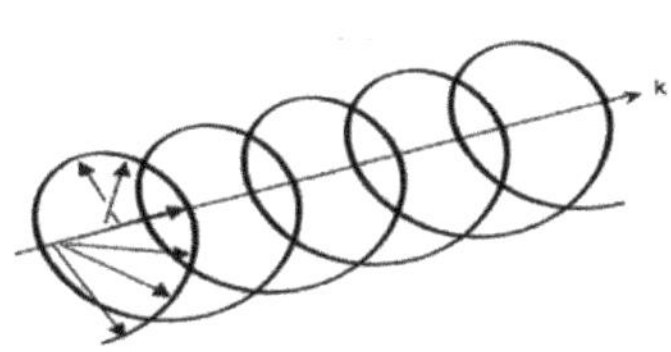

Figure 3.9. Schematic representation of circularly polarized light.

It has been shown that CD spectra between 260 and approximately 180 nm can be analyzed for the different secondary structural types (figure 3.10): alpha helix, parallel and antiparallel β-sheets, turn, and others. For example, α-helices display large CD bands with negative ellipticity at 222 and 208 nm, and positive ellipticity at 193 nm; β-sheets exhibit a broad negative band near 218 nm and a large positive band near 195 nm, while disordered extended chains have a weak broad positive CD band near 217 nm and a large negative band near 200 nm.

The chromophores of the aromatic amino acid in proteins are often in very dissymmetric environments, resulting in distinctive CD spectra in the near UV (250–300 nm), which can serve as useful probes of protein tertiary structure (figure 3.11). CD is an excellent spectroscopic technique for following the unfolding and folding of proteins as a function of temperature.

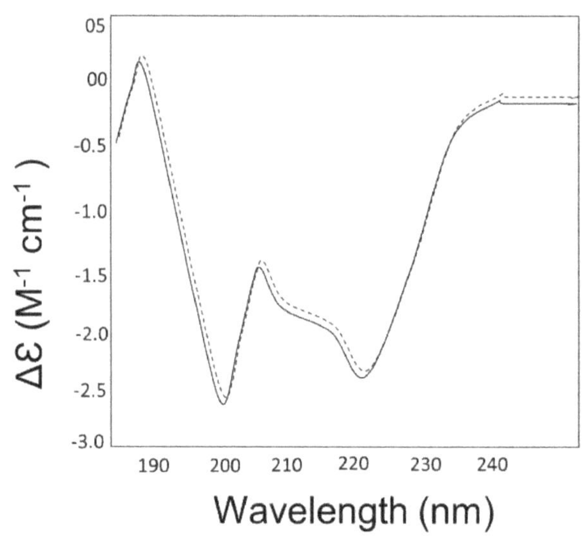

Figure 3.10. Far-UV CD spectra for unfractionated rhBMP-2 (—) and one of its homodimers, T66/T66 (------). The analysis demonstrates that the unfractionated BMP-2 and the isolated isoforms exhibit comparable secondary structure.

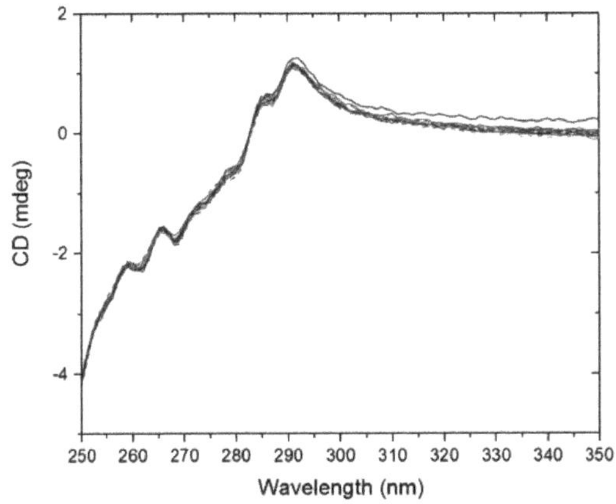

Figure 3.11. UV-near circular dichroism spectroscopy of a monoclonal antibody, showing a similar spectroscopic fingerprint for the mab in six different buffers.

How is the Complex Three-Dimensional Structure of a Protein Attained?

The simplest answer is that the sequence of amino acids in a protein determines its three-dimensional structure. The first evidence of the role of the amino acid sequence in dictating the three-dimensional structure of proteins was obtained by Christian Anfinsen in 1950 using the enzyme ribonuclease. He showed that ribonuclease could be refolded after denaturation (loss of three-dimensional structure) while preserving enzyme activity, thereby suggesting that all the information required by proteins to adopt its final conformation is encoded in its amino-acid sequence (Anfinsen 1973). In other words, the native structure is the thermodynamically most stable structure.

The problem is for large proteins, where it is difficult to predict the three-dimensional structure from the amino acid sequence. Ribonuclease is a relatively small protein made of a single polypeptide chain of one hundred and twenty-four amino acid residues; of the twenty natural amino acids, ribonuclease has nineteen of them, excluding tryptophan. If we assume that amino acids in a protein have a molecular mass of 110 Da, then ribonuclease should have a molecular mass of 13,700 Da (13.7KDa). This is a relatively small protein compared with most; for example, IgG has a molecular mass that is more than ten times larger, namely 150 KDa. Proteins are called *macromolecules* due to their large size. This contrasts with *molecules* like water, which have a size of 18 Da.

Proteins that are larger than 50 KDa form domains that are independently folded. Some proteins are elongated and stabilized by coiled-coil structures of triple helix, like collagen, found in connective tissues. The interior of protein subunits and domains consists of closely packed atoms with almost no water molecules. A consequence of the close-packed interior of proteins is that all proteins have approximately the same density, about 1.37 g/cm^3. The term commonly used is the partial specific volume ($\bar{v}_2$), which is the reciprocal of the density as a practical approximation; $\bar{v}_2$ varies from 0.70 to 0.76 cm^3 /g for different proteins, with an average value of $\bar{v}_2$ = 0.73 cm^3 /g (Murphy et al. 1998).

The amino acids found in the interior of proteins are, in general, hydrophobic. Hence, by examining a protein sequence, one can predict, in general, that hydrophobic amino acids will be buried in the protein's interior. The "hydrophobic effect" is one of the drivers leading protein to fold and adopt a given conformation, as was predicted by Charles Tanford (see Tanford 1980). The principle was first explicitly introduced to protein chemists in 1938 by Irwing Langmuir (Langmuir 1938a, b), an outsider to the protein world, working in surface chemistry.

The first person with a primary interest in proteins to become an enthusiast of the hydrophobic factor was the Irish crystallographer John D. Bernal (1901–1971). Bernal was a major driving force in the genesis and early development of biomolecular crystallography. The interest of Bernal in hydrophobicity was a sideshow regarding the dispute between Bernal and Langmuir on a bizarre chemical theory named the "cyclol hypothesis." The emphasis was on "cyclol," and hydrophobicity was inserted into the debate as a weapon.

The "cyclol" theory had been proposed by the British mathematician Dorothy Wrinch (1894–1976). The controversial idea of cyclol protein structure theory was based on ideas of mathematical symmetry. Although the theory was attractive for the mathematical symmetry that it possessed and its ability to explain many properties of proteins, it came under attack almost immediately by British X-ray crystallographers, who asserted that the theory did not conform to their experimental data. Wrinch suggested that polypeptide chains collapsed into a globular structure because their structure was not linear (as had already been demonstrated at that time), but cyclical.

Although Wrinch's theory was incorrect, many years after her death, peptides were found in fungi that adopted the structure proposed by Wrinch (Wang et al. 2018). Nature always gives us surprises.

The study of protein conformation in solution continued, and Bernal proposed in 1939 that the hydrophobic factor was related to the incapacity of the hydrophobic groups of forming hydrogen bonds and therefore their repulsion from the aqueous media. The role of hydrophobicity in protein conformation was mentioned in a seminal review by Walter Kauzmann (Kauzmann 1959), where he does not provide references, but the idea of the hydrophobic factor appeared to have been floating around in the community of protein chemists, and he summarized it in his review.

Walter Kauzmann is considered one of the two great thinkers about protein science in the twentieth century, together with Linus Pauling. Kauzmann hypothesized the overall organization of complex protein structures, based on what he called "the hydrophobic bond." When the first protein structures were later revealed (in 1965), the discoverers called their findings "a remarkable vindication of the predictions of Kauzmann in 1959." The word hydrophobicity has been employed as a chemical term to denote the antipathy between water and many kinds of nonpolar chemical entities. Charles Tanford stated that "the word hydrophobic is really a misnomer. It implies that the dissolved substances dislike water, whereas, in fact, it is water that dislikes the dissolved substance" (Tanford 1961).

We know that proteins are formed by amino acids held together that ought to be a long and flexible chain, where a globular protein conformation is not due mainly to the hydrophobic effect but also to other forces that stabilize the structure in an aqueous solution, such as hydrogen bonds (figure 3.12).

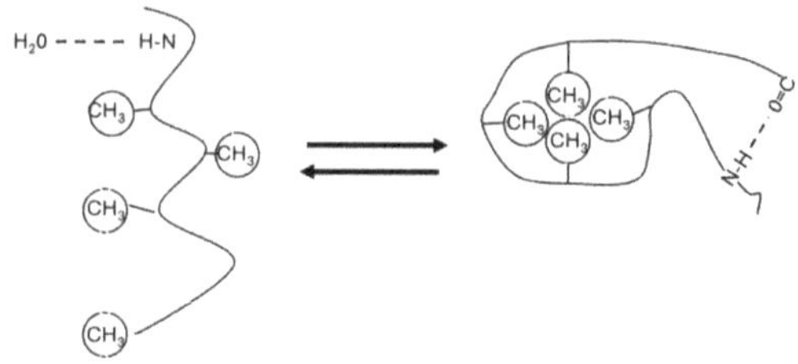

Figure 3.12. Schematic representation of protein folding with the contribution of the hydrophobic effect and hydrogen bonding.

How Big Is a Protein?

In general, we can say that a protein is approximately one millionth of a millimeter. For practical purposes, if we approximate a protein as a sphere, we can calculate the minimum radius of the protein by several hydrodynamic techniques (La Verde et al. 2017; Parker and Lollar 2021). The hydrodynamic value that can be experimentally derived is the Stokes radius (Rs), which is the radius of a sphere with the same hydrodynamic properties (i.e., frictional coefficient) as the biomolecule. Hence a protein of 10 KDa will have a minimum radius of 1.4 nm, and a protein of 250 KDa will have a radius of 5.2 nm. Since proteins are not true spheres, the apparent radius could be larger than the approximation to a sphere. A gel filtration column can also be used to determine the Rs of a sample protein by comparison with the Rs of standard calibration proteins (Le Maire et al. 1986, La Verde et al. 2017).

Table 3.0. Some protein standards for calibrating a gel filtration column.

Protein	**Molecular Weight (KDa)**	**Stokes Radius (Rs) mm**
Thyroglobulin	669	8.6
Catalase	250	5.2
IgG1	150	4.3
Ovalbumin	43	2.8
Albumin	66	3.5
Cytochrome c	12	1.7

How Far Apart Are Proteins in a Solution?

The cytosol is a very crowded and viscous place, with large proteins, sugars, ions, RNA, and water molecules circulating and bumping among each other in a "molecular storm." Hence, the average distance separating molecules in solution is small. In a 1 molar solution, there are 6×10^{23} molecules/liter = 0.6 molecules/nm^3, or, inverting, the volume per molecule is V = 1.66 nm^3/molecule at 1 M. For a concentration C, the volume per molecule is V = 1.66/C. We can take the cube root of the volume per molecule as an indication of the average separation.

$$d\,(\text{nm}) = V^{1/3} = 1.184 / C^{1/3} \qquad (3.27)$$

C is in molar, and d is in nanometers.

Based on the equation above, we can calculate that in a solution that is diluted to 1 mM, the approximated distance (d) between molecules is around ~11.8 nm. In a more concentrated solution (1 M), or a thermodynamically non-ideal solution, where molecule A may know of the presence of molecule B, the distance (d) between molecules may be just 1.18 nm. Protein sizes are around 4 to 100 nm. In a crowded and viscous cytosol, this could be like a full theater, with no room to move around.

The space between proteins is filled with water molecules, ions, lipids, and other small molecules. The water molecules can move around and get closer to the proteins up to a certain distance. Thus, every large molecule (proteins) is surrounded by an exclusion zone (a personal space), namely the depletion zone (Zhou and Hall 1996; Minton 2006; Miklos et al. 2010), which cannot be penetrated by small molecules, limiting the space available for the small molecules because proteins are not static; they are "kicking and alive" as we will see later.

If two large proteins stick together, the exclusion zone merges. This has the consequence of freeing more space for the water molecules to run around and therefore increases the entropy of the overall system. The increase in *order* by the association of protein molecules is compensated by increasing *disorder* by giving the water molecules more space to move around; we say this is an *entropic force.* Cells take advantage of the entropic forces to create different molecular structures, such as in microtubules, collagen, and actin. (See more details in Chapter 8).

How Does a Protein Fold and Why?

There are several possible orientations for each of the amino acids in the backbone of a protein. It cannot be folded randomly; there must be a route. Proteins know how to fold themselves. The answer to the problem of protein folding can be addressed by studying folding kinetics (the how) and using thermodynamics that is based on universal laws (the why). In the following sections, we discuss this approach.

As we have mentioned, proteins are classified as macromolecules due to their large size. We have also seen that proteins adopt helical structures (secondary structures) and superstructures (tertiary structures). The question is how these large and complex macromolecules adopt their final biologically active structure. Since the proteins come out of the ribosome and fold very quickly, the only possibility for this to happen is that there are intermediates that occur during the folding of certain protein regions, which serve as nucleation centers in the folding process.

One possible answer to the question is that all possible conformations are tried out to find the conformation that could minimize the energy. The work of Cyrus Levinthal in 1969 (Levinthal 1969; Durup 1998) indicated that given a protein's degree of freedom, finding the final conformation will take an astronomical number of conformations to try; that process could take an exponentially long time ($\sim 2 \times 10^{27}$ years). Nevertheless, we know that proteins are synthesized as linear chains that must adopt the right folding in seconds. Hence, it would be too long for a rather small protein to find the right conformation if all possible conformations were to be tried. The difference between the calculated and the actual folding is known as Levinthal's paradox.

The solution to Levinthal's paradox, despite almost thirty years of research, has not been found yet. Nevertheless, one possible solution is the assumption of the existence of a funnel-like landscape (see figure 3.13). From the 1980s to the 1990s, Ken Dill devised a theory to explain protein folding in terms of funnel-shaped energy landscapes. Today, that theory shows how amino acids in a protein chain can glom together in different ways and still arrive at the same protein shape, and it has emerged as the accepted view of how proteins fold to reach the lowest state of free energy of the protein (see, for example, Dill and Chang 1997; Bryngelson and Wolynes 1997; Onuchic et al. 1997).

A relatively smooth landscape, in this modeling, leads to the relatively short time that characterizes the overall protein folding process (see fig. 3.13).

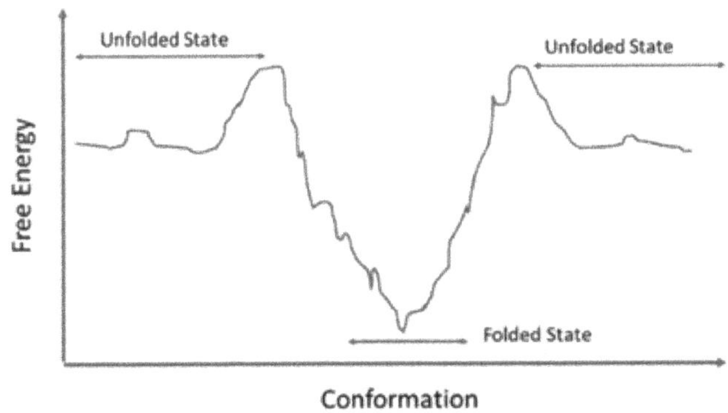

Figure 3.13. Folding funnel showing how proteins fold into their native structures by minimizing their free energy.

The basic idea of this approach is that there is no single path of folding but that the shape of the energy landscape resembles a funnel with wavy walls (funnel-like) and some traps and partial intermediates that are slightly more stable than unfolded proteins. The basic assumption is a minimum of frustration, i.e., a relatively smooth landscape near the native structure.

Wolynes examined the evolutionary aspects of the funnel-like landscape (Wolynes 2015). Anfinsen's refolding experiment itself suggests that folding can be thought of as the search for a *minimum free energy* arrangement in the space of a heteropolymeric chain of residues. The solvent-averaged interactions between amino acids depend on the residue types; some residues will interact more strongly with others. According to Ken Dill, the control of protein folding is thermodynamic; the final conformation is the most stable; however, there also exists a kinetic component, meaning that reaching the final conformation can be obtained by different pathways and not just one. Nevertheless, the first step to obtaining the final conformation is the hydrophobic collapse, where regions that repel water interact among themselves to protect themselves, forming a state named *molten globule.*

During their journey to find the final conformation, proteins can find themselves in zones of local minimal energy-yielding conformational intermediates of a short or long half-life. Those conformational intermediates could be locked, leading to anomalous proteins acting in different disease states.

The cell has a very clever system to prevent the formation of those aberrant conformations and facilitate the finding of the minimal energy or the most stable and active conformation. Nature has designed some proteins to prevent nonspecific aggregation by binding to nonnative proteins known as *molecular chaperones*. The term molecular chaperone was coined by Ronald Laskey, a British scientist (Laskey et al. 1978). There are several families of chaperones, and each possesses different functions. They assist in protein folding by binding to and stabilizing folding intermediates until the polypeptide chain is fully translated. It is for this reason that many chaperones, but by no means all, are heat-shock proteins, that is, proteins expressed in response to elevated temperatures or other cellular stresses.

The tendency to aggregate increases as proteins are denatured by stress. Thus, the role of heat-shock proteins is to prevent protein aggregation. Heat-shock protein chaperones are classified based on their observed molecular mass into Hsp60, Hsp70, Hsp90, Hsp104 (60KDa, 70KDa, 90KDa, and 104KDa of molecular mass respectively) and small Hsps (Bascos et al. 2019).

After discovering the heat-shock proteins and their role as molecular chaperones, some scientists believed that the concept of molecular chaperones had challenged the paradigm of protein self-assembly proposed by Anfinsen. According to Anfinsen, proteins fold by themselves only based on the amino acid sequence. Nevertheless, it looks like the molecular chaperones only provide a favorable environment to facilitate the folding, without altering the tendency of a protein to adopt a given conformation that is dictated by the sequence (primary structure) as was postulated by Anfinsen. If we think of the cell's cytosol as a very viscous mixture (further away from a true solution), where all proteins constantly interact nearby, the tendency to aggregate and unfold is high. Therefore, another role of the molecular chaperones is to prevent protein aggregation, and they also participate in tagging proteins for degradation when they are no longer needed by the cell.

Prediction of Protein Conformation

Chou and Fasman (Chou and Fasman 1975,1978) were the first to develop a method of secondary structure prediction that depends on assigning a set of prediction values to a residue and then applying a simple algorithm to those numbers. Peter Y. Chou was a postdoctoral fellow at Gerald Fasman's laboratory at the time they developed the method. The method takes into consideration the relative frequencies of each amino acid in α-helices, β-sheets, and turns based on known protein structures solved with X-ray crystallography. They assigned amino acids as formers, breakers, and indifferent for α-helical and ß-sheet regions in proteins. From these frequencies, a set of probability parameters were derived for the appearance of each amino acid in each secondary structure type, and these parameters were used to predict the probability that a given sequence of amino acids would form an α-helix, a β-strand, or a turn in a protein (Chou and Fasman 1975, 1978).

The method is at most about 50–60 percent accurate in identifying correct secondary structures (Kabsch Sander 1983). Several improvements were made during the years of the Chou-Fasman method (Nishikawa1983; Chen et al. 2006), but nowadays machine learning and faster computers have improved the predictions, giving birth to the era of computational biology. The method developed by Jean Garnier, David Osguthorpe, and Barry Robson (Garnier et al. 1978; Garnier 2017), known as the GOR method, is a computer program to predict secondary structures in proteins when the amino acid sequence is known, using the theory of information. This is another advance in the area of prediction of protein folding.

Following Anfinsen's principle, if a protein of known sequence acquires a conformation *Z,* then a protein with a similar sequence will generate a conformation close to *Z.* The first protein that was deduced using this approach was α-lactalbumin in 1969, in a publication entitled "A possible three-dimensional structure of bovine α-lactalbumin based on that of hen's egg-white lysozyme" (Browne et al. 1969).

More complex can be the prediction of the tertiary structure (figure 3.14). Several methods have been developed, taking into consideration the most thermodynamically stable protein conformation by finding the minimum free energy of stabilization. Harold Scheraga's (1921–2020) combination of experimental and theoretical approaches showed how amino acid sequences influence the three-dimensional structure, folding pathway, thermodynamics, and biological activity of proteins (Ripoll et al. 1996; Lee et al. 1998; Arnautova et al. 2009). Most of the predictive methods for tertiary structure are based on homology to sequences already known and the X-ray structure of protein from crystals.

Back in 1992, as a side project, I was trying to obtain crystals of tubulin, a protein that tends to self-association. Hence, our approach was to create crystals of the protein bound to colchicine to avoid the self-association. Several conditions were tried, with only the presence of microcrystals not enough for X-ray diffraction. The thirty-year quest in the field of tubulin research was solved in 1998 by the group formed by Eva Nogales, Sharon Wolf, and Kenneth Downing, biophysicists with Berkeley Lab's Life Sciences Division. At a resolution of 3.7 *Å*, the model provides the first highly detailed three-dimensional look at tubulin (Nogales et al. 1998).

Historically, X-ray crystallography has been the dominant technique for elucidating protein structures. The rate-limiting step in elucidating the three-dimensional structure of proteins is the growing of the crystals, where researchers could spend years trying to grow the proper crystals, some molecules cannot be crystallized (Callaway 2015), and the intensive analysis needs to be done to obtain the final structure. This led researchers to develop other techniques, like cryo-electron microscopy (Cryo-EM). Briefly, two-dimensional microscopic images are produced by shining a beam of electrons at a flash-frozen protein solution. The emerging scattered electrons are then detected, and an image is produced. The images are collated into image- processing software to reconstruct the three-dimensional structure of the protein (Callaway 2015, 2020).

Cryo-EM has helped to speed up the resolution of four thousand proteins listed in the Protein Data Bank (Benjin and Ling 2020). Most of the development of the cryo-EM is based on the work done by German physicist Ernst Ruska, together with his doctoral advisor Max Knoll, who on March 9, 1931, presented the very first prototype electron microscope. Capable of four-hundred-power magnification, the apparatus was the first demonstration of the principles of electron microscopy. The primitive electron microscope was used to analyze samples of viruses and bacteria.

If the wavelength used by the electron microscope is of the order of 0.02 Å, then it's perfectly possible to use the technique to solve the structure of proteins. However, the utilization of electron microscopy for the elucidation of protein structure has been a slow process due to technical reasons. This was mainly related to keeping the sample to analyze in an aqueous environment to preserve its stability, but also, radiation can damage the sample. Several decades of research have shown that radiation damage is reduced significantly at liquid nitrogen (Hayward and Glaeser 1979) or helium (Chiu et al. 1981) temperatures. Typically, a high-resolution structure is degraded through bond breakage and rearrangements, as well as interactions with highly reactive free radicals from the surrounding solvent (Glaeser 2008). It is believed these effects are minimized in cryo-EM by preserving structure in vitreous ice of either cells by high-pressure freezing or molecules by plunge freezing (Dubochet et al. 1988).

Obtaining the three-dimensional structure of membrane proteins is particularly challenging using X-ray crystallography. Hence, cryo-EM has been employed to obtain a resolution of 17 Å of the nicotinic acetylcholine receptor (Toyoshima and Unwin 1990; Unwin 1995, 2003). In 2021, significant advances in the prediction of the three-dimensional structure of proteins were obtained thanks to advances in artificial intelligence and machine learning. Deep Mind Technologies (London, UK) made a protein prediction software named AlphaFold (Varadi et al. 2022) that successfully predicted the sequences of proteins with a difference of one atom, which relies on pattern recognition.

The question is how precise the prediction of the novo proteins is. It looks like the AlphaFold is not as precise yet. Some have claimed that with the birth of AlphaFold, the problem of protein folding has been solved. I think we are not there yet.

Figure 3.14. Three-dimensional conformation of a protein, where α-helix, β-structure, and turns are kept together by several weak interactions.

Water Prefers to Socialize with Other Water Molecules

Let us examine some of the properties of water. Most biological reactions occur in water. Water loves to interact with itself because of its polar nature (Stillinger 1980; Lu et al. 2008). This means there is a slightly transient negative charge in the oxygen due to its electronegativity (the tendency to attract an electron from the hydrogens). The consequence is that the hydrogens in water molecules will have a transient positive electric charge (figure 3.15). The great capacity of water as a solvent is due to this polarity, allowing strong interactions with other water molecules or other polar molecules in the solution. If there are nonpolar molecules or hydrophobic molecules in the solution, water will not interact with those molecules. The structure of liquid water consists of molecules connected by short-lived hydrogen bonds. In ice, the hydrogens of water act as hydrogen-bond donors to the ion-pairs of oxygen, which act as hydrogen-bond acceptors. The consequence is the formation of a hexagonal lattice of hydrogen-bonded water molecules, with each H_20 molecule having four neighbors arranged in a tetrahedron. The open structure of ice causes its density to be less than that of the liquid state. Water is chemically a very simple molecule, but its behavior in solution still giving us surprises. In a paper appearing in *Nature Chemistry* (Litman et al., 2024), British and German researchers show that ions and water molecules on the surface of most salt water solutions, known as electrolyte solutions, are organized in a surprising and completely different way from what was thought until now, a finding that could lead to better models of atmospheric chemistry and have other important practical applications. The results show that both cations and anions are depleted at the water/air interface. And that simple electrolyte cations and anions are capable of orienting water molecules both up and down. This does not coincide with textbook models, which teach that ions form an electrical double layer and orient water molecules in a single direction.

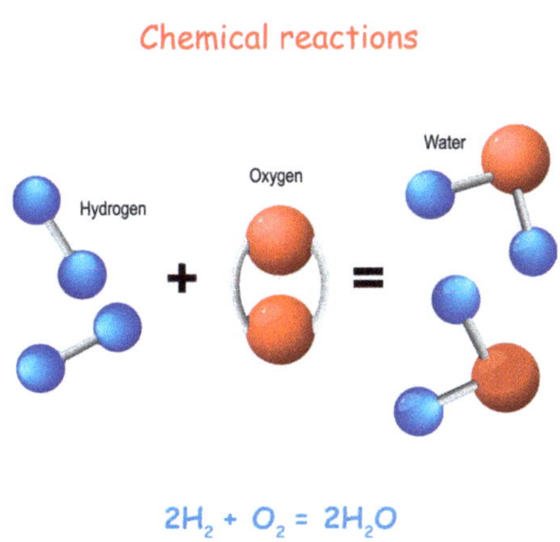

Figure 3.15. Chemical reactions forming water.

Source: Alamy Stock Vector/Tetiana Zhabska. Image ID: 2DJ654T. order reference: OY72287376. 27 august, 2022.

As we indicated before, one of the leading forces in stabilizing the structure of proteins is the hydrophobic effect. This results from two phenomena: 1) The tendency of hydrophobic regions to exclude water and 2) the strong ability of water to form a special type of bonds among itself, named hydrogen bonds. Any substance that is introduced into the water will form hydrogen bonds with the substance. If the substance is hydrophobic, such as oil, for example, it will introduce a cavity in the water network, decreasing the ability of water to form hydrogen bonds. Hence, to decrease the hydrophobic surface exposed to the water molecules, the hydrophobic substances tend to interact among themselves to minimize the surface-to-area ratio exposed to water.

In our example of the oil drop in water, oil-water will become larger to minimize the surface exposed to water and avoid the breaking of the hydrogen bonds between water molecules.

Hydrogen bonds are responsible for the unique properties of water. Hydrogen bonds also contribute to the stabilization of protein structure. A hydrogen bond is a type of interaction between polar molecules. It is weaker than an ionic bond or a covalent bond but stronger than nonpolar interactions (van der Waals forces). Hydrogen bonds are formed by the attraction between two atoms that already participate in other chemical bonds. The bonds may form between atoms within a molecule or between two separate molecules. One of the atoms is hydrogen, while the other may be any electronegative atom, such as oxygen, chlorine, or fluorine. The energy of hydrogen bonds has been estimated to be between 6–9 kcal/mol. These are relatively weak forces compared with covalent bonds (60 kcal/mol), which hold the primary structure of proteins together.

Another type of stabilizing force in proteins is electrostatic interactions. These, in general, are interactions that occur between charged or dipolar atoms and molecules. The interaction energy depends on a variable called the dielectric constant D of the medium. Calculating the D for proteins is extremely complicated due to the heterogeneous structure of the proteins forming several subdomains and microenvironments. Water has a very high D value; hence it will solvate the ions in solutions. In an organic solvent, the D value is low. Therefore, the strength of the electrostatic interactions is stronger. However, it is impossible to put a protein in an organic solvent because the unfolding of the protein will take place, exposing the hydrophobic solvent to the hydrophobic amino acids in the core of the protein, with the consequential loss in tertiary and secondary structure and therefore biological activity.

The electrostatic interaction energy (*E*) between two-point charges, namely q1 and q2, separated by a distance r in a medium of a dielectric constant D, is given by:

$$E = \frac{q1q2}{Dr} \qquad (3.28)$$

Long-range interactions are significant at low protein concentrations, whereas long- and short-range interactions are significant at higher protein concentrations (Chari et al. 2009). Nonpolar interactions also contribute to stabilizing the secondary and tertiary structure of proteins. These forces are called van der Waals or London dispersion forces. They act as a short distance when atoms are very close together so that there is an interpenetration of their electronic clouds, creating electrostatic dipoles that could be attractive or repulsive.

Although the attractive forces of the dipole-dipole interactions are weak and short-range, they are additive and may significantly contribute to the stabilization of the tertiary structure of proteins as well as disulfide bonds that may be present. The change in pH of the solution will lead to the ionization of amino acids in a protein and change the shape and structure of proteins. Thus, proteins have an optimal pH when structure function is preserved (figure 3.16). The figure shows the solubility as function pH and ionic strength. The solubility at low pH and low ionic strength is better than at the range of pH 5.5 to 6.5. Increasing the ionic strength improves the solubility at higher pH values for the model protein (figure 3.16).

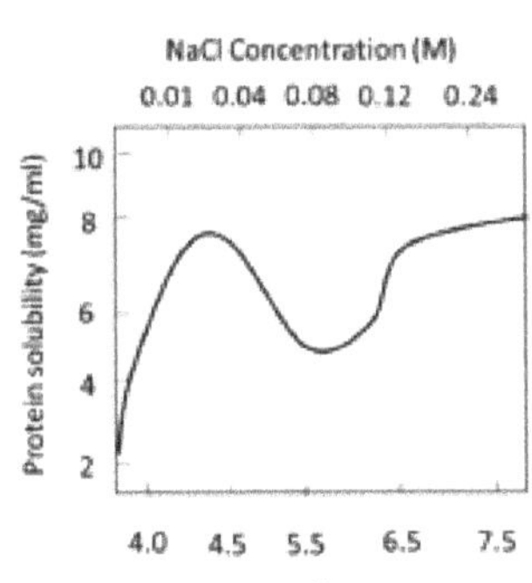

Figure 3.16. Solubility of a model protein as function of pH and ionic strength (NaCl concentration).

pH is a fundamental solution variable that changes the protonation state. The change of protein surface properties mediated by solution pH is related to protein solubility and aggregation. Figure 3.17 shows a contour plot for a protein as a function of pH and storage stability on the degree of solubility. The protein has a limited pH area where stability is optimal. Increasing the ionic strength improves the solubility at higher pH values.

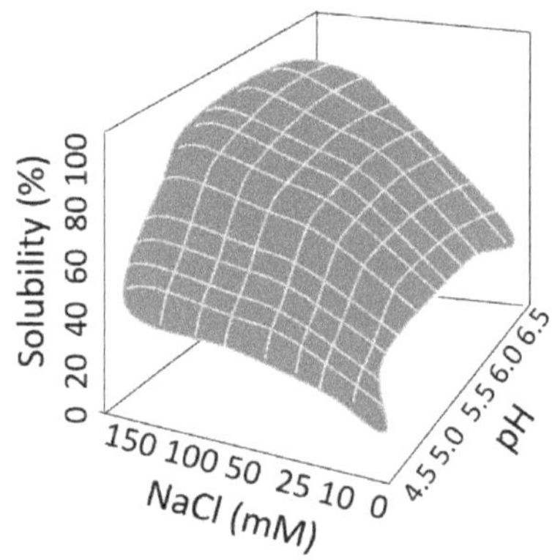

Figure 3.17. The protein has optimal biological activity between pH 6.0 to 6.5 and 50-100 mM NaCl.

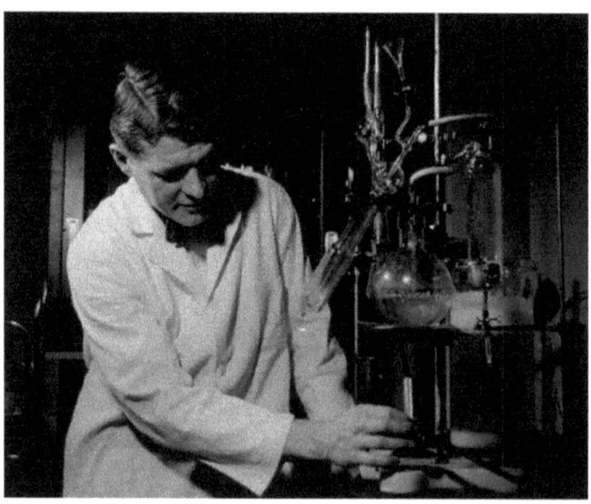

Figure 3.18. Christian Anfinsen, wearing white lab coat, working with chemical apparatus in the laboratory.

Source: Alamy Stock Photo. Image ID: HRNPKR. Science History Images. Reference: OY72287264. Christian Boehmer Anfinsen, Jr. (1916–1995) was an American biochemist. He shared the 1972 Nobel

Prize in Chemistry with Stanford Moore and William Howard Stein for work on ribonuclease, especially concerning the connection between the amino acid sequence and the biologically active conformation. Anfinsen published more than 200 original articles, mostly in the relationships between structure and function in proteins. He was also a pioneer of ideas in nucleic acid compaction.

The Right Balance Between Stability and Flexibility of Structure: "Kicking and Screaming"

In the preceding sections, we have learned that proteins have a defined three-dimensional structure formed by different types of interactions among amino acid residues. Nevertheless, they are not rigid; their dynamic and conformational flexibility permits the protein to carry many of its biological functions. They bind other molecules (metals, organic molecules, nucleotides, proteins, DNA, RNA) and catalyze reactions (enzymes) and movement (muscle contraction, kinetic rotors). They switch to different conformations to accommodate ligand binding. Proteins have several motions due to backbone vibrations and the side-chain motion of amino acid residues. Some of the aromatic amino acid residues (tryptophan, tyrosine, phenylalanine) can have aromatic ring flipping.

The dramatic fluctuations of the tertiary structure of proteins have been described as *breathing* motions. These motions have been observed in diluted solutions. It is unclear if the same type of movement (breathing) takes place inside living organisms, where the environment inside the cell is viscous and the movement appears to be greatly inhibited at high protein concentrations. Motion in the three-dimensional protein structure increases with decreasing protein concentration and increasing temperature and varies widely among the proteins.

The chemical composition of a protein (primary structure) also plays a large role in dictating the degree of motion of proteins. For example, lysozyme and bovine serum albumin, which are stiffened by multiple disulfide bonds, show relatively little increase in breathing in dilute solutions (Makowski et al. 2008). Gregorio Weber (1916–1997) was one of the first to realize that the rigid structure of proteins provided by X-ray diffraction was far from reality, and he initiated the field of the dynamics of proteins. In a seminal paper in 1975, he described the proteins as "kicking and screaming" (Weber 1975).

Demonstration of the breathing motion of proteins has been shown by nuclear magnetic resonance spectroscopy relaxation (NMR) to visualize movement in the range of *pico*seconds and *nano*seconds, up to hours. Depending on the NMR method used, the radio waves spectrum like the ones used in magnetic resonance images measures the relaxation from an excited state. The accessibility of backbone NH groups to solvent has been measured by rates of isotopic exchange. The amide hydrogen atoms exchange with tritium from $^3H_20$ or deuterium from D_20 (Englander 1972). Also, computational methods have been used to calculate electrostatic interactions (Newton's laws of motion of atoms). Those measurements have indicated the movement of proteins in the nanoseconds, motions on the order of 10 ns–1 μs, as was suggested by Weber.

Weber was also responsible for many of the more important theoretical and experimental developments in modern fluorescence spectroscopy. In particular, Weber pioneered the application of fluorescence spectroscopy to the biological sciences, including the synthesis and use of dansyl chloride as a probe of protein hydrodynamics; the extension of Perrin's theory of fluorescence polarization to fluorophores associated with random orientations with ellipsoids of revolution and to mixtures of fluorophores; the first spectral resolution of the fluorescence of the aromatic amino acids and of intrinsic fluorescence of proteins; the first demonstration that both FAD and NADH make internal complexes; and the first report on aromatic secondary amines, which are strongly fluorescent in nonpolar solvents, but hardly in water, the most spectacular case being the anilino-naphthalene sulfonates (ANS). Fluorescence spectroscopy has evolved from a narrow, highly specialized technique into an important discipline widely utilized in the biological, chemical, and physical sciences thanks to the scientific contributions of Gregorio Weber.

A clear correlation between protein dynamics and function has been demonstrated with the enzyme ribonuclease A (Rasmussen et al 1992). Protein motion is a characteristic of life. Some conformational flexibility is required, which explains why the free energy of stabilization of the native state compared with the unfolding state is low, just around 10–15 kcal/mole, much less than the energy of a covalent bond. "Kicking and screaming" requires flexibility; together with flexibility comes instability induced by several factors. Proteins contain multiple functional groups, many of which can undergo some type of chemical modification or damage. To stabilize proteins, one must have a detailed understanding of the degradation profile.

Fluorescence Spectroscopy as a Probe of Conformational Changes

Fluorescence in protein chemistry is a phenomenon by which a protein emits light when exposed to radiation, such as ultraviolet light (Lakowicz 1999 Michalet et al. 2006, Kobiilka and Gether 2002). It is applied in the studies of protein conformation and intermolecular contacts, such as in protein-ligand and protein-protein interactions, allowing qualitative, quantitative, and structural data elucidation. The absorbed radiation (the human eye is not able to see ultraviolet light) is transformed into light with a wavelength longer than the incident radiation (Lakowicz 1999; Michalet et al. 2006).

Fluorescence usually occurs in real time, with the fluorescent light emission occurring just after the protein is exposed to the primary energy source (phosphorescence is the emission of light long-lived). The excited molecules do not always fluoresce. The quantum yield describes the probability of fluorescence. Fluorescence originates from electronic transitions from aromatic amino acids (Wahba et al. 2015; Zhao and Zastrow 2022; Sillen et al. 2003), such as tryptophan (W), phenylalanine (F), and tyrosine (Y), which take electrons previously displaced by incident radiation to excited states in the protein structure to ground states[5] (see figure 3.20).

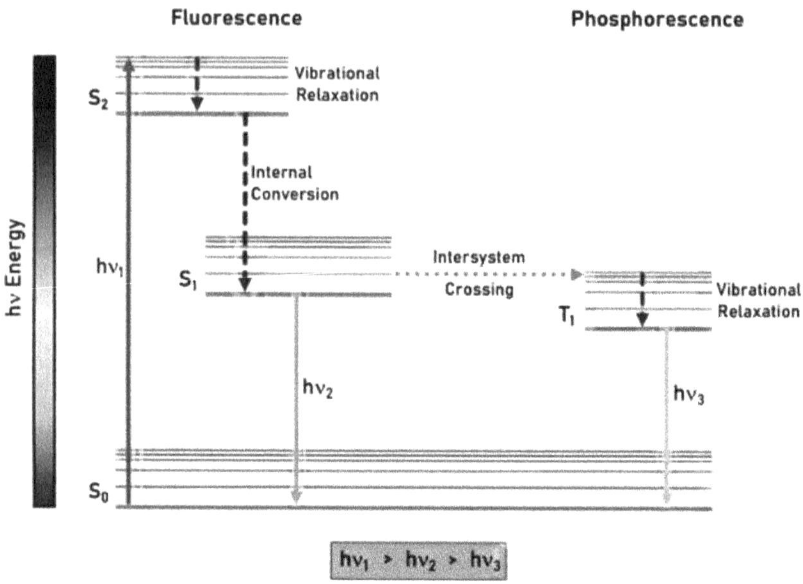

Figure 3.20. Jablonski diagram, showing the different stages of fluorescence and phosphorescence. The diagram illustrates the electronic states, the vibrational levels of a molecule, and the transitions between them.

The aromatic residues undergo π–$\pi*$ transitions under excitation with light of a particular wavelength and then release photons, emitting fluorescence. The protein fluorescence emission maximum is always observed at longer wavelengths than the excitation (λem > λex). This phenomenon occurs due to vibrational relaxation in the excited singlet state and is called Stokes Shift (Gehlen 2020). Conformational change can be monitored by applying intrinsic protein fluorescence, especially if fluorophores (tryptophan, tyrosine, and phenylalanine) have changed their original locations and, consequently, their intermolecular interactions. Being very sensitive to the polarity of the local nano environment, protein fluorophores, such as tryptophan, show maximums of the emission light (λ_{max}) in a wide range from ~308 nm (azurin) to ~355 nm (glucagon). Many biological substances emit characteristic fluorescence. The chlorophyll in leaves fluoresces red.

Some soluble receptors (SR) are usually expressed in mammalian cells as hybrid molecules consisting of the extracellular domain of the cell-surface receptor fused to the Fc portion of a human immunoglobulin (Ig). Such fusion proteins, when expressed, are designed to dimerize, in a manner analogous to the dimerization of the immunoglobulin heavy chain, to create a bivalent, soluble receptor capable of blocking the action of cytokines and growth factors.

When examined in a PBS buffer (10 mM sodium phosphate, 150 mM NaCl, pH 7.4), the SR-BP dimer and SEC fractions (see figure 3.21) were found to yield intrinsic tryptophan fluorescence spectra with different spectral characteristics (figure 3.22a). The emission maximum of the dimer is positioned at 347 nm, indicating that the dimer contains surface-exposed tryptophan residues in contact with bound, immobilized water molecules (Permyakov 1993).

Another notable difference between the spectra is that the SEC fractions exhibit higher fluorescence intensities at their emission maxima than the dimer. The fluorescence intensities of the SEC fractions increase over the dimer in the same order as the extent of blue shifting of their emission maxima; that is, with a decreasing SEC fraction number. When viewed together, the spectral differences evident in figures 3.22a and b indicate that the local environment of the tryptophan residues in the SR-BP dimer are different from those of the tryptophan residing in the aggregates in the SEC fractions. This is consistent with the view that the SR-BP dimer comprising the aggregates exhibits different tertiary structure from that of the free dimer. Also, fluorescence is useful when the ionization of a group perturbs the spectrum of a neighboring aromatic residue (Timasheff 1970).

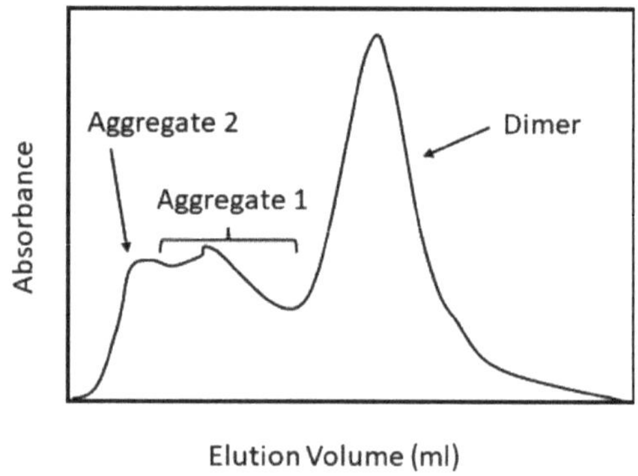

Figure 3.21. SEC fractions of SR-BP.

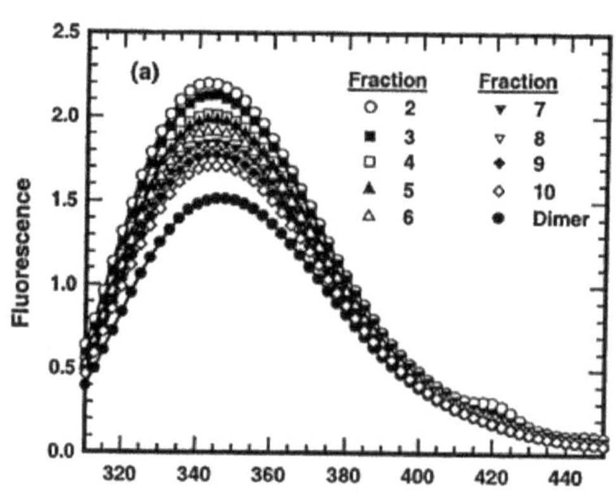

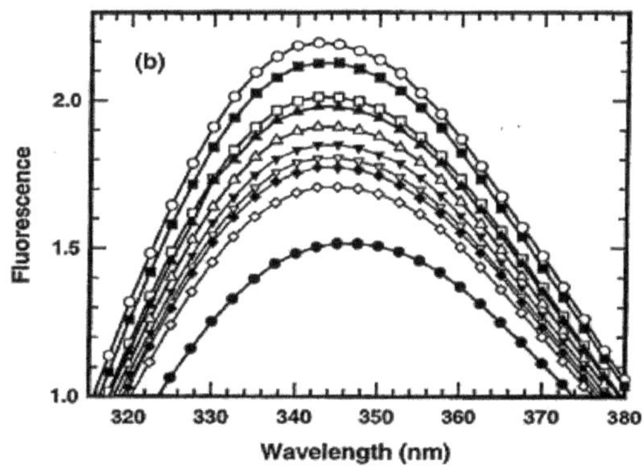

Figure 3.22. Intrinsic tryptophan fluorescence spectra of SR-BP in PBS.

The tryptophan fluorescence was selectively excited at 300 nm, and the emission was scanned from 450 to 310 nm using a wavelength step size of 0.5 nm and a data- averaging time of one second. The excitation and emission bandwidth were 3 and 4 nm respectively. Samples placed in the cuvette were gently stirred with a Teflon-coated stir bar during fluorescence data collection. Protein spectra were corrected by subtracting a solvent baseline. The spectra shown in figure 3.22a are replotted on an expanded scale in figure 3.22b.

Extrinsic Fluorescence

In many cases, information can be obtained from certain fluorescent species introduced into proteins. Thus, enzymes, other proteins (Ovádi et al. 1983; Buckler et al. 1993; Han et al. 1996), and receptors (Martinez-Carrion and Raftery 1973; Antollini and Barrantes 2007) can be marked with fluorescent species or covalent unions. Fluorescent analogs of substrates to study enzyme active sites, such as ε(etemol)-ATP, are frequently used or the ANS probe (1.8-anilinonaphthalene sulfonic acid), the latter being an extrinsic fluorescent probe that exhibits weak fluorescence in aqueous solution but stronger fluorescence when bound to nonpolar surfaces in proteins to detect hydrophobic regions in proteins (Stryer 1965; Daniel and Weber 1966; Slavik 1982).

Extrinsic dyes can be covalently attached to proteins, e.g., via the ε-amino group of lysine, the α-amino group of the N-terminus, or the thiol group of cysteine. For a review of the available dyes and their use, see Hawe et al., 2008.

Polarization

Fluorescence polarization (FP) is used to monitor binding reactions in solution. It can be used to assess biomolecular interactions, such as protein-protein interactions and enzyme activity. When a small, fluorescently labeled molecule (tracer) is excited with plane-polarized light, it will emit mostly depolarized light because the tracer tumbles rapidly during the time between excitation and emission[6]. However, when the tracer binds a much larger molecule, it rotates more slowly, and the emitted light remains largely polarized (figure 3.23).

To observe depolarization due to motion, it is necessary that the lifetime of the excited state be sufficiently long enough that the reorientation can occur before emission.

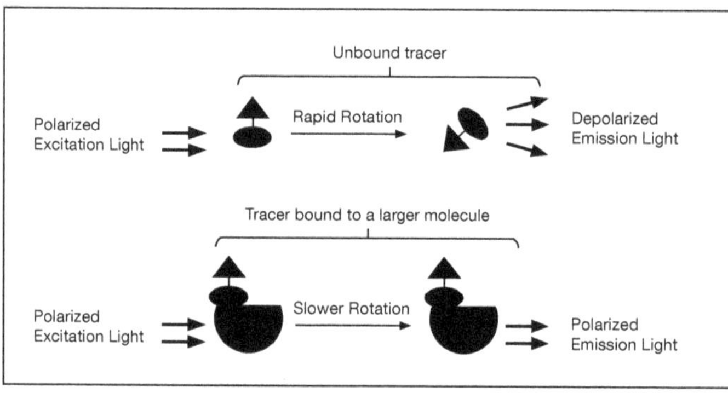

Figure 3.23. Schematic showing that with increasing mobility, polarization decreases.

If a fluorophore is bound to a macromolecule and then the macromolecule either forms dimers, aggregates, or binds to a large molecule, the polarization increases due to the reduced mobility of the structure. If a macromolecule undergoes a conformational change, for example, a helix-coil transition, the polarization decreases as the molecular structure becomes more disordered. Conversely, if the structure becomes more ordered, polarization increases (Weber 1960; Lundblad et al. 1996; Jameson and Ross 2010).

Quenching

Several conditions can diminish the fluorescence intensity of a fluorophore, denominated fluorescence quenching. This reduction of fluorescence intensity can occur through several processes, some via dynamic quenching, others via static quenching (Lakowicz 1999). A molecule that causes the reduction of fluorescence intensity of a fluorophore is called a quencher. In a dynamic quenching, a quencher collides with the fluorophore, deactivating it and therefore reducing the fluorescence intensity. This type of quenching, based on a collisional effect, can be described mathematically by the Stern-Volmer equation (Lakowicz 1999).

$$1/T = 1/T_o + K_q[Q] \qquad (3.29)$$

K_q is the bimolecular Stern-Volmer quenching rate constant, and T_o and T are the singlet lifetime in the absence and presence of some quencher concentration [Q], respectively. K_q approaches zero at low accessibilities and approaches the diffusion-controlled upper limit at high accessibilities.

The fluorescence intensity of an emitter is decreased by both the collisional and the static quenching process (Eftink and Ghiron 1976a), according to:

$$F_o/F = (1 + K_{sv}[Q])\, e^{v[Q]} \qquad (3.30)$$

F_o and F are the fluorescence intensity in the absence and presence of a quencher, respectively, at a quencher concentration [Q]. K_{sv} is the Stern-Volmer quenching constant, and v is the static quenching parameter (Eftink and Ghiron 1976b). K_{sv} is related to K_q by $K_{sv} = T_oK_q$.

The initial slope (IS) of a F_o/F vs. [Q] plot is given by Tao and Cho (Tao and Cho 1979).

Figure 3.24 shows the fluorescence quenching of phosphopyridoxylated polylysine[7]. The increase in the slope of the Stern-Volmer plot with the increase of the temperature at which the quenching reaction was carried out is typical of a collisional type of quenching.

$$\left[\frac{dF_0/F}{d[Q]}\right]_{[Q]=0} = K_{sv} + v \tag{3.31}$$

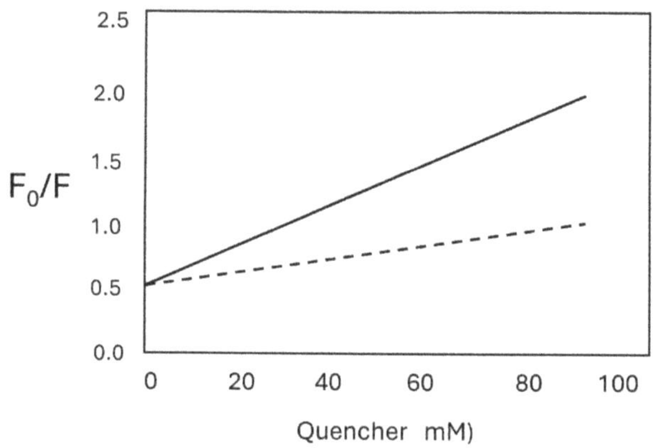

Figure 3.24. Nitromethane quenching of phosphopyridoxylated-poly-lysine. Quenching was carried out using a concentration of 0.2 μM Schiff's bases as determined from the absorbance at 325 nm, using an extinction coefficient of 7,400 $M^{-1}cm^{-1}$ (adapted from Pérez-Ramírez 1990). The temperature of the sample cuvette was 15 °C (----) and 25 °C, (—), respectively.

Self-quenching is a phenomenon that occurs at high fluorophore concentrations. At low concentrations, the fluorescence signal is directly proportional to the fluorophore concentration until it reaches a critical concentration with maximum fluorescence intensity. Above this concentration, the opposite is observed, with fluorescence intensity being inversely proportional to fluorophore concentration (Bae et al. 2021). A proposed explanation for this behavior lies in the collision between fluorophore molecules that cause the inner filter effect, which involves the depletion of excitation light and reabsorption of emission light.

Fluorescence Resonance Energy Transfer (FRET)

FRET is a distance-dependent physical process by which energy is transferred nonradiatively from an excited molecular fluorophore (the donor) to another fluorophore (the acceptor) by means of intermolecular long-range dipole-dipole coupling. The rate of energy transfer from a specific donor to a specific acceptor (K_T) is given by

$$K_T = 1/T_d\,[R_o/r]^6 \qquad (3.32)$$

T_d is the lifetime of the donor in the absence of the acceptor, r is the distance between the donor and acceptor, and R_o is a characteristic distance, called the Főrster distance (figure 3.25; see also Főrster 1959), at which the efficiency of transfer is 50 percent (Lakowicz 1999).

This dependence of the transfer rate on distance has resulted in numerous energy transfer applications in biochemical research, especially since the critical distances range from 20 to 40 Å. Any phenomenon that affects the donor-acceptor distance will affect the transfer rate, allowing quantitation.

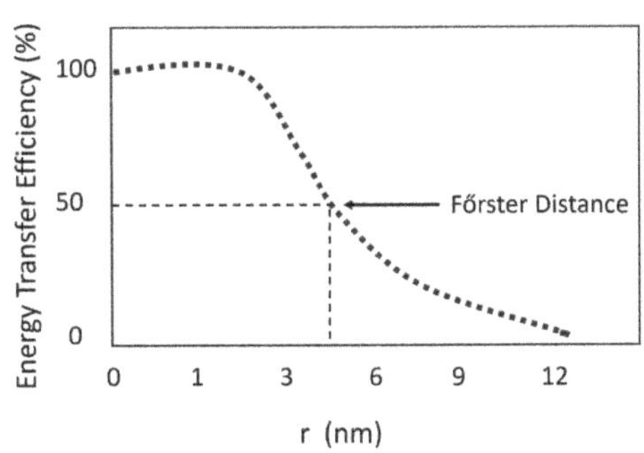

Figure 3.25. Distance and energy transfer efficiency.

FRET has been used to measure the active sites in dimeric enzymes (Martinez-Carrion and Boettcher 1975). Aspartate aminotransferase is a dimeric enzyme with identical subunits. Each subunit contains one pyridoxal phosphate at the active site that shows fluorescence properties. It has been shown that there is migration of electronic excitation energy from the aromatic residues (donor) to phosphopyridoxyl residues, the acceptor (Martinez-Carrion and Boettcher 1975). Also, based on FRET, enzyme-cleavable sensor molecules (Takakusa et al. 2002) have been designed, as well as clinical assays (Carmona et al. 2006).

Several variables can induce protein instabilities, and fluorescence spectroscopy is a powerful technique that allows the identification of conformational changes in proteins induced by several factors (see figure 3.26). Intermolecular contacts can also be studied using fluorescence.

Elevated Temperature	**Freezing and Thawing**
Mechanical Stress	**Light**
Oxidative Stress	**pH changes**
Dehydration	**Interfaces**

Figure 3.26. Factors that cause protein instability.

Infrared Spectroscopy

There are many spectroscopic methods that have been used in biochemical research. We do not intend to illustrate them all here. Infrared is associated with heat and is used in various applications, including thermal imaging, remote controls, and some forms of spectroscopy. Infrared spectroscopy allows the analysis of the conformation of macromolecules in the liquid and solid state. This is an advantage in the biopharmaceutical industry, where it is often required to preserve proteins in the solid state by the process of lyophilization.

However, the process of lyophilization, if not carried out in optimal conditions, can also produce instability in proteins (the discussion on lyophilization is outside the scope of this book).

Infrared spectroscopy uses long wavelength radiation (low energy) to excite local vibrations, rotations, and changes of functional groups in molecules. It is vibrational energy that keeps the covalently bonded atoms of a molecule constantly oscillating about an equilibrium position. These can be interpreted as an exchange of interactions between such groups and their environments (e.g., presence of hydrogen bonds). Thus, when exposed to an infrared radiation, energy is absorbed by the molecule, which increases in vibrational motion.

Infrared spectroscopy is the analysis of infrared light interacting with a molecule. Infrared spectroscopy exploits the fact that molecules absorb radiation at specific frequencies characteristic of their structure. The range of the infrared region is 12800 ~ 10 cm^{-1} and can be divided into the near-infrared region (12800 ~ 4000 cm^{-1}), mid-infrared region (4000 ~ 200 cm^{-1}), and far-infrared region (50 ~ 1000 cm^{-1}).

The first generation of infrared spectrometers was developed in the late 1950s. It utilized a prism optical splitting system. The prisms were made of NaCl. This setup had a lot of problems. It had a narrow scan rate, and reproducibility was very poor. The modern infrared spectrometer uses an interferometer. With this replacement, infrared spectrometers became exceptionally powerful. The accuracy of wave number is high, and there is high resolution, higher signal-to-noise ratio of spectrum, and reduced interference from straight light.

A common FTIR spectrometer consists of a source, interferometer, sample compartment, detector, amplifier, A/D convertor, and computer. The source generates radiation, which passes the sample through the interferometer and reaches the detector (figure 3.27). Then the signal is amplified and converted to a digital signal by the amplifier and analog-to-digital converter respectively. Eventually, the signal is transferred to a computer in which the Fourier transform is carried out. The Michelson interferometer is used to split one beam of light into two so that the paths of the two beams are different. Then the Michelson interferometer recombines the two beams and conducts them into the detector, where the difference of the intensity of these two beams is measured as a function of the difference of the paths.

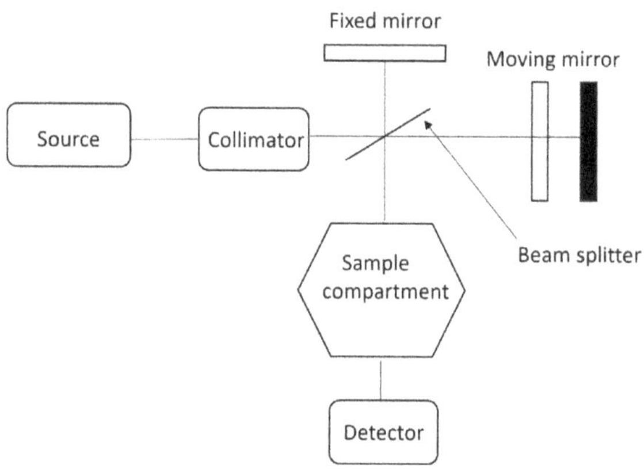

Figure 3.27. Diagram of an FTIR spectrometer.

The addition of the Fourier transform in the 1960s to infrared analysis resulted in higher resolution and decreased noise, increasing the attractiveness of this technique. The Fourier transform is a mathematical function that decomposes a waveform, which is a function of time, into the frequencies that make it up.

The infrared portion of the electromagnetic spectrum is divided into three regions: the near, mid, and far infrared, named for their relation to the visible spectrum (see Chapter 2, figure 2.8). In infrared spectroscopy, the x-axis represents wave numbers in cm-1 = $1/\lambda$ instead of wavelength= λ. Characteristic bands found in the infrared spectra of proteins and polypeptides include amide I and amide II. These arise from the amide bonds that link the amino acids. The absorption associated with the amide I band leads to stretching vibrations of the C=O bond of the amide, and absorption associated with the amide II band leads primarily to bending vibrations of the N—H bond.

C=O and the N—H bonds are involved in the hydrogen bonding between the different elements of secondary structure (see figure 3.28). The locations of both the amide I and amide II bands are sensitive to the secondary structure content of a protein. β-sheets have a characteristic signal at 1620–1640 cm^{-1}. The average frequency of the main component is about 1629 cm^{-1}, with a minimum of 1615 cm^{-1} and a maximum of 1637 cm^{-1} (Nevskaya and Chirgadze 1976). For α-helical structures, the mean frequency is found to be 1652 cm^{-1} for the amide I and 1548 cm^{-1} for the amid II absorptions (Chirgadze and Nevskaya 1976).

The half-width of the α-helix band depends on the stability of the helix. For the most stable helices, the half-width of about 15 cm^{-1} corresponds to a helix-coil transition-free energy of more than 300 cal/mole. Other helices display half-widths of 38 cm-1 and helix-coil transition-free energies of about 90 cal/mole (Nevskaya and Chirgadze 1976).

The 3_{10}-helix differs from the α-helix in that the internal hydrogen bonding occurs between residues i and i+3 instead of i and i+4 in α-helices. Assignment of beta (β) turns by means of a normal mode analysis for insulin demonstrates a strong overlapping of the different types of β-turns with the α-helical absorption (Krimm and Bandekar 1986). However, an absorption near 1680 cm-1 is now clearly assigned to β-turns.

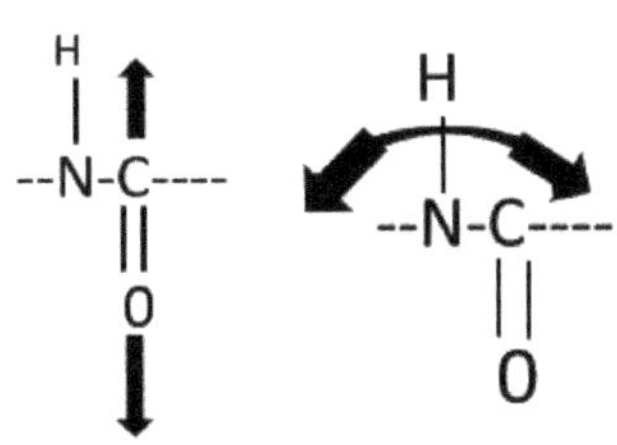

Figure 3.28. C=O stretching (amide I vibration, 1600-1700 cm^{-1}) and NH bending vibrations (amide II ~1549 cm^{-1} vibration).

The frequencies of the main absorption bands from synthetic polypeptides that adopt an antiparallel chain structure have been compiled (Chirgadze and Nevskaya 1976). From these data, it follows that the amide I absorption is primarily determined by the backbone conformation, independent of the amino acid sequence, its hydrophilic or hydrophobic properties, and charge. The average frequency of the main component is about 1629 cm^{-1}, with a minimum of 1615 cm^{-1} and a maximum of 1637 cm-1. The average value for the second frequency is 1696 cm^{-1} (the lowest value is 1685 cm^{-1}). The parallel β-sheet structure that is not common in synthetic polypeptides leads to an amide I absorption near 1640 cm^{-1}.

Water is a very strong infrared absorber, with significant bands centered at wave numbers of ~3800-3900 cm^{-1} (H-O stretching band), 2125 cm^{-1} water association band, and H-O-H bending vibration at 1640 cm^{-1}, overlapping with the amide I vibrations band. Despite the high signal-to-noise ratio of modern FTIR spectrometers, it is often necessary to perform measurements in aqueous solution around 1640 cm^{-1} in D_20, where no strong bands are close to the amide I frequency.

FTIR, like circular dichroism spectroscopy, provides a means of estimating the secondary structure present in proteins with the advantage that with FITR, information can also be obtained in the solid state, as in lyophilized proteins.

In the infrared spectrum of cytosolic holoaspartate aminotransferase, the band arising from the symmetric stretching of the dianionic phosphate monoester of bound pyridoxal phosphate has been identified (Sanchez-Ruiz and Martinez-Carrion 1986). FTIR has also been used as a tool to monitor the degree of phosphorylation of several proteins, such as ovalbumin and hen egg phosphovitin (Sanchez-Ruiz and Martinez-Carrion 1988).

CHAPTER 4

THERMODYNAMIC AND PHYSICAL STABILITY

Applying the laws of thermodynamics to understanding the physical stability of proteins is a powerful tool to ascertain if a process will indeed occur or not. *Physical stability* refers to structural stability, and its main concern is about changes in secondary, tertiary, and quaternary structures without affecting the primary structure. This contrasts with *chemical stability* (biochemical stability), which includes cleavage, oxidation, deamidation, racemization, and beta elimination, which we will cover in Chapter 5. Physical and chemical stability are linked; thus, a physical instability, like denaturation, can expose an amino acid residue that could be chemically modified by exposure to the solvent at a given pH, leading to chemical instability. A chemical instability, like deamidation, could impact the conformation of the protein and, therefore, its physical stability.

All the favorable processes that occur in nature (spontaneous processes) tend to occur naturally and release energy. In general, a process occurs because it will minimize the *free energy.* Energy is the ability to do work, and we can classify the energy as kinetic energy or the energy of motion, which is involved in i) movement, and ii) thermal energy, the energy that objects have due to their motion. It depends on the object's mass and velocity.

Another type of energy is potential energy (figure 4.0), based on position or stored energy (chemical, electrical, gravitational). *Free energy* is energy minus unusable energy. The product of temperature and entropy gives unusable energy, as we will see in the following section.

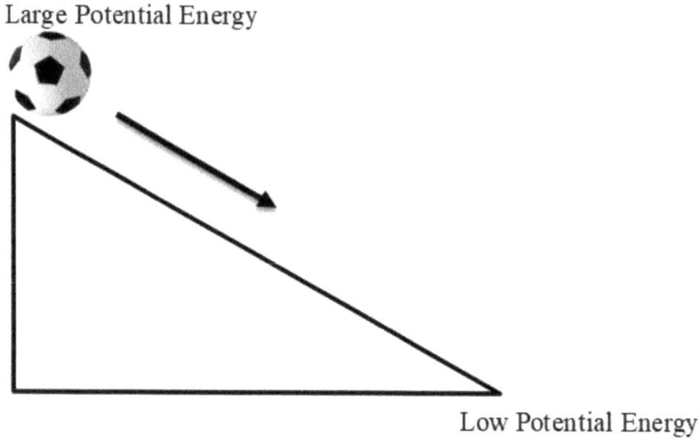

Figure 4.0. Energy on a given system.

Thermodynamics deals with the relationships between heat and other forms of energy. It describes how thermal energy is converted to and from other forms of energy and how it affects matter. The fundamental principles of thermodynamics are expressed in laws. This book is concerned only with systems at equilibrium (equilibrium thermodynamics). Thus, equilibrium thermodynamics deals with *closed systems* and provides information about the energetics of isolated systems that form the building blocks of biological systems. The complete systems, organelles, cells, etc., are *never in true equilibrium* while they are functional. They are continuously exchanging matter and energy to and from other systems and to chemical reactions.

Equilibrium means death for the cell. Therefore, *nonequilibrium thermodynamics* deals with *open systems*, with physical systems that are not in thermodynamic equilibrium but can be described in terms of macroscopic quantities (nonequilibrium state variables), which represent an extrapolation of the variables used to specify the system in thermodynamic equilibrium (averaging the microscopic variables over a "local equilibrium ensemble").

Thermodynamic Laws

The first principle of thermodynamics establishes that *energy can neither be created nor destroyed*. It can, however, be transferred from one location to another and converted to and from other forms of energy. In equation form, the first law of thermodynamics is:

$$\Delta \mathbf{U} = \mathbf{Q} - W. \quad (4.0)$$

Here, ΔU is the change in internal energy U of the system. Internal energy is defined as the energy associated with molecules' random, disordered motion. This "storm" is happening inside our bodies, where molecules are colliding among themselves. It is separated in scale from the macroscopic ordered energy associated with moving objects. Thus, a glass of wine sitting on a table at room temperature has no apparent kinetic or potential energy.

Q is the net heat transferred into the system—that is, Q is the sum of all heat transferred into and out of the system. W is the net work done by the system—that is, W is the sum of all work done on or by the system. We use the following sign conventions: if Q is positive, there is a net heat transfer into the system; if W is positive, there is net work done by the system.

The second law of thermodynamics indicates that *spontaneous processes proceed toward lower energy*. There is also a *third law of thermodynamics;* according to the third law, the entropy of a perfect crystal is zero when the temperature of the crystal is equal to absolute zero (0 kelvin). The fourth law of thermodynamics refers to far-from-equilibrium theories applied to living systems. Ilya Prigogine (1917–2003) developed theories on "far-from-equilibrium" thermodynamics and named the dissipative structure theory, according to which biological life is argued to have self-assembled from inorganic nonlife through the nonequilibrium thermodynamic processes of fluctuations and bifurcations (Goldbeter 2003; Grandy 2012). "We are led, therefore, to regard energy dissipation as the 'driving force of evolution'" (Nicolis and Prigogine 1977).

The second law of thermodynamics can be expressed in several ways, the simplest being that heat (Q) as a function of temperature (T) will naturally flow from a hotter to a colder body. It is a property of thermodynamic systems called entropy, represented by "S"—a measure of the disorder within a system or the *degree to which energy is dispersed.*

As a state function (independent of the pathway), entropy change, ΔS, depends only on initial and final states. ΔS has a positive value when the disorder increases and a negative value when the disorder decreases. The following conditions increase entropy: a change of phase: solid→liquid→gas, an increase in the number of gas molecules, or a solid dissolving to form a solution.

Now, equating entropy with disorder is convenient, but it is incorrect. The creationists held to this by indicating that life violates the second law of thermodynamics due to high organization. However, entropy can be reduced locally if it is increased globally, which is what life processes do. Hence, we must analyze the system's entropy and the surrounding entropy. Heat is a form of energy, and we want to analyze if heat (and entropy) can flow to the environment. This analysis requires that we know the temperature. We can experiment using a water bath at a controlled temperature. If the system begins to heat up from a heat-releasing reaction (exothermic), the heat will flow into the surrounding bath until the system's temperature equals that of the bath. Conversely, if the system cools down (endothermic reaction), it will take heat from the surrounding bath. If the water bath is large enough, the heat flow from the system will not change the temperature of the water bath. The water bath then serves as an energy container and temperature control.

Similar situations are encountered in living organisms; they also act as heat baths. Our cells are immersed and exposed to a constant temperature of 37°C, except for seminiferous tubules, which require an optimal temperature of 32°C (Pérez-Ramírez and Castells 1991). All the chemical reactions in our cells occur at a constant temperature. However, to examine how the energy will flow in that system, we need to introduce the concept of usable energy or the so-called *Gibbs free energy* (ΔG) that we briefly mentioned before.

If we examine the example of the soccer ball on the top of the slope (figure 4.0), it has high potential energy and spontaneously will roll down the hill to minimize the potential energy. The process can be described by ΔG, where:

$$\Delta G = \Delta G_{Final} - \Delta G_{Initial} \tag{4.1}$$

Thus, a process is spontaneous if *ΔG has a negative value.* If the value of ΔG is positive, the process is non-spontaneous from left to right (reactants to products). When the value of ΔG = 0, the process is at equilibrium.

We can express ΔG as a function of the total energy of the system, named enthalpy (ΔH), including chemical and potential energy and the degree of disorder/energy distribution, also called entropy (ΔS) or the number of ways to arrange a molecule or macromolecules; in our case here, as a function of temperature (T) that affects kinetic energy. The enthalpy indicates the amount of thermal energy of a thermodynamic system, i.e., a defined chemical process. The reaction enthalpy, ΔH (change in enthalpy), indicates the amount of heat absorbed or released in a chemical reaction. Thus, using the enthalpy change (ΔH) of a chemical reaction, one can make a statement about whether it is an endothermic or exothermic reaction.

Enthalpy change > 0: the reaction is endothermic
Enthalpy change < 0: the reaction is exothermic

Entropy (ΔS) measures the irreversibility of chemical and physical processes. The greater the irreversibility of a process, the more significant the change in entropy. Thermodynamic systems always strive for greater entropy, i.e., a more significant *distribution of energy.* If the ΔG of a chemical reaction is positive, the reaction is said to be endergonic. Endergonic reactions are *not* spontaneous. On the other hand, if ΔG is negative, this reaction is called exergonic. These reactions are spontaneous. Whether the respective reaction is exergonic or endergonic, i.e., occurs voluntarily or not, depends partly on the temperature.

Free energy (ΔG) is then the total energy (ΔH) minus the product of temperature (T) and entropy (ΔS) of the system. Because entropy represents how much energy has become dispersed and useless, free energy represents that part of the useful energy because we subtract the useless part, namely, TΔS. Below is the summary equation for the first law of thermodynamics.

$$\Delta G = \Delta H - T\,\Delta S \tag{4.2}$$

In order that we may compare the energetic changes associated with different reactions, we must first standardize the conditions of the reaction. Thus, note that sometimes ΔG is also written as ΔG°. The former refers to the change in Gibbs free energy when the reaction takes place at a constant temperature and pressure under arbitrary but reported conditions. Meanwhile, ΔG° is ΔG under standard conditions of 25°C (298K), the standard atmospheric pressure of 101325 Pa (and 1 Atm), with components in a standard state throughout at 1 molar concentrations. The same is true for variables ΔH and ΔS.

Let us go back and examine the structure of proteins, considering some of the thermodynamic principles we have introduced.

Primary structure (random coil)
Unfolded [U]
Higher enthalpy
Higher entropy

Tertiary structure
Native [N]
Lower enthalpy
Lower entropy

Figure 4.1. A protein in its nonnative (unfolded) and folded state respectively.

The protein's primary structure is a string exposed to the solvent, in this case, an aqueous solution; it has few stabilizing interactions with the constituent's amino acid residues; hence, the enthalpy is higher than the folded molecule (tertiary structure). Furthermore, hydrophobic residues are exposed, making the interaction with molecules a destabilizing factor. Nevertheless, the entropy or the degree of randomness is higher (favorable) than in the folded protein.

Recall that the change in ΔG must be negative for a reaction to proceed (that is, to favor the products). Therefore, either the change in enthalpy must be negative (and dominant over an unfavorable entropy term), or the change in entropy must be positive (and dominant over an unfavorable enthalpy term). The entropy term is controlled by temperature. A higher entropy effect at high temperatures leads to the protein unfolding. The enthalpy accounts for noncovalent bonding (hydrogen bonds, van der Waals, and ionic interactions), whereas the entropy term accounts for the hydrophobic effect (desolvation).

In the case of proteins, the *negative Gibbs free energy of folding* is dictated primarily by the negative enthalpy despite the decreased entropy. Let us examine the folded form of the protein, the so-called conformational stability, or the Gibbs free energy of stabilization ($\Delta G°$native – $\Delta G°$unfolded). It is a minimal value and accounts for no more than 10–15 Kcal/mol between the folded and unfolded state. Therefore, the free energy given by the energy (enthalpy) minus entropy time and temperature is lower for the folded protein, and because the system tends to minimize its free energy, it will tend to an ordered state at equilibrium.

The concept of ΔG is the coin that allows us to see if a process will occur despite a decrease in entropy. Can a process occur if the free energy and the entropy decrease simultaneously? At a constant temperature, the process is possible if the total energy of the system ($\Delta H°$) decreases more than the entropy term. Events in the "messy" natural world arise from lawful physical and chemical processes. Thus, free energy captures the battle between deterministic forces (chemical bonds), the molecular storm, necessity, and chance.

We can calculate the *apparent free energy of stabilization* using the two-state approximation, assuming equilibrium between unfolded [U] and folded [N] conformations. Many proteins appear to undergo completely cooperative folding and unfolding, as their folding reactions meet both the kinetic and thermodynamic criteria for two-state folding. The criteria for two-state folding are, however, also met by a multistate folder whose folding reaction is populated by intermediates that are too unstable to be detected experimentally (Nosaka et al.1978; Feng et al. 2005). Nevertheless, the utility of two-state models is so great that even when the folding kinetics may be more correctly describable as multistate, the two-state model is invariably extended to account for the kinetics (Malhotra and Udgaonkar 2016). In this analysis, we can equate unfolded to nonnative proteins since the probability of having a completed unfolded protein in solution is unlikely. There is a quantitative calorimetric examination for a two-state state unfolding (Privalov 1979) that we will review later in the chapter.

A critical aspect of the thermodynamic parameters $\Delta G°$, $\Delta H°$, and $\Delta S°$ is that they are *state functions,* and we just need to know the final and initial state independent of the pathway to reach those states.

If [N] and [U] are at equilibrium, we can approximate the equilibrium constant K_{Native} as:

K_{native} = Products/Reactants = ([Unfolded])/([Native])

Where $\Delta G° = G_{unfolded} - G_{native} = - RT \text{ in } K_{native}$ (4.3)

T is the temperature in Kelvins, and R is the gas constant (1.9872 cal/mol-K): the more significant and positive $G_{unfolded}$, the more stable the protein to denaturation.

If the protein evaluated is multi-domain, which unfolds irreversibly, then Gibbs free energy has no meaning. However, the folding/unfolding transitions of many proteins occur in multiple discrete steps associated with the formation of intermediates, which is indicative of reduced cooperativity (Feng et al. 2005). Hence, the two-state model is only applicable to small proteins that unfold reversibly. In the case of irreversible or slowly unfolding proteins, kinetic stability or the unfolding rate is essential. A kinetically stable protein will unfold more slowly than a kinetically unstable protein; we will return to kinetic stability later in the chapter.

Thermodynamic analysis of therapeutic proteins has been used as a screening tool in biopharmaceutical development. However, those large proteins unfold irreversibly. Thus, the general thermodynamic parameters obtained during those experiments, such as melting temperature and the onset of melt, are typically called *apparent rather than absolute values.* However, despite the apparent nature of those measurements, the information is helpful if used comparatively with similar proteins evaluated under *similar scan rates* (Guziewicz et al. 2007; Pérez-Ramírez et al. 2010).

Several studies have been performed varying the protein environment and evaluating the impact on the Gibbs free energy of stabilization. Those studies measure the fraction of protein unfolding by adding a protein denaturant, such as urea or guanidine hydrochloride. This is an indirect method where a curve is constructed, and extrapolation to zero concentration of denaturant can give information on the free energy of stabilization of the native state and compare the protein under different solution conditions as described by Nick Pace and coworkers (Pace and Vandenburg 1979; see also Chapter 6). ΔH and other derived thermodynamic functions are obtained by differentiating the van Hoff equation.

$$d\ln K/dT = \Delta H/RT^2 \quad (4.4)$$

The occurrence of a chemical reaction will be associated either with the liberation or absorption of heat. A calorimetry technique could detect this heat exchange in a closed system. The temperature dependence of protein unfolding is a critical parameter that has been extensively studied. Direct measurements of the heat required to unfold proteins were initially done by Peter Privalov (1932–2020) with a sensitive calorimeter. Julian Sturtevant in 1958 built the first calorimeter in the United States. In 1964, Privalov and his group developed the first differential scanning calorimeter (DSC) to study the energetics of proteins and nucleic acids, essentially founding the field of biological microcalorimetry (Privalov et al. 1964).

Having become convinced that it was possible to measure the very small amount of heat released during the folding of a protein, Privalov developed a supersensitive calorimetric technique, microcalorimetry, which continues to be widely used in many laboratories.

I remember meeting Professor Privalov at the biocalorimetry conference held in July 2006 in Zaragoza, Spain. We were introduced by a common friend,

Lucia Bianconi (from the University of Rio de Janeiro-Brazil), while having a cocktail *al fresco*. He was a very humble man and a deep thinker. He invited me to sit with him at the dinner table to continue our conversation on the thermodynamics of proteins and our shared interest in self-associating proteins.

Differential Scanning Calorimetry (DSC)

A differential scanning calorimeter (DSC) is commonly used in the biopharmaceutical industry. A DSC measures the heat changes associated with the thermal denaturation of the therapeutic protein of interest to evaluate i) different biological constructs to the same target and ii) different formulations during pre-formulation (Pérez-Ramírez et al. 2010, 2015). The isothermal titration calorimeter (ITC) provides thermodynamic data valuable to confirm the binding models (see Chapter 5) as well as quantitation of binding enthalpy (ΔH), entropy ($T\Delta S$), free energy (ΔG), and equilibrium binding constant (K_b).

A DSC can measure the *calorimetric enthalpy* (ΔH), and a parameter named heat capacity ΔC_p (the amount of energy necessary to raise the temperature of a gram of substance one degree Celsius), or in other words, the temperature derivative of the enthalpy function, [C_p= (dH/dT)p]. The change in heat capacity reflects the fact that the amount of heat required to raise the temperature of a solution of unfolded protein is greater than that required for a solution of folded protein of the same concentration. This increase in heat capacity upon unfolding results primarily from the restructuring of solvent (Edsall 1935; Madan and Sharp 1996).

While ΔC_p is slightly temperature dependent (Gill et al.1985), the assumption of a constant ΔCp does not lead to significant errors in any other parameter (Privalov and Gill 1988). It is usually assumed that ΔC_p is constant or changes by a minimal amount as a function of a limited T range and expressed as a derivative ($d\Delta C_p/dT$). Typical values of ΔC_p for protein unfolding are in the range of 0.71–4.78 cal/mol. DSC experiments are done at constant pressure, as the volume changes in protein unfolding are very small (<5 percent), then C_p (T) $\cong$ C_v (T) (Seelig and Seelig 2021). ΔG for the unfolding reaction is obtained by integration:

$$\Delta G_{N\text{-}D} = \Delta H^* \left(1 - T/T^*\right) - \int \Delta C_p dT + T \int (\Delta C_p/T) dT \qquad (4.5)$$

ΔH^* is the heat of denaturation at the transition temperature T^*.

To integrate the equation above, it is necessary to know the temperature dependence of ΔC_p.

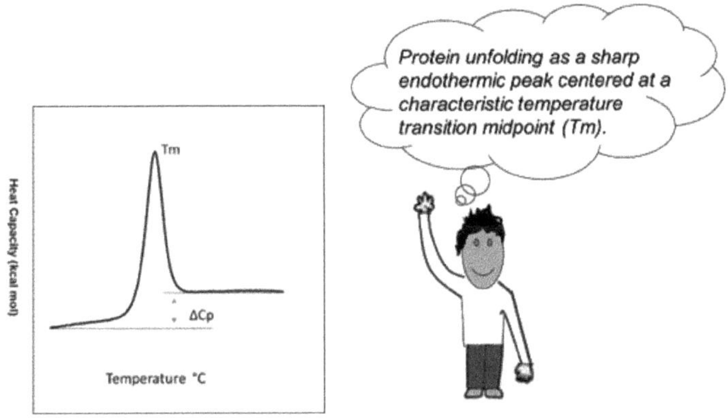

Figure 4.2. A DSC signal obtained for a model protein 1 mg/ml. Scan rate of 60°C h^{-1}. The diagram shows the definition of Tm and ΔC_p. The area under the calorimetric transition is the calorimetric enthalpy.

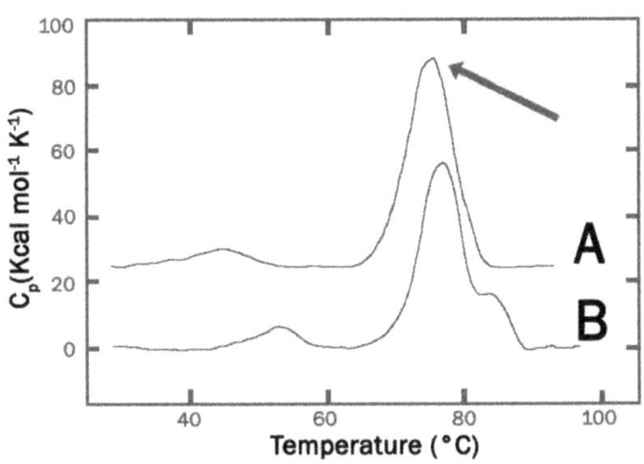

Figure 4.3. A DSC profile for a monoclonal antibody under two different solution conditions (A and B). In A, the monoclonal antibody is in 10 mM histidine pH 6.0, and in B, it is in the same buffer but with 2 percent sucrose. The peak indicated by the arrow corresponds to the melting temperature (Tm) for the Fab portion of the monoclonal. The area under the curve is the calorimetric enthalpy (ΔH_{cal}). In B, the Tm has slightly shifted to the right, suggesting that sucrose has a stabilizing effect.

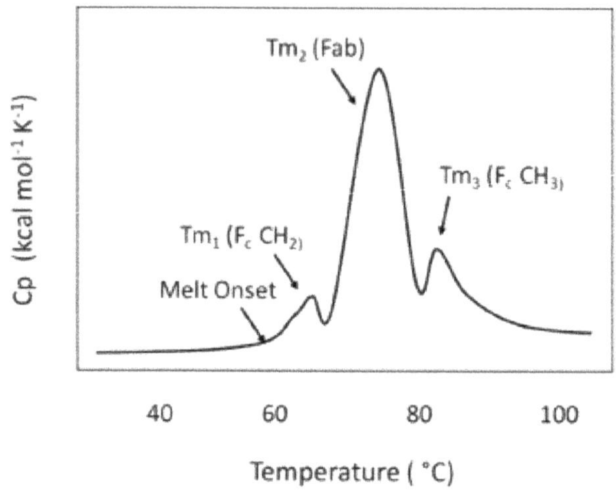

Figure 4.4. A DSC thermal unfolding profile of a humanized IgG1 antibody, showing the different transitions for the domains.

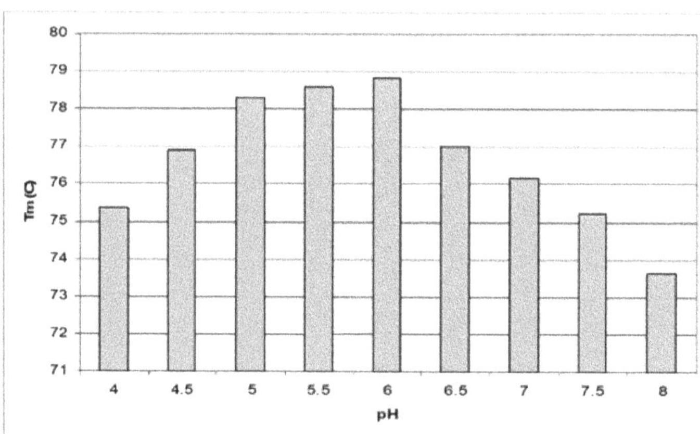

Figure. 4.5 pH screen: apparent T_{m2} for an IgG as function of pH, showing improved thermal stability at pH 5-5-6.0.

Buffer	Mab1			Mab2			
	T_{onset} (°C)	T_{m1} (CH2+Fab)	T_{m2} (CH3)	T_{onset} (°C)	T_{m1} (CH2)	T_{m2} (Fab)	T_{m2} (CH3)
A5.2	64.6	70.5	83.2	43.9	49.9	71.2	84.0
A5.5	65.3	71.1	83.5	43.4	50.2	71.7	84.1
H5.5	62.9	69.0	82.0	43.1	49.6	69.7	82.5
H6.0	65.0	70.4	83.0	43.1	49.9	71.1	83.9
H6.5	66.1	71.2	83.6	41.5	50.1	71.5	84.1
KP6.5	66.4	71.8	83.2	39.9	49.2	72.1	83.7
KP7.0	66.0	71.6	83.2	41.4	48.0	71.8	83.4

Table 4.1. Mab1 and Mab2 differ by one amino acid in the Fab region. A5.2 and A5.5 are 10 mM sodium acetate buffers at pH 5.2 and 5.5 respectively. H5.5, H6.0, and H6.5 are 10 mM histidine chloride buffers at pH 5.5, 6.0, and 6.5 respectively. KP6.5 and KP7.0 are 10 mM sodium phosphate buffers at pH 6.5 and 7.0 respectively.

Table 4.1 shows the comparison of the thermal stability of two monoclonal antibodies that differ in a single amino acid in the Fab region. Examining the onset of melt for the two constructs, it can be observed that Mab1 has a better thermal stability behavior at all pH tested.

The energetics of globular proteins folding and unfolding show remarkable regularity (Robertson and Murphy 1997), reflecting the common nature of the stabilizing forces in protein structure and function. When the ΔH change for different globular proteins is compared at a given temperature, a correlation emerges between the ΔH change and the size of the protein (Robertson and Murphy 1997). This suggests that proteins of similar size, which exhibit differences in ΔH, might have unstructured regions.

Upon protein unfolding, the major contribution to the ΔH change arises from the disruption of intramolecular interactions (hydrogen bonds, van der Waals, ionic, etc.) and the parallel solvation of the interacting groups. The change in ΔCp is also correlated to the size of the protein (Robertson and Murphy 1997). The heat capacity change is mainly related to the changes in hydration of groups that are varied from the solvent in the native protein and become solvent exposed upon unfolding. Before we continue with the description of the DSC applications, we need to introduce some concepts on the relationships between the changes in free energy with temperature.

The Gibbs-Helmholtz Equation

Frequently, we are interested in the way that ΔG, ΔH, and ΔS vary with temperature at constant pressure. If we know how G, H, and S vary with temperature for each of the two states of interest, we can find the temperature dependence of ΔG, ΔH, and ΔS. The Gibbs-Helmholtz equation relates the change in Gibbs energy to its temperature dependence and the position of equilibrium to change in enthalpy. Hermann von Helmholtz (1821–1894) was a German physiologist and physicist who made significant contributions to science. He postulated that everything was connected through energy conservation as a universal law and could be mathematically proven.

For a reaction where the temperature is held constant (eq. 4.2), and from the first law of thermodynamics, we know that:

$$H = U + PV \qquad (4.6)$$

U is internal energy, and PV is pressure and volume, respectively. If we have an infinitesimal process:

$$\Delta H = \Delta U + P\Delta V + V\Delta P \qquad (4.7)$$

Since:

$$\Delta U = T\Delta S - P\Delta V \qquad (4.8)$$

We find that:

$$\Delta G = V\Delta P - S\Delta T \qquad (4.9)$$

This shows the dependence of ΔG on temperature and pressure. To observe the change in Gibbs energy due to temperature change alone with constant pressure, like most of the biochemical reactions in the laboratory, the equation becomes:

$$\Delta G = -S\Delta T \qquad (4.10)$$

Solving for S and plugging it into eq. 4.2, the Gibbs-Helmholtz equation is found:

$$(\partial(\Delta G/T)\partial T)p = -\Delta HT^2 \tag{4.11}$$

A commonly written form of the Gibbs-Helmholtz equation for the free energy of unfolding *(ΔG_u)* is:

$$\Delta G_u(T) = \Delta H_{\mathrm{m}}(1 - T / T_{\mathrm{m}}) - \Delta C_p\left[(T_{\mathrm{m}} - T) + T\ln(T / T_{\mathrm{m}})\right] \tag{4.12}$$

The positive nature of the change in heat capacity, ΔC_p, leads to a parabolic relationship between ΔG and temperature (see figure 4.6), where there is a precise temperature of maximum stability, and there is a high- and low-temperature boundary where the free energy of the protein favors the unfolded state (Kumar et al. 2003). The shape and curvature of this free energy profile is unique to each protein and based on its specific thermodynamic parameters.

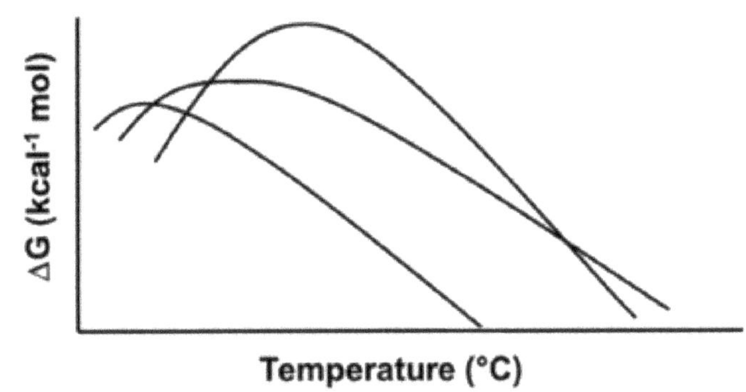

Figure 4.6. The free energy of folding for a typical protein is also temperature dependent at low temperatures. The shape of the curve is protein-dependent and determined by the protein's enthalpy, entropy, and heat capacity. At low temperatures, other factors (kinetic effects) help to prevent or slow aggregation.

DSC Scan Reversibility as a Temperature Function

To obtain accurate thermodynamic parameters from a DSC experiment, we should test for the scan's reversibility as a temperature function. As the equilibrium is being perturbed by scanning in temperature, thermodynamic reversibility in the DSC experiment is better demonstrated by showing that the Cp function is independent of scan rate. The quantitative evaluation for unfolding to test the two-state model (reviewed by Privalov 1979; Privalov and Khechinashvili 1974) can be expressed as $\Delta H_{cal}/\Delta H_{VH} \geq 1$, where ΔH_{cal} is the calorimetrically measured heat of unfolding (area under the scanning curve), and ΔH_{VH} is the apparent enthalpy change found from the temperature dependence of the apparent equilibrium constant for two-state unfolding (the Van't Hoff reaction).

If unfolding is two-state, $\Delta H_{cal} = \Delta H_{VH}$; otherwise, ΔH_{cal} is greater. Studies of single-domain proteins have given values of $\Delta H_{cal}/\Delta H_{VH}$ close to 1.0, whereas papain (a small protein whose X-ray structure has two well-defined domains) shows $\Delta H_{cal}/\Delta H_{VH} = 1.8$ (Tiktopulo and Privalov 1978). For more "complex" proteins that undergo irreversible thermal denaturation, scan rate–dependent DSC may lead to a detailed characterization of *protein kinetic stability* in terms of the corresponding denaturation free energy barrier as has been proposed by José Sanchez-Ruiz from the University of Granada in Spain (Plaza del Pino et al. 2000; Rodriguez-Larrea et al. 2006; Sanchez-Ruiz 2010).

The kinetic stability of proteins has tremendous implications for the development of therapeutic proteins by the biopharmaceutical industry. In addition, emerging molecular approaches to the inhibition of the formation or growth of amyloid structures (amyloid genesis is implicated in many human diseases, including Alzheimer's disease (AD), Parkinson's disease (PD), and type II diabetes) focus on the increase of the kinetic stability of protein native states (Plaza del Pino et al. 2000; Petrasi et al. 2005). For small, fast-folding proteins, deviations from the equilibrium two-state behavior detected by DSC thermograms can be interpreted as marginal thermodynamic free energy barriers for folding/unfolding (Muñoz and Sanchez-Ruiz 2004; Naganathan et al. 2005). This may allow the fast and convenient identification of potential instances of global downhill folding (Garcia-Mira et al. 2002).

DSC Practical Considerations

A correct DSC experiment consists of at least three runs: a buffer scan and two scans of the sample under study. The second scan of the sample is typically used to assess the reversibility of the transition. The reversibility is usually expressed as the percentage of the original signal recovered in the second scan. This is critical since an analysis based on equilibrium thermodynamics cannot be applied to a sample that undergoes an *irreversible transition.* Protein samples for calorimetric studies need to be prepared following very rigorous and exact procedures that guarantee the purity and integrity of the protein under study. Protein concentrations must be accurately determined since any error in concentration will be reflected directly in the measured thermodynamic parameters. For example, a 3 percent error in the protein concentration will introduce a 3 percent error in the enthalpy change.

Samples for calorimetric analysis need to be at the same chemical potential. This can be accomplished by extensive dialysis against the buffer of interest. Samples should be free of particles and aggregates and degassed before loading the sample in the calorimeter. The degassed buffer used in the last dialysis during sample preparation must be loaded into the reference cell of the instrument to ensure that the solvents in the two cells are identical. This is critical if absolute values of the heat capacity are needed.

The reference and sample cell in a DSC operates in a differential mode, such as the difference in heat capacity between them being the quantity measured. This is accomplished by continuously monitoring the differential electrical power required to maintain the temperature difference between the two cells equal to zero as the temperature of the entire setup is scanned at a constant rate (for experimental details, see Cooper et al. 2001). The differential electrical power (e.g., µcal/s) after normalization by the scanning rate yields the difference in heat capacity between the two cells. If the two cells were identical, a single scan would be enough to determine the difference in heat capacity between the sample and the reference solution. However, the two cells are never identical, and two separate scans are required to subtract and eliminate instrument defects (Cooper et al. 2001).

The exclusive association of the observed high Tm with optimal stability or low Tm with poor physical stability could be misleading in certain situations. A classic example is polyethylene glycol (PEG), which, at a low storage temperature, can stabilize proteins yet exhibit lower melting temperatures (Tm) because of protein hydrophobicity and the molecular size of the PEG used. At high temperatures, high molecular weight PEGs preferentially interact with the denatured state, giving a low Tm, but at low temperatures, they are preferentially excluded from the native state (Timasheff's mechanism, see Chapter 5). If lower Tm means lower stability at lower temperatures, it would be erroneous in the case of PEG.

Therapeutic proteins developed by pharmaceutical companies need to be stable for at least twenty-four months in the liquid state without a change in their potency and purity. These products are usually stored refrigerated between 2–8°C. Hence, maintaining the conformational stability at the recommended stored conditions is crucial for marketing and distribution. However, the drug product, once it reaches the market, may be subjected to several excursions or variables that could affect the conformation of a protein, such as extreme temperatures, agitation, exposure to different light, freezing, presence of oxidants, detergents, etc. Conditions need to be found by protein formulators to maintain the stability of proteins.

In a calorimeter experiment, heat is given to the protein at a rate of 1°C/min, yielding a positive and significant ΔC_p, making Gibbs free stabilization energy very sensitive to temperature. Single amino acid mutations in proteins have provided another way to probe the stabilization of proteins (Matthews et al. 1987; Matthews and Fersht 1995). The rationale for those studies is that the contribution of a particular amino acid to protein stabilization can be evaluated by adding or removing a side chain involved in those interactions. The difference between the wild type and the mutant stability, $\Delta\Delta G$, is typically taken as the contribution of the given amino acid substitution. Similarly, values of $\Delta\Delta H$ and $\Delta\Delta S$ can be obtained to map the thermodynamic stabilization of a given protein (Shirley et al. 1992).

Several DSC studies have revealed several regular features regarding the thermal denaturation of globular proteins. A large positive ΔC_p always accompanies denaturation. The specific heat capacity change, normalized to the number of residues in the protein, $\bar{\Delta} C_p$ (the amount of heat that must be added to one unit of mass of the substance to cause an increase of one unit in temperature), is directly proportional to the number of nonpolar contacts in the protein interior (Privalov and Gill 1988).

Although exposure of the polar surface makes a significant negative contribution to $\bar{\Delta}Cp$ (Murphy and Gill 1990, 1991; Privalov and Makhatadze 1990), globular proteins generally bury a constant amount of polar surface area per residue (Murphy and Freire 1992; Murphy et al.1992), and $\bar{\Delta}Cp$ can be considered to some degree as a measure of the hydrophobicity of the protein.

DSC is generally accepted as the technique of choice to determine the energetics of protein folding/unfolding transitions and the thermodynamic mechanisms underlying those reactions. DSC is a technique used in the biopharmaceutical industry to identify solution conditions that could stabilize a therapeutic protein during formulation development (Guziewicz et al. 2007).

In the preceding chapter, we discussed that proteins are *alive and kicking*. They are dynamic entities that constantly sample different partially folded states as a function of temperature and other solution variables. These variables dictate the standard free energy between the native, unfolded, and partially folded aggregation-prone states, leading to oligomerization. As a result, not all oligomerization events in proteins are similar. Altering these variables, such as temperature, salt, pH, and ligands, can induce a protein to aggregate due to "unnatural folding" to balance the thermodynamically unfavorable interactions between solvent and exposed hydrophobic residues in proteins. These variables may induce a protein to self-associate in mostly the native state to counteract the unfavorable interactions with the solvent. It is possible to have cold instability without inducing cold denaturation, destabilizing the native state of a protein and making it more prone to aggregation events (Pérez-Ramírez and Simler 2019). Therefore, stability studies at higher temperatures may not always predict protein behavior under refrigerated conditions.

Thus, protein aggregates can form from partially nonnative species (A*) that could be in equilibrium with the native species (N). A* can be part of the native state assembly and recruit other A* to form aggregates, as shown in figure 4.7 below. Thus, under conditions greatly favoring the native state, aggregation can proceed (figure 4.8).

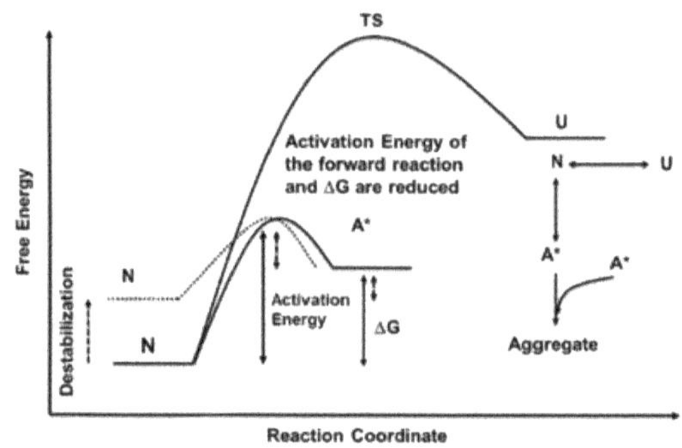

Figure 4.7. Illustrates the aggregation results from a repeated population of an aggregate-prone partially folded state (A*).

(Adapted from Pérez-Ramírez and Simler 2019).

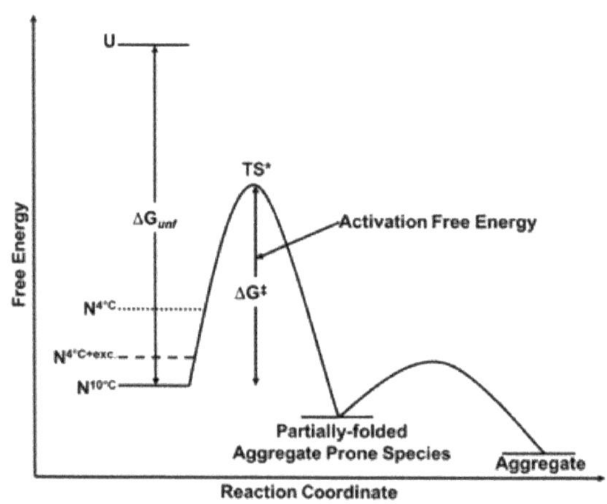

Figure 4.8. Schematic free energy diagram for the aggregation of a model glycoprotein at 10°C (N10°C, solid line), 4°C (N4°C, dotted line) and 4°C in the presence of a stabilizing excipient (N4°C+ exc.). ΔGunf and ΔG‡ are the free energy of unfolding and the activation free energy respectively. TS* is the transition state between the native form of MG and the partially folded aggregate-prone species

(Pérez-Ramírez and Simler 2019).

Privalov was also the first to describe the phenomenon of cold denaturation of proteins, where they lose their tertiary and quaternary fold at low temperatures (see review by Privalov and Makhatadze 1990). Privalov showed that cold denaturation is a general phenomenon (also reported in monoclonal antibodies, Lazar et al. 2010) caused by the specific and strongly temperature-dependent interaction of protein nonpolar groups with water. Privalov and coworkers were able to demonstrate that myoglobin, which has the largest ΔCp value known, can be cold-denatured in acetate buffer at acidic pH by supercooling the protein solution to −8°C. Consistent with the prediction, cold denaturation of myoglobin proceeded with the release of heat.

For all known proteins, cold denaturation occurs below the freezing point of water (273 K) and is difficult to observe experimentally. Water effectively becomes more hydrophobic as it cools, with fewer hydrogen bonds to the protein at a lower temperature, furthering the solvation of hydrophobic regions, which drives the unfolding of the protein. In contrast to expectations, hydration of these groups is favorable thermodynamically, i.e., the Gibbs energy of hydration is negative and increases in magnitude at a temperature decrease. As a result, the polypeptide chain, tightly packed in a compact native structure, unfolds at a sufficiently low temperature, exposing internal nonpolar groups to water.

Kinetic Versus Thermodynamic Stability

Protein stability can be defined in two ways. 1) *Thermodynamic stability*—related to a low amount of unfolded and partially unfolded states in equilibrium with the native, functional protein. Thermodynamic stability is strictly a function of the change in free energy (ΔG°), which is a state function, meaning that its value is determined exclusively by the difference between the initial state and the final state (or the free energy of the initial products and the final products). In other words, thermodynamic stability is independent of the pathway between reactants and products.

2) *Kinetic stability*—which is related to a high free energy barrier "separating" the native state from the nonfunctional forms (unfolded states, irreversibly

denatured proteins). Such a barrier may guarantee that the protein's biological function is maintained, at least during a physiologically relevant time scale, even if the native state is not thermodynamically stable concerning nonfunctional forms.

In general, kinetics treats things that impact the rate at which a reaction occurs. One implication is that reaction kinetics are partially a function of the stability of the intermediate species (and *transition states*) that the pathway of a chemical reaction must necessarily pass through in the journey to the formation of the final product. The *transition state*, or in other words, the intermediate state between the initial state (reactants) and the final state (products), is where the molecules do not have their original shape and are highly unstable (i.e., highly energetic) relative to the initial reactants. Then the initial barrier of activation energy that must be surpassed to set the reaction in motion will be relatively high, and the reaction will tend to proceed slowly (if at all).

The energy required to cross over the *transition state* is called the *activation energy (Ea)*. It is the minimum amount of energy required to activate a chemical reaction. This minimal energy is obtained from the collisions of the water molecules constantly bumping into each other in the cytosol and colliding with the reacting molecules. However, not all interactions (bumping) are productive; chemical reactions take time.

Enhancement of kinetic stability is essential for many biotechnological applications of proteins. All that is required is a sufficiently high free energy barrier "separating" the native state from the nonfunctional forms. Assuming transition-state theory applies, the rate of irreversible denaturation is related to the free energy barrier ($\Delta G^{\neq}$) (see Idakieva et al. 2005 and references quoted therein). If the activation free energy value is large enough, the rate of irreversible denaturation will be low and the protein will remain in the native, functional state during a sufficiently large time scale. In this case, we say that the protein is kinetically stable.

Let us set reactants X + Y. There are two different pathways through X + Y, both of which lead to final products that are lower in free energy than the reactants X + Y. Pathway A has much higher activation energy than Pathway B, but the final products at the end of Pathway A are lower in free energy (i.e., more thermodynamically stable) than those at the end of Pathway B. This means that Pathway A is thermodynamically favored, but Pathway B is kinetically favored (that is, it poses a lower barrier in terms of activation energy, and the reaction will proceed more rapidly via this pathway). The consequence of this is that the kinetic pathway, Pathway B, will be the favored one unless environmental conditions (temperature, pressure, concentrations of catalysts, etc.) are tuned in such a way that the higher activation energy of Pathway A can be surpassed.

An example of kinetic versus thermodynamic control is shown by the influence of N-linked glycans on the stability of glycoproteins, which has been studied using horseradish peroxidase isoenzyme C (HRP), which contains eight asparagine-linked glycans. HRP was deglycosylated (d-HRP) with trifluoromethanesulfonic acid and purified to an enzymatically active homogeneous protein containing (GlcNAc)2 glycans.

The thermodynamic stability of horseradish peroxidase isoenzyme C (HRP) in equilibrium studies at various temperatures is indistinguishable from that of chemically deglycosylated HRP (d-HRP), whereas the unfolding rate constant of d-HRP in guanidinium hydrochloride (GuHCl), and hence the kinetic stability, decreases two to threefold. The results are compatible with a glycan-induced decrease in the dynamic fluctuation of the polypeptide chain (Tams and Welinder 1998).

Figure 4.9 summarizes the main differences between thermodynamic and kinetic stability. Thus, when performing experiments using DSC (e.g., for excipient screening) at a high scan rate, a change in the *kinetics* and not the thermodynamics of unfolding could lead to the identification of false changes in thermal stability (Guziewicz et al. 2007). Therefore, when characterizing a protein, several orthogonal techniques based on different principles must be employed to avoid pitfalls and cross-verify conclusions drawn from within different data sets, adding to the robustness of the package. This will also serve to provide a robust overall package of information on the nature of the characteristics of the protein. Examples of orthogonal techniques used in the characterization of protein aggregates are size exclusion chromatography (SEC) combined with a multi-angle light scattering (MALS) and a concentration detector (either UV or differential refractive index) or the use of analytical ultracentrifugation.

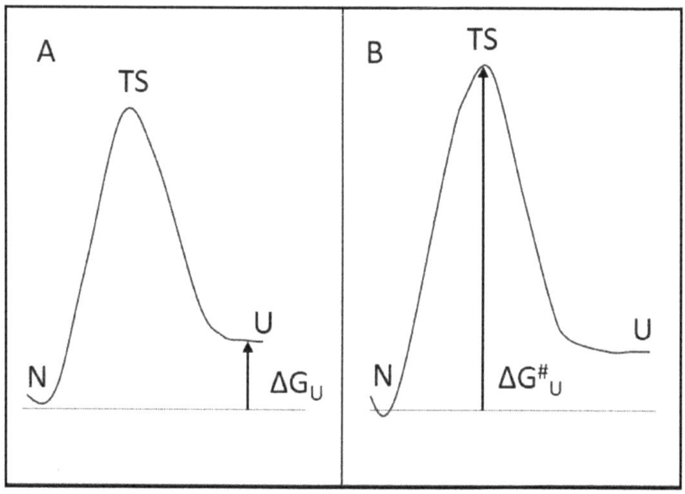

Figure 4.9. Free energy diagrams illustrating the different types of protein stability. *(A) The thermodynamic stability of proteins is defined as the difference in free energy (ΔG_U) between the native (N) and unfolded (U) states. The low transition state (TS) free energy for unfolding causes the N and U states to exist in equilibrium, which favors the N state. (B) Kinetically stable proteins (KSPs) are characterized by a high activation energy ($\Delta G^{\#}{}_U$) for unfolding that "sequesters" them in their native state (N). The stability of such proteins is under kinetic control as their unfolding rate is very slow, which results in a longer half-life.*

Table 4.1 provides a list of the most common analytical assays used in characterizing physical stability of proteins.

Table 4.1. Analytical assays to characterize the physical stability of proteins.

Category	Technique	Application
	Circular dichroism (CD)	Secondary and tertiary structure (aqueous)
	Fourier transform infrared spectroscopy (FTIR)	Secondary structure (aqueous, solid)
	Fourier transform Raman spectroscopy	Secondary structure (aqueous, solid)
Structural characterization	Second-derivative ultraviolet spectroscopy	Tertiary structure (aqueous)
	Fluorescence spectroscopy	Conformational changes, ligand binding (aqueous)
	Differential scanning calorimetry (DSC)	Structural stability, unfolding (aqueous)
	Hydrogen deuterium exchange spectroscopy	Local structural dynamics
	Analytical ultracentrifugation (AUC)	Matrix independent size, association behavior (soluble aggregates)
	Dynamic light scattering	Hydrodynamic size
	Static light scattering	Molar mass, size
	Size exclusion chromatography (SEC)	Size estimation and quantitation of soluble aggregates (non-dissociable)
Degradation product detection	SDS-PAGE	Size estimation, covalent vs. noncovalent aggregates
	Light obscuration	Size estimation (particles/insoluble aggregates)
	Coulter counter	Size estimation (particles/insoluble aggregates)
	Microflow imaging	Size estimation (particles/insoluble aggregates)

Figure 4.10. Josiah Willard Gibbs (1839–1903) was an American scientist elected to the National Academy of Sciences in 1879. Together with James Clerk Maxwell and Ludwig Boltzmann, he created statistical mechanics (a term that he coined), explaining the laws of thermodynamics as consequences of the statistical properties of ensembles of the possible states of a physical system composed of many particles.

CHAPTER 5

SOCIALIZING: BINDING, EXCLUSION, STABILITY

We have now discussed proteins based on their conformation dictated by their amino acids' constituents. The conformation is not static; proteins constantly interact with their environment. The environment surrounding a protein may have a stabilizing or destabilizing effect. Some conformational flexibility is required, which explains why the difference in the free energy of stabilization of the native state compared with the unfolding state is very small (see Chapter 3). This loose conformational stability permits proteins to exert their biological function as enzyme-substrate interactions, ligand-receptors, or the interaction protein-protein to form organized structures like collagen, microtubules, multi-enzymatic complexes, etc. Hence, proteins like to socialize among themselves and with other molecules present in their immediate surroundings (water, chemicals, etc.).

Protein interactions with the ligands in solution components can modulate protein conformation, stability, and their intrinsic biological role. Protein self-association can affect the secondary structure of the assembling subunits. However, the effects are variable and depend on the nature of the protein undergoing self-association. For example, the low molecular weight 4S aggregate of wild-type tobacco mosaic virus coat protein (TMVP) self-assembles into a high molecular weight 20S aggregate at pH 7.0 upon warming from 5 to 20°C. In addition, a 4S aggregate of recombinant E. coli-expressed TMVP self-assembles into a 28S aggregate under similar conditions.

The 4S, 20S, and 28S aggregates were found to exhibit very similar far-UV CD spectra, indicating that the gross secondary structure of these proteins was not affected by self-assembly (Shire et al. 1990). On the other hand, far-UV CD revealed a significant change in protein secondary structure upon the formation of the hexamer $12S_H$ subunit of bacterial transcarboxylase from 2.5 S_H monomers (Hennessey 1982). The studies on transcarboxylase provide a clear example where quaternary interactions affect the secondary structure of the assembling protein subunits. Thus, the results of the studies on TMVP and transcarboxylase indicate that the effect of protein self-association on secondary structure cannot be generalized.

Ligand Binding

A typical event in living systems is the noncovalent binding of one or more molecules to a single macromolecule. A molecule bound to a macromolecule is called a ligand. A molecule may have a single binding site or there be many sites (identical or different). Multiple sites may be interdependent. Occupation of one site by a ligand may affect the affinity of other sites. A well-established principle of binding by proteins is that ligands must compete with water and other small molecules to form interactions with protein surfaces. It has also been shown that ligands must compete with interactions among residues at protein surfaces. Results from structural surveys, a variety of experimental studies, and computations suggest that intramolecular interactions are present at protein surfaces, and the energetics of these interactions can change when proteins bind to other molecules (Robertson 2002).

Why Study Ligand Binding to Proteins?

Many parts of the compact protein structure can contribute toward the shift in energy and structure taking place on ligand binding. The energies of ligand-ligand interaction, although small, are sufficient to shift the covalent equilibria in which proteins take part to a significant extent, and this may be found responsible for the interconversion of chemical and osmotic energies in metabolism (Weber 1975). Understanding the mechanism of action of enzymes and regulatory systems usually requires knowledge of the number of binding sites and the strength of binding. Purification and process development may require a protein to be bound to a particular ligand for separation or to maintain stability. Biopharmaceutical formulation may require a protein to be bound to a ligand to maintain physical and chemical stability.

Most protein stabilizers are either bound or excluded from the protein. Since all biological macromolecules also bind multiple ligands, the effects of one ligand are generally influenced by or "linked" to the binding of other ligands.

For example, all proteins and nucleic acids bind ions as well as protons (Lipfert et al. 2014; Yang et al. 2019). Proton fluctuations may contribute to protein interactions (Kirkwood and Shumaker 1952a, b). Hence, the effects of pH and ion concentration on a protein-protein interaction are generally linked.

Jeffries Wyman developed the theoretical foundations for linked function analysis of macromolecular-ligand interactions over the past fifty years (see Wyman 1964; Wyman and Gill 1990). Wyman made fundamental contributions to our understanding of the binding of oxygen by hemoglobin and the linkage of this process with the binding of hydrogen ions and other species in the solution.

Wyman (1901–1995) was born in West Newton, Massachusetts. His ancestors were New Englanders, and he was the third in successive generations to bear that name. The first Jeffries Wyman (1814–1874) was a great comparative anatomist and the first director of the Peabody Museum of American Archeology and Ethnology at Harvard. In 1960 Eraldo Antonini, from the University of Rome's Biochemical Institute and the Instituto Regina Elena, invited Wyman to visit his department, which was headed by Alessandro Rossi Fanelli. When they offered Wyman a position as guest scientist, he accepted for a "trial period" that lasted twenty-five years. While working in Rome, Wyman fully developed the theory of linked functions and reciprocal effects in his landmark 1964 paper in *Advances in Protein Chemistry*. Using straightforward thermodynamic principles embodied by the first and second laws, Wyman showed that the responses of a macromolecular system to chemical and physical variables like chemical potentials, pHs, and temperature, are mutually dependent.

Wyman's logic of linked function analysis provided the basis for explaining allosteric effects. In simple terms, the Wyman linkage analysis indicates that the binding of a ligand will drive a conformational change only if the ligand binds more to one conformation than the other.

Wyman met Jacques Monod while he was in Paris from 1952 to 1955. Monod and his group were working on regulatory aspects of metabolism, and especially feedback inhibition of metabolic pathways. Monod and Jacob had introduced the term "allosteric" in 1961 to describe enzymes that can bind effector molecules at sites quite distinct from the catalytic site (reviewed by Alberty and Di Cera 2003). The binding of the effector induces conformational changes that promote or inhibit catalytic activity. When Wyman visited Paris in 1964, Jacques Monod invited him to give a seminar on his views about the effects of conformation changes in hemoglobin on ligand binding and their relation to cooperativity and effectors. These discussions led to the famous paper by Monod, Wyman, and Changeux (1965) proposing that allosteric proteins are oligomers composed of several subunits (protomers), the oligomer being capable of existing in two distinct conformations, denoted as T and R, which differ in their affinity for ligands and effectors.

Homotropic interactions are always cooperative, and heterotrophic interactions, which are caused by displacement of the R/T equilibrium by the effector, can involve either activation or inhibition at the catalytic site for an enzyme or at ligand-binding site for hemoglobin. Thus, binding studies and linkage effects aim to provide information on those critical aspects of macromolecular function that can be related to structural features.

Figure 5.0. The equilibrium constant for a given reaction is the first step to get to the Gibbs free energy.

Figure 5.1. Jeffries Wyman, biochemist. Photograph from Wellcome Collection. Public Domain Mark.

Let us look at the effect of ligands. Take reaction:

$$E + S \rightleftarrows E^*S \; ; \quad K_{overall} = \frac{[E^*S]}{[E][S]}$$

Apparent reaction:

$$E \overset{S}{\rightleftarrows} E^*S \; ; \quad K_{app} = \frac{[E^*S]}{[E]}$$

$$K_{overall} = \frac{K_{app}}{[S]}$$

And by Wyman:

$$\left[\frac{\partial \ln K_{app}}{\partial \ln [S]}\right]_{T,P} = +1 \qquad (5.0)$$

Complete reaction is:

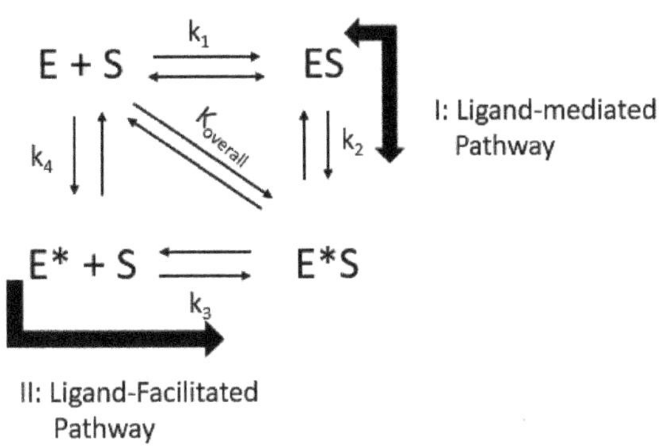

Path I: KI app = $k_1\, k_2\, [S]$
Path II: KII app = $k_3\, k_4\, [S]$

Then $k_1 k_2 = k_3 k_4 = K_{overall}$
and $k_3/k_1 = k_2/k_4$

If k_2 is different from k_4, then k_3 is different from k_4.

Consequence: If a ligand affects a reaction, then the ligand's binding constant must change during the reaction. Specifically, if a reaction is pH dependent, then the pK of an ionizable group must change (i.e., be different in the reactant and the product).

The first steps in any binding measuring reaction are (1) the determination of possible linkage between binding and protein self-association; (2) the establishment of specificity; and (3) the evaluation of the rate of re-equilibration of the binding reaction, since a true equilibrium constant can be obtained only if the re-equilibration is rapid relative to the length of the binding experiment.

Protein self-association is an intermolecular event and is, therefore, dependent on the total protein concentration. Consequently, a ligand binding linked to protein self-association must depend on protein concentration. However, the association state of the protein must be established directly in the presence of the ligand by a suitable hydrodynamic (analytical ultracentrifugation) or scattering technique at varying ligand activity.

The establishment of specificity should be examined at different levels. First, does the ligand bind exclusively to the protein being analyzed, or can it bind to a variety of proteins? This can be investigated by binding experiments to various proteins. Specific binding always involves the formation of discrete contacts with specific regions on the protein molecule. These frequently result in well-defined perturbations in protein and ligand spectroscopic characteristics that can give important insight into the interactions' nature. A test of the rate of re-equilibration can be performed if the binding induces a spectral change either in the ligand or the protein, and the reversibility could be established with a competitive ligand that could displace the first ligand.

In the analysis of binding data, a reasonable approach would be to first try the *Scatchard* equation:

$$\frac{r}{[A]} = \frac{n}{Kd} - \frac{r}{Kd} \qquad (5.1)$$

r is equal to the number of moles of ligand bound to one mole of macromolecules, which is defined as:

$$r = \frac{[A]bound}{[P]total} = \frac{[PA]}{[P]+[PA]} \qquad (5.2)$$

A is the ligand, and P the protein.

With Eq 5.1 (the Scatchard equation), a plot of r/[A] versus r gives a line with a slope of -1/Kd and x-intercept of n/Kd.

George Scatchard (1892–1973) was born in Oneonta, New York. In his insight into the thermodynamic relations of systems containing many components, he was directly in the line of descent from Willard Gibbs. He also resembled Gibbs in his resolutely brief and austere presentation of his results. It was often not easy to follow the path by which he traveled from basic principles to results, but those who made the effort could always verify the fact that the path was there and that it provided a firm footing for the traveler (Edsall and Stockmayer 1980).

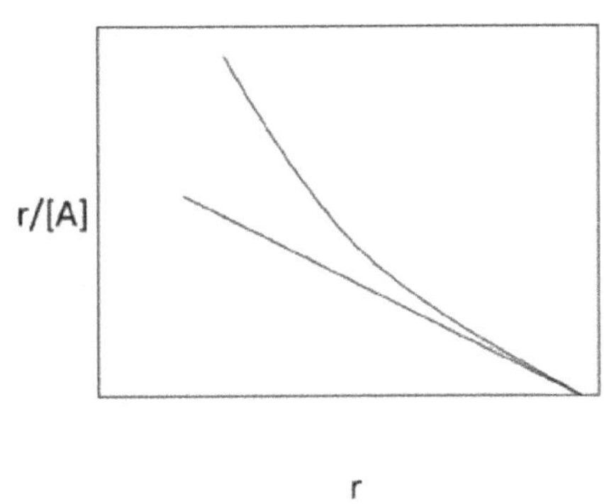

Figure 5.2. Scatchard plot for ligand binding to a protein with two binding sites of identical affinity (linear plot). The curvilinear plot represents the binding of a ligand to a protein with several binding sites for the same ligand with different affinities. The same curvilinear plot could be obtained if there is ligand-induced protein self-association.

Depending on the macromolecular concentration, the intrinsic binding constant, and the mechanism of the equilibrium linkage, the Scatchard binding curves can be concave either upward or downward, and those obtained at different protein concentrations could either converge to a single point on the y-axis or diverge (Can and Hinman 1976; Nichols and Winzor 1976). These characteristics are like those exhibited by allosteric (multiple ligand-binding sites on a protein that shows cooperativity between binding) and heterogeneous binding (multiple ligand-binding sites on a protein that are independent and with different affinity) and can be confused as such. The shape of the Scatchard plot can be used effectively in diagnosing and analyzing ligand binding linked to self-association (see Na and Timasheff 1985).

1. Ligand binding that *induces* self-association is defined as one in which the ligand binds to the monomer, causing it to self-associate.
2. Ligand binding that *facilitates* self-association is defined as one in which the ligand binds only to the polymer. The protein can undergo self-association in the absence of the ligand, but the association is enhanced by the binding of the ligand to the polymer.
3. Ligand binding that induces and facilitates self-association is defined as the combination of the above two. Ligands can bind to both monomers and polymers, and the self-association can take place in the absence of ligands but is enhanced by the binding of the ligand to the polymer.

The most abundant natural molecule surrounding proteins is water, which is essential for protein function. Nevertheless, proteins can work in an ambience with limited water. Some enzymes are dehydrated and maintain function (Tuena de Gómez-Puyou and Gómez-Puyou 1998; Karan et al. 2012). Several studies indicate that proteins require only a few water molecules to help maintain their structure and function fully.

Water that immediately surrounds the protein appears to be single and densely packed compared with bulk water and with less mobility than bulk water. If a substance is dissolved in a mixture of two solvents, one generally forms a more favorable interaction with one substance than the other. The result is that the solvent layer in immediate contact with the molecule tends to be richer in one of the two components, such that the solvent composition of the nearby solution bathing the molecule differs from that of the bulk solution. In biological molecules, the preferred solvent is most often water. Hence, the term preferential hydration is commonly used, which we will discuss later.

If we think of the interior of a cell as a very crowded environment, where many proteins and other structures are in close contact, the amount of water immediately surrounding the protein is minimal. There is much debate about the size of the water layer surrounding the proteins, but it appears to be not more than two water molecules thick. However, the charge of the protein plays a critical role in the magnitude of the water layer. Crystal structure analysis of proteins has shown that proteins contain water buried inside. Homologous proteins tend to have buried water in the same places.

As discussed in the preceding sections, the interaction between two proteins or the interaction of a protein with ligands such as water can be analyzed using the mathematical expression of the first law of thermodynamics that links the enthalpy and the entropy to the Gibbs free energy (Chapter 4). From the Gibbs energy, we obtain the strength of binding. Nevertheless, in several instances, when comparing the strength of different ligands to a protein, the magnitude of the change is small or nonexistent. Carefully inspecting the contribution of the enthalpy and entropy to binding may help elucidate the mechanism of the binding of a particular ligand to a protein, or the type of protein-protein interaction taking place, either with a significant conformational change in the protein or favoring hydrophobic interactions, etc.

Isothermal Titration Calorimetry to Elucidate the Mechanism of Binding

The best way to elucidate the mechanism of ligand-protein interactions is to use tools that can provide insight into the binding thermodynamics; in that regard, the isothermal titration calorimeter (ITC) is best suited for the task. ITC does not only provide the affinity of the ligand for the macromolecule but also a complete thermodynamic analysis. ITC measures the heat absorbed or generated when molecules interact. These heat changes are usually small, typically sub-millionths of a degree, but are universal and can be detected by these sensitive instruments. Briefly, a biomolecule solution is traditionally placed in the sample cell, and a ligand solution in a matching buffer is placed in the syringe (figure 5.3).

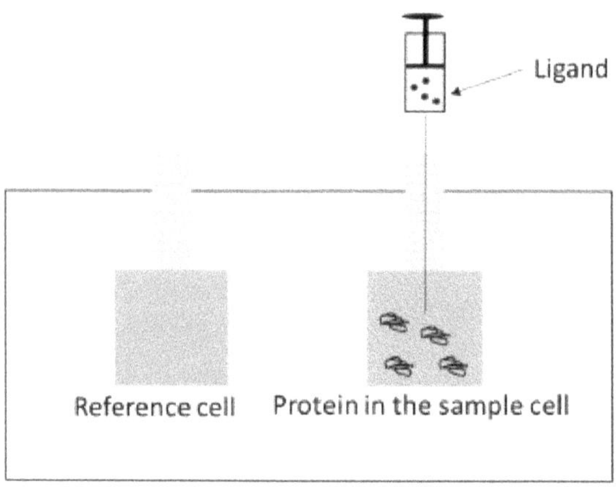

Figure 5.3. Isothermal titration calorimeter.

When the ligand solution is injected into the cell, the ITC instrument detects heat released or absorbed due to the interaction. This is done by measuring the changes in the power needed to maintain isothermal conditions between the reference and the sample cell. Injections are performed repeatedly in small increments, resulting in smaller peaks, indicating that the biomolecule has become saturated (figure 5.4A). Eventually, the peak sizes remain constant and represent only the heat of dilution. A critical aspect of an ITC experiment is choosing the proper concentration for protein and ligand. The choice of protein concentration depends on the "c" value as given by:

$$c = nP_t/K_b \quad (5.3)$$

Pt is the protein concentration in the measurement cell; *n* is the number of binding sites per protein molecule. The shape of the binding isotherm depends on the c value. The c value should be between twenty and a hundred to obtain a sigmoidal shape of the binding isotherm in order to get a good fit to estimate K_b, ΔH, ΔS, and n. If the K_b and *n* are not known, assume that the K_b is between 1×10^3 and 1×10^6 M^{-1}.

The heat of each injection is integrated with respect to time and plotted against the molar ratio of components. These are then fitted to the appropriate binding model to calculate the equilibrium association constant (K_b) or dissociation constant (K_d), enthalpy, and stoichiometry. If the equilibrium constant (K_b or K_d) and enthalpy are known, then the free energy (ΔG) and the entropy (ΔS) can be determined (see Velazquez Campoy and Freire 2005).

There are other ways to measure the interactions of a given ligand with a protein, such as ELISA (enzyme-linked immunosorbent assay) and surface plasmon resonance (SPR); the latter is commonly used in the biopharmaceutical industry to characterize interactions of proteins of interest in early research. SPR measures the change in the reflective angle of the incident light caused by surface waves called "plasmon polaritons." Since plasmon polaritons are sensitive to binding events occurring on a surface, they proportionally affect the angle of the reflected light (Drescher et al. 2009; Stabenfeldt et al. 2010).

Figure 5.4B shows representative ITC binding isotherms for two interactions with the same affinity but with different binding mechanisms. Polar interactions such as hydrogen bonds, van der Waals interactions, and London dispersion forces tend to contribute favorably to the enthalpic component, whereas entropically favored interactions tend to be more hydrophobic and associated with a conformational change. The entropy of binding arises due to changes in restriction/freedom of the backbone and side chain atoms and rearrangement or release of solvent water molecules and ions (Ladbury et al. 2010).

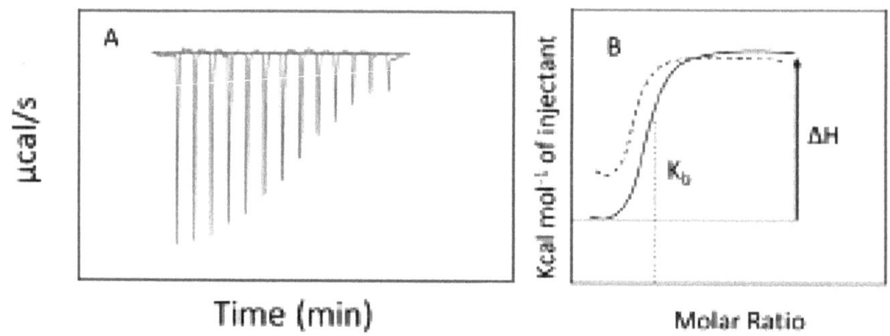

Figure 5.4 (A) ITC ligand peptide titrated into a binding model protein. (B) Binding isotherm for two binding peptides to the model protein that show similar affinity (K_b) but different mechanisms of interaction (dashed line). The dashed line is binding driven by hydrophobic interactions (entropy-driven interaction) compared with an enthalpy-driven interaction (continued line), which tends to be driven by hydrogen bonding and van der Waals interactions.

Apart from water, some other molecules or solvents stabilize proteins; some have importance in vivo, and others are used in the pharmaceutical industry to keep proteins stable for further analysis. Of the substances used commonly to stabilize proteins in vitro are sugars such as sucrose, sorbitol, and trehalose. Certain salts, such as ammonium sulfate, have also been known for a long time to stabilize biological activity.

Proteins are constantly interacting with different components in the molecular storm that takes place in the crowded cell environment. The thermodynamic properties of the system are a function of all the variables in the system. Often, we want to know how a given property changes when one of the other variables are kept constant. Mathematically, this is accomplished through partial differentiation and multicomponent solution thermodynamics. Given a function of two or more variables f(x1,x2, . . . ,xn), the partial derivative of f with respect to x1 gives the rate of change of f with respect to x1. It is computed by taking the first derivative of f with respect to x1 while holding all other variables x2, . . . ,xn fixed, i.e., treating them as constants.

Analysis of Complex Mixture Solutions

Many practical applications of protein thermodynamics are found in processes that involve mixtures of different components, such as the formulation development process in the biopharmaceutical industry to stabilize proteins. We call these mixtures multicomponent systems. The reason for developing the thermodynamics of multicomponent systems is the discovery of a series of phenomena not presented in the study of pure substances. The variation of the proportions of the components through mass transformation or chemical reactions gives the systems the possibility of reaching new types of equilibrium phases, which are governed by the degrees of freedom of the system.

The fact of not being able to explain the properties of mixtures as the average value of the properties of pure components leads us to the need to generalize the fundamental equations of thermodynamics for systems formed by more than one component or several phases (see Casassa and Eisenberg 1960, 1961,1964). The reason for jointly treating the states of matter (liquid, gas) with the same principles of thermodynamics lies in the fact that in a mixture of ideal gases, each of the components behaves as if it were in a state of purity, that is, with no interaction between the different gases. However, in a dissolution, this is not what happens. The presence of one of the solution species, for example, the solvent, affects the thermodynamic properties of the solute. Thus, the volume resulting when a liter of ethyl alcohol is mixed with a liter of water is less than two liters (1.93 liters) and not two as expected. This result is due to the intense forces of attraction of water and ethyl alcohol. In this case, the variable volume is a function of each component's relative number of moles *in the solution.*

The partial molar properties depend exclusively on the extensive nature of the mixing properties from which they come, but at no time do they depend on the principles of thermodynamics, nor do they prejudge the existence of entropy and the internal energy of the systems; therefore, it could have been studied long before the postulates of thermodynamics.

The equilibrium states of a thermodynamic system composed of m components are determined by m+2 variables. Of all the sets of independent variables capable of characterizing the state of equilibrium of the system, there is one that stands out for its practical value; this is the one that is described by the variables p, T, n_1 ..... n_m where p is pressure, T is temperature, and n_m is the number of moles of the component m. This set of variables has the following advantages:

1. Can be measured in any system.

2. *Extensive variables* (depends on the size or mass of the system, i.e., volume, total mass, heat capacity, internal energy, etc.) are functions of all variables in this set. For example: $U = U(p,T, n_1 .... n_m)$, $H = H(p,T, n_1 .... n_m)$, $S = S(p, T, n_1 .... n_m)$, $G = G(p,T, n_1 .... n_m)$... etc.

3. *Intensive variables* (do not depend on the size or mass of the system. i.e., temperature, pressure, density, specific heat capacity, etc.) are only a function of the m+1 first variables. For example: $Cp = Cp(p,T, x_1 ..... x_{m-1})$, $S = S(p,T, x_1 ..... m_{-1})$, $G = G(p,T, x_1 .... x_{m-1})$, etc., since it is verified that $\sum_m x_m = 1$ for each of the components in each phase.

The *intensive property* does not depend on the quantity of the substance taken; for example, the boiling point of water. *Extensive property* depends on the amount of substance taken; for example, the volume of water or heat required to boil water.

The most studied systems are those of two components, which will be described by the variables p, T, and x1. However, many times the concentration used is the molarity or molality, the latter being the preferred option; in such circumstances, the above set shall take the form *p, T, c, n* or *p,T, m, n*.

Partial Molar Properties

A partial molar property describes how an extensive property varies with changes in the molar composition of a solution. Technically, it is a *partial de-*

rivative of the extensive property with respect to the number of components of interest, assuming constant temperature, pressure, and components not of interest. The property that depends upon the quantity of matter in the system is known as an *extensive property.* The volume of the system is extensive. But the volume of one mole of gas or matter in the system is fixed and independent of the amount of matter present in the system.

When two substances are combined into a mixture, we expect the properties of the mixture to be some combination of the properties of the two pure substances. For some properties, we can simply take a weighted average of the two pure substances. Mass works like this; atoms of one substance do not suddenly become massive or less massive when they are combined with another substance. We can write:

$$m = m_A X_A + m_B X_B \tag{5.4}$$

However, not all properties behave in this way. When salt is dissolved in water, the volume of the water does not increase by the volume of the salt added. The volume increase is described by a quantity we call the *partial volume* of salt in water. As you might expect, the partial molar volume of salt in water is the increase in the volume of water caused by adding one mole of salt. In general, the partial molar property of substance X in a given mixture is the increase in that property of the mixture when one mole of X is added.

The modulation of biochemical reactions and biological processes by solvent additives, whether ligands, cosolvents, osmolytes, or denaturants, was first described by Jeffries Wyman in his theory of linked functions (Wyman 1948, 1964). The reciprocity between two linked processes, a biological reaction and the thermodynamic binding of a ligand, is stated by the Wyman linkage equation (Wyman 1964) in the form of (i) equilibrium constants and (ii) free energy perturbations at any concentration of the ligand X, which modulates the reaction:

$$(\partial \ln K/\partial \ln a_x)_{T,P,a_{j\neq x}} = v_x^{\mathrm{Prod}} - v_x^{\mathrm{React}} = \Delta v_x \tag{5.5}$$

and

$$(\partial \Delta G^\circ / \partial \mu_x)_{T,P,n_{j \neq x}} = [(\partial \mu_{React} / \partial n_x)_{T,P,n_{j \neq x}} - (\partial \mu_{Prod} / \partial n_x)_{T,P,n_{j \neq x}}] / (\partial \mu_x / \partial n_x)_{T,P,n_{j \neq x}}$$

K is the equilibrium constant of a reaction, React ⇌ Prod; v_x is the dialysis equilibrium binding of ligand X to the reacting entity; T, P, a_j, and n_j are the Kelvin temperature, pressure, activity, and concentration of component j, respectively; ΔG° is the standard free energy change of the reaction; and μ_i is the chemical potential of component i, $\mu_i = \mu_i^\circ + RT \text{ in } a_i$.

In simple terms, if ligand binding between two states is different, then the equilibrium between those states will be shifted to the state with higher affinity.

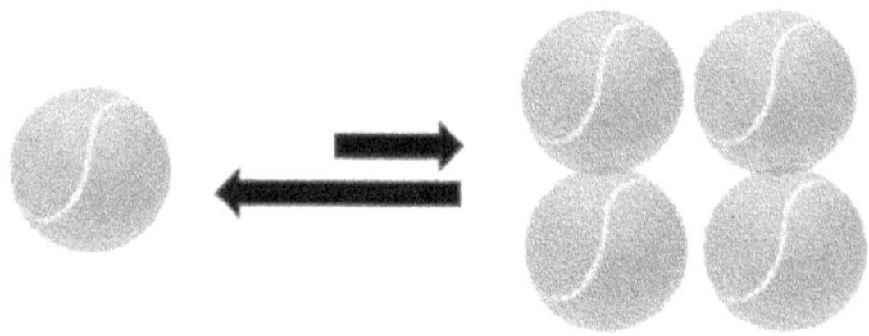

Figure 5.5. Equilibrium between a monomer and a tetramer modulated by a ligand preferentially bound to the monomeric form.

If a reaction is modulated by a destabilizer, as in a native to nonnative protein equilibrium, the destabilizer will be preferentially bound to the nonnative form (figure 5.6).

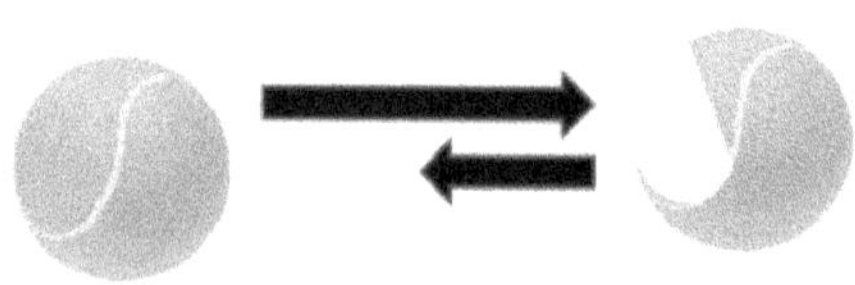

Figure 5.6. Equilibrium between the native and nonnative form of a protein modulated by a ligand preferentially bound to the nonnative form of the protein.

Timasheff's Mechanism of Protein Stabilization by Solvent Components

Serge Timasheff had a unique mind to apply multicomponent solution thermodynamics to understand the behavior of proteins in complex mixtures, and he has been called the man with a genius for solutions in biology (Schellman and Somero 1996). Now we know the principles of protein stabilization by solvent components due to the work of Serge Timasheff. The long-term goal of Timasheff's research was to arrive at an understanding of the thermodynamic and molecular levels of the nature of the interactions by which solvent additives and cosolvents used at a high concentration ($\geq$ 1M) affect the stability, state of aggregation, and solubility of proteins. Such cosolvents are used i) to maintain proteins stable in solution (e.g., sucrose, glycerol, various osmolytes), or ii) to destabilize protein structure in solution (e.g., urea, guanidine hydrochloride, alcohols), and to modulate the processing of expressed proteins, i.e., solubilize inclusion bodies, control the rate of refolding, and prevent nonspecific aggregation.

In addition to the rational development, based on thermodynamic and molecular understanding, the development of additivity and compensation would permit the formulation of complex solvent systems, consisting of mixtures of strong and weak stabilizers and destabilizers, to use systematically in the extraction, folding, and stabilization of proteins newly synthesized by recombinant techniques and their long storage in active forms.

Figure 5.7. Serge Timasheff (1926–2019) in his office at Brandeis University, ca 1996.
Source: Biophysical J. 1996, 71 (4):1985-1993. Reproduced with permission of the copyright holder.

Serge Timasheff was a physical biochemist, a towering figure in solution thermodynamics, and a gentleman of his time (Aune et al. 2021). He was introduced to the study of the thermodynamics of three component systems, especially solutions of macromolecules in water, by Kirkwood (Kirkwood and Buff 1951; Timasheff et al. 1957), with whom he completed his PhD. Kirkwood and Goldberg developed a theory of light scattering in which they showed that the presence of a third component could perturb the scattering and give misleading results (Kirkwood and Goldenberg 1950). Kirkwood also showed how the scattering could be corrected for the presence of the third component.

Timasheff applied the same theoretical idea to the scattering of X-rays from solutions. This work put the interpretation of the X-ray scattering of nucleic acids on a firm theoretical basis and eliminated systematic errors affecting the credibility of the method (Schellman and Somero 1996). Timasheff became interested in why certain substances, when added at high concentrations (>1M), stabilize protein structure. For a summary of Timasheff's many contributions to protein biochemistry, see his recollections of the work performed in his laboratory (Timasheff 2004).

Let us look at the destabilization reaction, that is, a protein denaturation, expressed as a simple two-state equilibrium between the native (N) and denatured (or unfolded or destabilized) (D) states. When the equilibrium is shifted toward the native, or stable, form of the protein by the solvent additive (S), then the reaction can be expressed by:

$$\mathrm{N} \underset{\underset{\text{Stabilizer}}{\longleftarrow}}{\overset{\mathrm{K(S)}}{\rightleftharpoons}} \mathrm{D} \qquad (5.6)$$

$$\mathrm{K} = [\mathrm{D}]/[\mathrm{N}]$$

The equilibrium constant of this reaction is the function of the concentration of the stabilizer, a linkage function. Hence, the effect of the stabilizer on the equilibrium constant can be expressed quantitatively by the Wyman linkage relation (Wyman 1964). It states that, upon changing only the cosolvent concentration, the change in the equilibrium constant is given by the difference in the number of cosolvent molecules bound by the denatured and native protein molecules:

$$\frac{d\log K}{d\log a_s} = \Delta\nu_S = \nu^D - \nu^N \qquad (5.7)$$

Where a_s is the thermodynamic activity of the additive S, related to its free concentration by the activity coefficient, Δv_S is the difference between the number of moles of solution component S (stabilizer) bound per mole of protein (as measured by dialysis equilibrium or a similar technique) between the denatured (v^D) and native (v^N) states of the protein. Figure 5.8 shows a hypothetical Wyman plot.

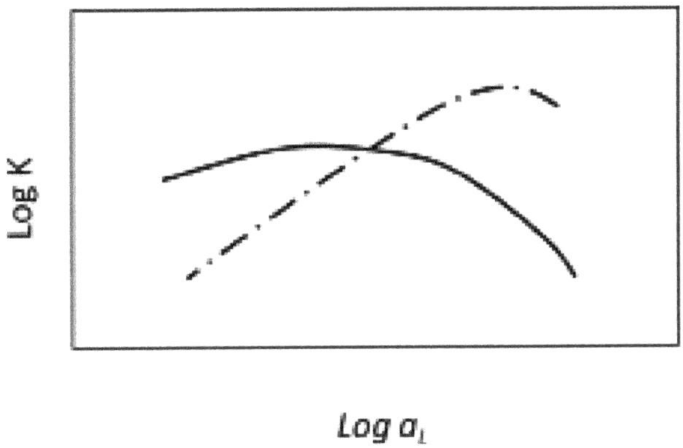

Figure 5.8. Hypothetical Wyman plot showing that the slope at any concentration of ligand (L) is the change in ligand binding at the given ligand concentration (it can be positive or negative). Hence, the measured Δv_L is, in fact, the difference between changes in the numbers of effective water molecules (r_w) and ligand molecules (r_L), which interact thermodynamically with the protein without necessarily forming direct contacts (Tanford 1969). $\Delta V_L = \Delta r_L - (m_L/m_w)\,\Delta r_w$ (where m_L and m_w are molal concentrations of ligand and water, and a_L is the activity of the ligand).

In the case of stabilization, the reaction is shifted to the left (Eq. 5.7), so the equilibrium constant must decrease with an increase in the concentration of the additive. In this case, ($d\log K/d\log a_S$) is negative, so Δv_S must also be negative.

This means there must be less binding of the stabilizer to the denaturant form of the protein than to the native one. In other words, if there is less binding of the stabilizer to the denatured form of the protein, then the protein will preferentially bind to water or be preferentially hydrated.

As we discussed previously, water does not like to solvate hydrophobic regions like the ones of a denatured/unfolded protein, increasing the partial molar free energy. In consequence, the equilibrium will be shifted to the native state in the presence of the stabilizer, where the stabilizer will still be preferentially excluded from the domain of the protein and the protein will be preferentially hydrated with an increase in the partial molar free energy, but less than in the denatured state.

If we return to the definition of *partial molar free energy,* it is the partial derivative of the Gibbs free energy with respect to the amount of the species (all other species' concentrations in the mixture remaining constant) and at a constant temperature. When pressure is constant, *the chemical potential* is the partial molar Gibbs free energy. Hence, partial molar free energy is the thermodynamic quantity, which indicates how the properties of a mixture vary with changes to their molecular composition when being kept at a constant temperature and pressure. This is one of the fundamental concepts of multicomponent solution thermodynamics.

The principles of preferential exclusion of excipients or osmolytes are quite extensive. Many prokaryotes, fungi, plants, and animals that live under osmotic stress accumulate small molecular weight compounds for adaptation. These organic substances, principally sugars, polyhydric alcohols, amino acids, and methylamines, are compatible solutes (Borowitzka & Brown 1974; Borowitzka 1985; Yancey et al.1982; Somero 1986). Their presence at physiological concentrations does not significantly disturb the structure and function of macromolecules within cells.

Timasheff's group was able to measure the binding of stabilizing solvents after dialysis equilibrium to different proteins in the native state (see Lee and Timasheff 1974, 1981; Lee et al. 1979; Arakawa, et al. 1990; Xie and Timasheff 1997). Thermodynamically, the stabilizing action of a cosolvent on the native structure may be defined for three reference states (Timasheff 1993; Timasheff 1994). 1) *The cosolvent solution of the given composition;* 2) *Pure water* (in practice this is taken as the dilute buffer free of cosolvent); and 3) *A cosolvent solution of a concentration different from that being measured.*

The three situations are explained below. In the analysis, component one is defined as water, component two as protein, and component three as cosolvent (Scatchard 1946; Stockmayer 1950).

The Cosolvent of a Given Composition

When the reference state is the cosolvent, the effect of the cosolvent on protein stability is expressed by the Wyman linkage relation (Wyman 1964; Wyman and Gill 1990), and equation 5.0 applies. Therefore, if the equilibrium constant K of the denaturation reaction is affected by the cosolvent, then at constant temperature and pressure:

$$-d\Delta G°/(RTd \log a_3) = d \log K/d \log a_3 = v_3^{products} - v_3^{Reactants} = \Delta v_3 \qquad (5.8)$$

a_3 is the activity of the cosolvent, i.e., its concentration; c3 is multiplied by the activity coefficient, f_3 ($a_3 = c_3f_3$); and v_3 is the extent of binding of the cosolvent to the protein. In the analysis, we will use the molal concentration scale m_i, i.e., moles of substance i/1000 g of water. As described by Timasheff (Lee and Timasheff 1974; Timasheff 1992, 1993), the extent of binding v_3 is identical to the value of the thermodynamic binding measured by a direct technique such as dialysis equilibrium. Thermodynamically, then:

$$V3 \equiv (\partial m_3/\partial m_2) = -[(\partial \mu_2/\partial m_3)_{m2}]/[(\partial \mu_3/\partial m_3)m_2] \qquad (5.9)$$

Equation 5.9 indicates that dialysis equilibrium binding is the change of concentration of free cosolvent caused by the addition of the protein at chemical equilibrium, i.e., at the constant chemical potential of the cosolvent μ_3. This in turn is equal to the perturbation of the chemical potential of the protein by the cosolvent at constant protein concentration. The denominator of the last term of Eq. (5.9) accounts for the thermodynamic nonideality of the cosolvent solution. We now take the definition of the chemical potential of any component *i*:

$$\mu_i = \mu_i^\circ + RT \ln m_1 + RT \ln \gamma_i \qquad (5.10)$$

μi° is the standard chemical potential, and γ_i, is the activity coefficient of component i, giving for the denominator:

$$(\partial\mu_3/\partial m_3)_{m2} \equiv RT[(1/m_3) + (\partial \ln \gamma_3/\partial m_3)] \qquad (5.11)$$

Thermodynamic dialysis equilibrium binding, as seen by the inspection of equation (5.9), specifies only that the concentration of ligand (cosolvent) in the free state is changed by introducing the protein into the system. It does not state whether this change is positive or negative. As such, this parameter can assume negative or positive values or zero (Na and Timasheff 1981; Timasheff 1992) whether the interaction of the protein with the cosolvent containing solution is thermodynamically favorable, unfavorable, or indifferent (Timasheff 1992). This is shown by the fact that the binding of a ligand molecule to the surface of a protein requires that water molecules be displaced. In other words, the binding process is an exchange reaction (Timasheff and Kronman 1959; Schellman 1987,1990; Timasheff 1992). The competition between ligand and water can be represented by the equation (5.12).

$$P \,.\, H_20_m + L \rightleftharpoons PL + m\, H_20 \qquad (5.12)$$

$$K_b = ([PL][H_20]^m/([P \,.\, H_20_m][L])$$

K_b is the binding (exchange) constant, depending on the relative affinities of ligand (cosolvent) and water molecules for a locus on the surface of the protein molecule, and the site will have a preference by occupancy by the cosolvent or by water. Hence, the interaction is called *preferential interaction,* and the *cosolvent binding* is called *preferential binding.* This is also referred to as preferential exclusion of the ligand. Therefore, the negative values of the binding measured mean that the protein has a higher affinity for water than for the cosolvent. Preferential binding and preferential hydration are related by a reciprocal equation (Timasheff 1993):

$$m_3(\partial m_1/\partial m_2)_{\mu 1,\mu 3} = m1\ (\partial m_3/\partial m_2)_{\mu 1,\mu 3} \qquad (5.13)$$

The term $(\partial \mu_2/\partial m_3)_{m2}$ from equation 5.9 is the preferential interaction parameter. Preferential binding is the manifestation of many weak interactions of the ligand with the protein (Timasheff and Xie 2003).

The Reference State Is Water

Using the reference state as pure water (Timasheff 1992), the effect of a cosolvent on protein stability is expressed by the difference in the transfer of free energy of the protein from water to the cosolvent solution of the given composition in the denatured and native states, i.e., $\Delta\mu_2{}^{D}{}_{Tranf} - \Delta\mu_2{}^{N}{}_{Tranf}$. The transfer free energy $\Delta\mu_{Tranf}$ is the total free energy of interaction (binding) of the protein with the cosolvent. It is related to experimentally measurable quantities by the integral under the $(\partial\mu_2/\partial m_3)$ curves from zero to the given concentration of the cosolvent. It is experimentally measurable by the determination of dialysis equilibrium binding at a series of cosolvent concentrations and the numerical integration under the plot of:

$$-RT\ (\partial m_3/\partial m_2)_{\mu 3}\ [(1/m_3) + (\partial \ln \gamma_3/\partial m_3)]\ \text{vs}\ m_3 \qquad (5.14)$$

Examples of the calculation using different cosolvents have been described by Timasheff (see Timasheff 1993).

A *third reference state* at any cosolvent concentration is the same as the last one, with the exception that the lower limit of the integral under the $(\partial\mu_2/\partial m_3)$ curves with the lower limit as the reference solvent concentration, and the transfer free energy refers to the difference in the free energy of the interaction of the protein with cosolvent at two concentrations of the cosolvent from zero to the given concentration of the cosolvent.

To identify the stabilizing/destabilizing capacity of a cosolvent, both the preferential binding and the transfer free energy must be measured for each of the two end states (Timasheff 1992; Kendrick et al. 1997) of the unfolding equilibrium. The first approach is to measure a cosolvent's stabilizing/destabilizing capacity relative *to water*. This is expressed by the difference between the transfer free energies of the protein in the denatured and native states from water to the cosolvent solution (Timasheff 1992, 1993). Another approach is to measure at a given *constant concentration of the cosolvent* whether a cosolvent is a stabilizer or destabilizer, defined by the difference between the preferential binding of the cosolvent to the protein in the denatured and native state.

This can be measured in two ways: from the slope of the double logarithmic plot of the denaturation equilibrium constant versus the thermodynamic activity of the cosolvent, or by determining in dialysis equilibrium the binding of the cosolvent to the protein in the denatured and native states and taking the difference. In the second approach, once again the unfolded form of the protein can be obtained by its chemical destabilization (e.g., reducing agents). These results can be used to calculate the transfer of free energies.

Experimental Approach to Measure Preferential Binding

Measuring the binding of extremely weak ligands as water is complicated experimentally. It involves the performance of very careful dialysis equilibrium experiments. In these experiments, the ligand (cosolvent) concentration is very high (>0.5 M), and its changes and interactions with proteins are extremely small (10–100 μM); a precision detection technique is required. Those techniques are precision densimetry and differential refractometry.

Timasheff's laboratory performed the measurements using a densimeter. Those measurements are complicated because they require instruments that can measure in the fifth decimal place of density due to the very weak interactions. I remember that the Anton-Paar density meter was set up in a constant temperature room at 20°C, and nobody could open the door of the room while the experiments were performed. A small air current could dramatically affect the results. Timasheff and coworkers have discussed the experimental approach in detail (e.g., see Lee and Timasheff 1974; Lee et al.1979).

In determining the preferential interaction, density measurements must be performed on two sets of protein solutions, one which has been dialyzed (constant chemical potential) and the other not dialyzed (constant composition of the solvent). The second type of measurement requires that the protein be dried carefully. It is then weighted out rapidly and dissolved in the experimental medium. Any extraneous water molecules will change the solvent composition sufficiently to affect the density measurement.

For each solvent composition, one must do measurements as a function of protein concentration, and extrapolate to zero. Usually, five to ten protein concentrations between 2 and 20 mg/ml are sufficient. The preferential binding is then calculated from the results of the density measurements. The equations employed are the ones from Cassasa, Cohen, and Eisenberg (Cassasa and Eisenberg 1964; Cohen and Eisenberg 1968):

$$(\partial m_3/\partial m_2)_{T,\mu1,\mu3} = (M_2/M_3)(\partial g_3/\partial g_2)_{T,\mu1,\mu3} \qquad (5.15)$$

$$(\partial g_3/\partial g_2)T,\mu_1,\mu_3 = [(1-\phi'_2{}^{\circ}\rho_{\circ}) - (1-\phi_2{}^{\circ}\rho^{\circ})]/(1-\bar{v}_3\,\rho^{\circ}) \qquad (5.16)$$

g_i is the concentration of component i in units of grams of i per gram of water, ρ° is the density of the solution in g/ml, $\bar{v}_3$ is the partial specific volume of the cosolvent at the experimental concentration, $\phi_2{}^{\circ}$ is the apparent partial specific volume of the protein measured under conditions at which the molal concentration of the diffusible component (cosolvent), m_3 is kept identical in the solvent, and the protein solution, ϕ'_2 is the apparent partial specific volume of the protein measured under conditions at which the chemical potential of the cosolvent μ_3 is kept constant between the protein solution and the reference solvent. This is attained within a close approximation by bringing the system to dialysis equilibrium.

The two parameters, ϕ_2 and ϕ'_2, are known as the isomolal and isopotential apparent partial specific volumes of the protein. The superscript ° denotes that this is the value extrapolated to zero protein concentration. The parameters ϕ_2 and ϕ'_2 at each protein concentration are obtained from density measurements (Kielley and Harrington 1960):

$$\phi = (1/\rho_\circ)\,[1 - (\rho - \rho_\circ)/C] \qquad (5.17)$$

ρ is the density of the protein solution in mg/ml, $\rho_\circ$ the density of the solvent, and C the protein concentration in mg/ml. These values are then plotted as a function of protein concentration. Their extrapolation to zero protein concentration gives the values to be used in Eq. (5.15 and 5.16) to obtain the preferential binding parameter $(\partial m_3/\partial m_2)T,\mu_1,\mu_3$

All the cosolvent concentrations are expressed in molal concentrations (moles of cosolvent/100 g of water). Thus, all solutions must be prepared gravimetrically. The derived parameter $(\partial \mu_2/\partial m_3)_{m2}$ requires knowledge of the activity coefficients of the cosolvent as a function of its concentration. This is necessary to evaluate $(\partial \mu_3/\partial m_3)_{m2}$ (see Eq. 5.9 and 5.11). Activity coefficients could be obtained from the literature for many of the cosolvents used.

Activity coefficients can be measured by vapor pressure osmometry. By studying preferential hydration as a function of concentration, it is possible to obtain the change in molar free energy of a macromolecule induced by adding cosolvents such as sucrose, a stabilizer, or urea, a destabilizer of proteins. If one is dealing with a molecule that can exist in two forms, native and unfolded, then the effect of an added cosolvent will be the stabilization of the native protein if the cosolvent is preferentially excluded, relative to water, from the protein surface (figure 5.9).

Destabilizing cosolvents like urea will preferentially interact with the unfolded protein and favor its unfolding. This is a modified form of Le Chatelier's principle. The measurements provide quantitative descriptions of enhanced or diminished stability and permit the a priori determination of the conditions of maximal stabilization. Timasheff's mechanism of protein stabilization by solvents can be summarized as follows:

1. Nonspecific stabilization by preferentially excluded solutes, i.e., negative ligand binding.
2. Protein chemical potential is increased, but the increase in chemical potential is higher in the denatured state compared with the native state for a stabilizing excipient.
3. Magnitude of exclusion varies directly with protein surface area, so it is greater for partially or fully unfolded species than for the native state.
4. As a result, equilibrium is shifted toward the native state.
5. The stability of the system increases because the unfolded protein has a higher surface area from which the solute must be excluded.
6. Also, multimers are favored over constituent monomers.

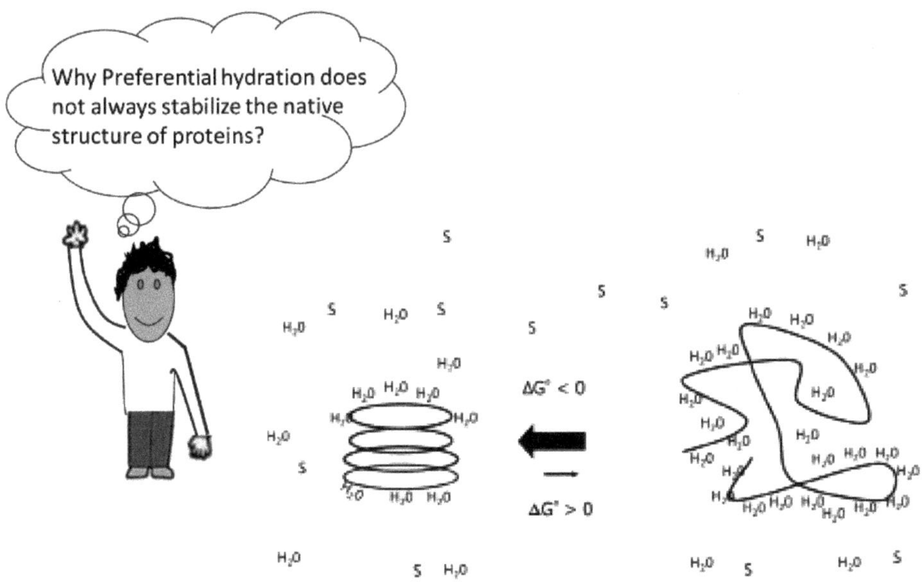

Figure 5.9. Schematic of Timasheff's mechanism of protein stabilization by a preferentially excluded excipient (preferential hydration) such as sucrose (S), which shifts the native and nonnative equilibrium to the left.

Timasheff developed a general relation that shows whether a cosolvent will be a protein unfolder or folder is defined by the relative affinities of cosolvent and water for loci on a protein molecule, summed over the entire protein surface (Timasheff 1992). In the analysis, these assumptions must be compared for the native and unfolded states of a protein. This can be done at two reference states: The first is relative to water, where the effect of the cosolvent is defined thermodynamically by the change in transfer free energy during unfolding; the second is relative to the solvent of the given composition, where the change in preferential binding defines the effect.

Sites on protein surface can be either 1) indifferent (occupancies by ligand or water are thermodynamically identical); 2) thermodynamically active, where the affinities for ligand and water molecules are non-equal, and an exchange equilibrium constant defines occupancy, K_{exc}; 3) those for which the ligand has no affinity (these will be occupied by water, i.e., they manifest themselves as cosolvent exclusion). If sites at which the ligand has a higher affinity than water predominate, the ligand will be a denaturant; in the opposite case will be a stabilizer. Thus, there is a continuum in water cosolvent competition, ranging from strong denaturants (urea, GuHCl) to strong stabilizers (sucrose, sorbitol, trehalose) with weakly acting agents ($GuS0_4$, $MgCl_2$, DMSO) in between. Thus, the global interaction of a compound with a protein can be one of the compensations between sites at which it strongly displaces water and acts as a denaturant and those from which it is excluded and acts as a stabilizer (see Timasheff 1992).

Timasheff's work has demonstrated that preferential hydration is a decisive factor in the determination of the stability and conformation of proteins and nucleic acids. Indeed, the concept provides the necessary theoretical context for answering one of the most fundamental questions in evolutionary biology: Why has natural selection favored the evolution of unique mixtures of solutes found in cells? (Schellman and Somero 1996). Sugars have been extensively used in the biopharmaceutical industry to stabilize protein drug products (figure 5.10), and they stabilize proteins by the Timasheff's mechanism, having as a consequence an increase in T_m. Lee and Timasheff (1981) investigated the increase in the thermal unfolding transitions of a-chymotrypsin, chymotrypsinogen, and ribonuclease as functions of sucrose concentration and correlated the observations with the preferential exclusion theory.

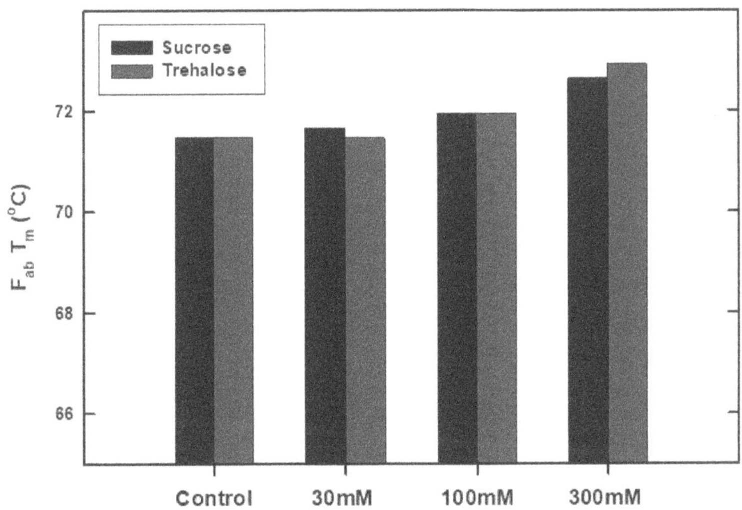

Figure 5.10. Stabilization of a mab by nonreducing sugars as measured by DSC, showing an increase in Tm2.

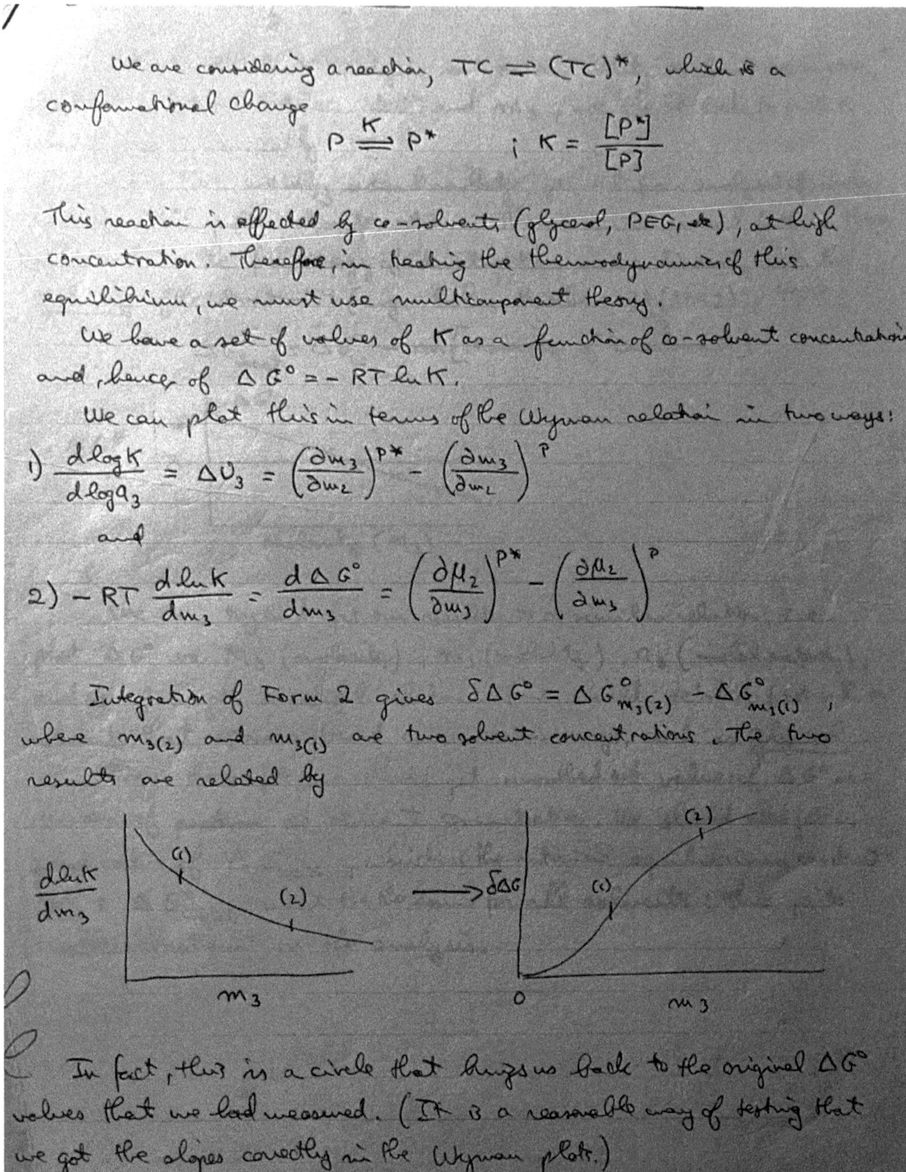

We are considering a reaction, $TC \rightleftharpoons (TC)^*$, which is a conformational change

$$P \overset{K}{\rightleftharpoons} P^* \quad ; \; K = \frac{[P^*]}{[P]}$$

This reaction is affected by co-solvents (glycerol, PEG, etc), at high concentration. Therefore, in treating the thermodynamics of this equilibrium, we must use multicomponent theory.

We have a set of values of K as a function of co-solvent concentration, and, hence, of $\Delta G^\circ = -RT \ln K$.

We can plot this in terms of the Wyman relation in two ways:

$$1) \; \frac{d\log K}{d\log a_3} = \Delta \nu_3 = \left(\frac{\partial m_3}{\partial m_2}\right)^{P^*} - \left(\frac{\partial m_3}{\partial m_2}\right)^{P}$$

and

$$2) \; -RT\frac{d\ln K}{dm_3} = \frac{d\Delta G^\circ}{dm_3} = \left(\frac{\partial \mu_2}{\partial m_3}\right)^{P^*} - \left(\frac{\partial \mu_2}{\partial m_3}\right)^{P}$$

Integration of Form 2 gives $\delta\Delta G^\circ = \Delta G^\circ_{m_3(2)} - \Delta G^\circ_{m_3(1)}$, where $m_{3(2)}$ and $m_{3(1)}$ are two solvent concentrations. The two results are related by

In fact, this is a circle that brings us back to the original ΔG° values that we had measured. (It is a reasonable way of testing that we got the slopes correctly in the Wyman plots.)

Figure 5.11. Handwritten notes from Serge Timasheff to the author on a protein-solvent project (December 18, 1992).

Stabilizing Osmolytes

Osmolytes are organic solutes that help cells adapt to dehydration or fluid excess. Osmolytes are generated within cells in response to osmotic stresses. Examples include glycine, glutamine, inositol, sorbitol, taurine, methyl sulfonium compounds, urea, and methylamine compounds, particularly trimethylamine oxide (TMAO), which are compatible osmolytes that commonly occur in the tissues of marine organisms. TMAO is best known as a "counteracting solute" that protects proteins against various destabilizing forces (Yancey et al. 1982, Yancey 2005). Methylamines (e.g., trimethylamine N-oxide (TMAO) can enhance protein folding and ligand binding and counteract perturbations by urea (e.g., in elasmobranchs and mammalian kidneys). Trehalose and proline in overwintering insects stabilize membranes at subzero temperatures. Trehalose in insects and yeast, and anionic polyols in microorganisms around hydrothermal vents, can protect proteins from denaturation by high temperatures (Yancey 2005).

Ions/solvents favoring strong water interactions are known as *kosmotropes*, or structure makers (stabilizers). Ammonium sulfate, sucrose, and trehalose are all kosmotropes and salting-out compounds. They remove protein water of hydration to release low entropy water. Kosmotropes cause water molecules to interact favorably, which also (in effect) stabilizes intramolecular interactions in macromolecules, such as proteins. Kosmotropic ions are depleted from the hydrate water, whereas chaotropic ions are enriched in this phase. Moreover, recent research has revealed that effects due to the partitioning of solutes scale linearly with salt concentration, which is in full agreement with the analysis by Schellman using Kirkwood-Buff integrals (Schellman 1987, 2005). By contrast, chaotropes disrupt hydrogen bonding between water molecules; induce weaker interactions with water than water itself: structure breakers and solubilizers (urea, guanidine hydrochloride), disrupt hydrogen bonding.

Increasing the ionic strength of a solution, let us call it salt concentration for now, increases the solubility of a solute, such as a protein (figure 5.12). This effect tends to be observed at lower ionic strengths. At higher salt concentrations, the solubility of a protein decreases (precipitation).

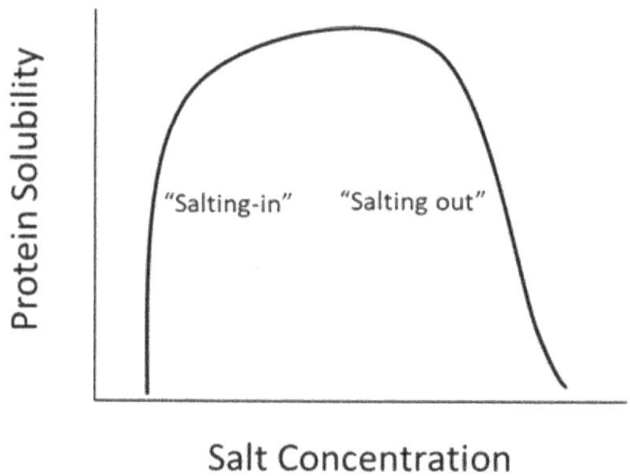

Figure 5.12. Solubility of a model protein as a function of salt concentration.

Many prokaryotes, fungi, plants, and animals that live under osmotic stress accumulate compatible solutes (Borowitzka & Brown 1974; Yancey et al. 1982; Somero 1986). TMAO has a demonstrated ability to counteract the perturbing effects of urea on enzyme activity when accumulated in a 2:1 ratio with urea (Somero 1986). Using methylamine alone might have the unfavorable consequence of imparting too much stability, i.e., too much rigidity, to the system. Their combined use, however, gives the desired osmotic effect while mutually neutralizing their perturbations of biological function (Lin and Timasheff 1994). At a 2:1 molar ratio of urea and trimethylamine N-oxide (TMAO), the effects of the two cosolvents on the transition temperature (T_m) were found to be essentially the algebraic sum of their effects when used individually. Preferential interaction measurements of urea, TMAO and urea in its 2:1 molar ratio mixture with TMAO, have shown that the presence of TMAO has no effect on the interaction of urea with the protein in either the native or the unfolded (reduced carboxymethylated RNase T1) state (Lin and Timasheff 1994).

The transfer free energies of native and unfolded protein from water to 1 M TMAO are estimated as 1.2 and 5.4 kcal/mol respectively (Lin and Timasheff

1994). The unfavorable transfer free energies of native and unfolded protein from water to TMAO provide a molecular-level rationale for the preferential hydration of proteins by osmolytes.

The analysis of the properties of the methyl amines (e.g., TMAO) and the realization that any cosolvent that has some affinity for protein can act simultaneously as stabilizer or destabilizer of different regions of a protein (Timasheff 1992, 1993) due to local interactions, have made methyl amines good candidates for simultaneously stabilizing and solubilizing components of a formulation. The principle has been demonstrated for urea which, while a general denaturant due to its binding to a large number of loci on the protein, is simultaneously a weak inhibitor of denaturation by being excluded from other loci (Timasheff 1992, 1993). In the case of urea, denaturation prevails.

Methylamines should function in the same way. In their case, stabilization over the entire surface due to their general exclusion will be stronger than destabilization due to the very weak binding to hydrophobic patches. This very weak binding, on the other hand, should lead to prevention of coalescence of these patches and solubilization of the protein. Hence, methyl amines should act simultaneously as protein solubilizers and stabilizers, which circumvent the problem with the use of non-ionic detergents whose binding is too strong and result in a true coating. Since the effects of cosolvents are additive, mixtures of methyl amines with sugars and polyols should boost solubilization while maintaining the proteins stable. In this way by proper choosing of stabilizing agents, or combination of stabilizers, as well as mixtures of stabilizers with destabilizers, it becomes possible to manipulate protein stability.

According to Timasheff's mechanism, all stabilizing cosolvents have been found in binding studies to be preferentially excluded from the immediate vicinity of protein molecules. While this exclusion is the thermodynamic source of the stabilization, different cosolvents are excluded by different mechanisms that depend on the nature of their interactions with surface groups in proteins, whether, polar, charged, or hydrophobic (Gekko and Timasheff 1981; Lee and Timasheff 1981; Arakawa and Timasheff 1982). Therefore, depending on their nature, patches on the protein surface can interact differently with each cosolvent (Timasheff 1992, 1993).

Solution Variables and Protein Stability

The buffer or formulation where the protein is embedded exerts a profound effect on the structure of the protein. The effect of solution variables could be either on the secondary, tertiary, or quaternary structure, the so-called *physical stability*. The three-dimensional structure of proteins in solution can be affected by ligands (stabilizers, destabilizers), pH, ionic strength, temperature, and pressure. In general, some of these variables show interrelations. The *chemical stability* of the protein could be affected by the same variables that affect the *physical stability* of proteins. Before continuing, we need to define what we mean by chemical stability. In simple terms, we will refer to chemical stability as any changes to the protein's primary structure, such as cleavages/hydrolysis of the peptide bond.

Usually, hydrolysis of the peptide bond requires very harsh conditions such as incubation in strong acids and high temperatures. However, the hydrolysis could also be carried out by enzymes named proteases, such as pepsin, trypsin, etc., that help in the hydrolysis of proteins during digestion. Several other chemical instabilities could happen on proteins, such as oxidation, deamidation, isomerization, and glycation (nonenzymatic covalent attachment of sugars to proteins). Buffer composition and pH have a strong effect on deamidation and isomerization reactions in proteins (Geiger and Clarke 1987; Wakankar and Borchardt 2006; Lu et al. 2019). Protein stability typically decreases sharply in acidic and basic pH ranges (Privalov 1979). This behavior could be explained as the participation of protons in the unfolding process where the pKs (the apparent ionization constant) of some ionizable residues are different in the native and unfolded species. A group such as a side chain COO^- or NH_3^+ in a protein shows an observed pK that depends on the local charge environment. Nearby positively charged groups repel proton ions in solution and cause the pK to decrease (Slebe and Martinez-Carrion 1976; Pace et al. 2009), whereas neighboring negatively charged groups attract protons and cause the pK to increase (Matthews and Guard 1986; Harns et al. 2009).

Charge interactions usually cause the pK of a group to be lower in the native than in the unfolded state at acidic pH (below the isoelectric point) but higher in the native than the unfolded state at alkaline pH (Baldwin and Eisenberg 1987). The reason is that the protein has a net negative charge above the isoelectric pH but a net positive charge below the isoelectric pH, and a protein molecule expands to occupy a larger volume when it unfolds, therefore decreasing charge density (Baldwin and Eisenberg 1987). When an ionizable group is part of a salt bridge, the sign of the pK shift for a COO^- group will be lowered due to the proximity of opposite charges. Conversely, the pK for the NH_3^+ groups will be higher, providing stabilization to the salt bridges.

The most common chemical instabilities in proteins are hydrolysis, oxidation, deamidation, isomerization, β-elimination, and disulfide exchange. One or a combination of the above mechanisms of chemical degradation can take place in a protein. There is extensive literature about chemical stability in proteins (see Manning et al. 1989; Schoeneich et al. 1997). Below is a summary of the main degradative mechanisms in proteins.

Hydrolysis: A general term for a class of chemical reactions in which bonds are broken by the addition of water. It is a common cause of degradation in proteins involving the cleavage of the peptide bond in an acid-catalyzed environment. Several solution factors and enzymatic proteolysis could induce these reactions. Therefore, aseptic processing in handling proteins is a must since terminal sterilization (e.g., gamma irradiation) cannot be used. Proteolysis that occurs near the N-terminal or C-terminal portion of a protein is called *clipping.* In the absence of proteolysis, the hydrolysis of the peptide bond can occur by direct attack of water on the peptide bond carbonyl. The attack is facilitated by protonation of the carbonyl. The reaction is acid catalyzed. Peptide bonds located on the c-terminal portion of aspartic residues are sensitive to proteolysis. The side chain of tryptophan is sensitive to hydrolysis.

The main degradation product of tryptophan is kynurenine, which fluoresces at much higher wavelength (450 nm) than tryptophan itself.

The structural basis for the cleavage in the hinge region of IgG seems to be largely the result of a single amino acid change relative to human IgG_1—the change of a proline in the core hinge of IgG_1 to serine.

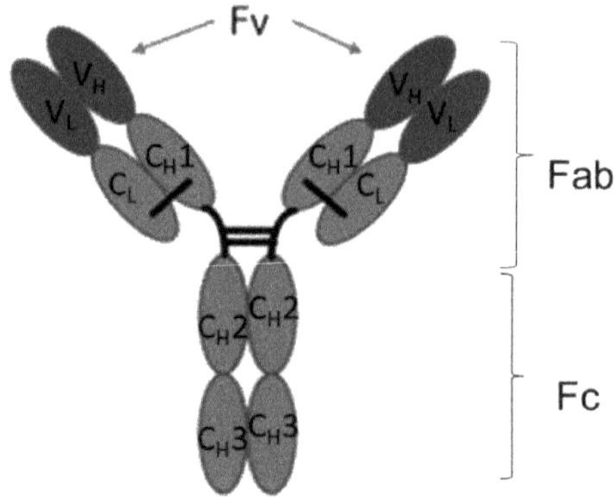

Figure 5.13. Schematic representation of an IgG_1. Covalent hetero tetramer (dimer of dimers). Heavy chains are: V_H, C_H1, C_H2 and C_H3. Light chain: V_L and C_L. There is a hinge linking C_H1 and C_H2 in heavy chains. There are inter-subunit disulfide bonds. Fab (antigen-binding fragment): variable domains V_H/V_L combine to form a unique binding pocket. Fc (crystallizable fragment): constant sequences. There are several isotypes/classes: IgA, IgD, IgE, IgG and IgM. Subtypes/subclasses: IgG_1, IgG_2, IgG_3 and IgG4.

Deamidation: Proteins may degrade via asparagine (Asn) deamidation and aspartate (Asp) isomerization, but the factors responsible for such degradation remain poorly understood. Deamidation goes through the formation of cyclic intermediates (Geiger and Clarke 1987) detected in neutral or alkaline solution conditions (pH 7–8.5), indicating general acid-base catalysis. Asn-Gly seem like the most frequent sites for deamidation, Asn-Asn and Asn-Asp are also common sites for deamidation.

Deamidation may be mitigated by decreasing the pH below 6.5. The rate of protein deamidation at acidic pH is often much slower. Deamidation can be detected by using molecular mass or charge differences. Isoelectric focusing may help in the visualization of deamidated forms. Charge variant quantitation may be obtained by ion-exchange chromatography (IEC). This technique allows the isolation of the different species for further characterization by peptide mapping or mass spectrometry.

Isomerization of aspartic acid residues to iso-aspartic is a common instability occurring at acidic pH (Geiger and Clarke 1987; Reissner and Oswald 2003). Deamidation of asparaginyl residues occurs via a succinimide intermediate, which subsequently hydrolyzes into a mixture of aspartyl and isoaspartyl residue-containing forms. The isoaspartyl residue is generally difficult to detect because isoaspartic acid is isomeric with aspartic acid, therefore showing no mass difference. Recent studies using electron capture dissociation mass spectrometry have demonstrated an approach to differentiate these two isoforms in synthetic peptides (Cournoyer et al. 2006, 2007).

The identity of the amino acid residues succeeding Asn and Asp influences the rate of succinimide formation in proteins. Thus, Asn-Gly (NG) and Asp-Gly (DG) motifs are by far the most prone to modification, accounting for 67 percent and 36 percent of hot spots observed respectively. The causes of deamidation are usually pH (typically > 7), temperature, time, solvent accessibility, conformation, and amino acid sequence. It has been postulated that deamidation may provide a signal for protein degradation, thereby regulating intracellular levels.

Amino acid	Residue Composition	Residue Average Mass	Delta Mass
Asparagine	$C_4H_6N_2O_2$	114.04	0
Succinimide Intermediate	$C_4H_3NO_2$	97.073	-17.031
Aspartic acid	$C_4H_5NO_3$	115.089	0.985
IsoAspartic acid	$C_4H_5NO_3$	115.089	0.985

Table 5.0. Identification of Deamidation

How Does One Identify the Sites of Deamidation?

- Isolation of intact protein variant by chromatography. Hint: If rate of deamidation is very low and therefore difficult to detect, incubation of protein samples in slightly alkaline buffer (i.e., 20mM sodium phosphate, pH ~ 8) or at elevated temperature (i.e., 37°C) can accelerate the rate of reactivity at potential deamidation sites (forced deamidation).

- Ion-exchange chromatography (IEC) is the most widely applied chromatographic technique for the isolation of deamidated proteins to detect charge changes.
- Separation is performed under nondenaturing conditions; therefore, native conformation is preserved, a requirement for the evaluation of bioactivity and antigenicity properties.
- Formation of either a succinimide from an Asn or an IsoAsp from Asp will not result in a change in net charge, and therefore resolution of the intact and variant forms by IEC would not be expected.
- The potential exists for distinguishing Asn from succinimide as well as Asp from IsoAsp by exploiting subtle differences in hydrophobicity and, hence, retention behavior associated with the structural modifications. Typical RPLC conditions can result in partial to complete protein denaturation, and therefore interpretation of in vitro/in vivo bioactivity results must be made with caution.

It is often not possible to directly characterize variants resulting from single-site amino acid modifications in large, glycosylated proteins (>20–30kD) due to the limits in mass resolution as well as the heterogeneity introduced by the carbohydrate structures. In general, cleavage of the deamidated protein variant into smaller fragments using specific proteases, i.e., trypsin, which cleaves at the C-terminal end of lysine and arginine residues, is necessary for the identification of the site of modification. The peptide fragments of varying hydrophobicity are separated by reversed-phase HPLC, resulting in a pattern of peaks or "fingerprint" diagnostic for a particular protein and used to monitor product identity and purity.

The effluent from the HPLC is directed into the electrospray mass spectrometer (quadrupole or ion-trap) in a technique referred to as LC-MS, allowing the identification of the various theoretical peptides as well as "new" peptides arising from post-translational modifications, such as deamidation and oxidation. Mass spectrometers work by ionizing molecules and then sorting and identifying the ions according to their mass-to-charge (m/z) ratios. Two key components in this process are the ion source, which generates the ions, and the mass analyzer, which sorts the ions. Several different types of ion sources are commonly used for LC/MS. Each is suitable for different classes of compounds.

Several different types of mass analyzers are also used. Each has advantages and disadvantages depending on the type of information needed.

Degradation hotspots can be characterized by:

- Conformational flexibility.
- The size of the C terminally flanking amino acid residue and secondary structural parameters.
- Not all sequences that deamidate/isomerize when exposed to stress become hot spots under refrigerated conditions (2–8°C).
- In-silico prediction methods have been developed to evaluate the degradation propensity of both Asn and Asp residues in the complementarity-determining regions (CDRs) of mAbs (Sydow, et al. 2014).

Oxidation is another common mechanism of degradation in proteins (Franson et al. 1996; Franson and Hagman 1996; Bobroski et al. 2007). Methionine residues are more susceptible to becoming oxidized. The specific position within the molecule affects the propensity to oxidation. Cysteine oxidation is temperature, pH, and buffer dependent. Also, the reaction could be affected by neighboring groups and active oxygen species. Other residues with a propensity to photo-oxidation in visible light are histidine, tyrosine, methionine, cysteine, and tryptophan. Histidine is most sensitive to oxidation at neutral pH and by metals (e.g., copper, iron). Tyrosine is oxidized at alkaline pH and can form di-tyrosine cross-linked complexes (Aeschbach et al. 1976). These can be formed inter or intramolecularly between two proteins from two free tyrosine or between free tyrosine and proteins.

In pathophysiological contexts, di-tyrosine cross-linking has been reported in low-density lipoproteins, arising from oxidative damage within human atherosclerotic lesions (Leeuwenburgh et al. 1997). Tryptophan is a major target for many oxidants, and many oxidation products have been characterized (Bellmaine et al. 2020). Many of these involve the initial formation of (nitrogen-centered) indolyl radicals (Tryptophan) which, depending on the properties of the protein and environment, can self-react to generate di-tryptophan crosslinks.

Racemization and *β elimination* are closely related to degradation modes. In both reactions, there is an abstraction of a proton from an α carbon of an amino acid, so the reaction is favored under alkaline pH. The proton can recombine with the resulting carbanion to reform the original amino acid in either the D or L form (racemization). Beta elimination can occur at cysteine, serine, threonine, phenylalanine, and lysine residues (Manning et al. 1989, 2010). These reactions could be accelerated by high temperatures, alkaline pH, or metals. Racemization and β-elimination are favored if there is:

- A good leaving group (e.g., Ser, Cys).
- An aromatic ring attached to a β carbon (e.g., Phe, Tyr) as the double bond will be stabilized by conjugation with the ring system.

β-elimination of cystine involves cleavage at the C-S bond to produce a vinyl bond between Cα and Cβ, forming dehydroalanine with expulsion of a persulfate anion at high pH and elevated temperature. Dehydroalanine and persulfate can contribute to cross-linking and protein aggregation.

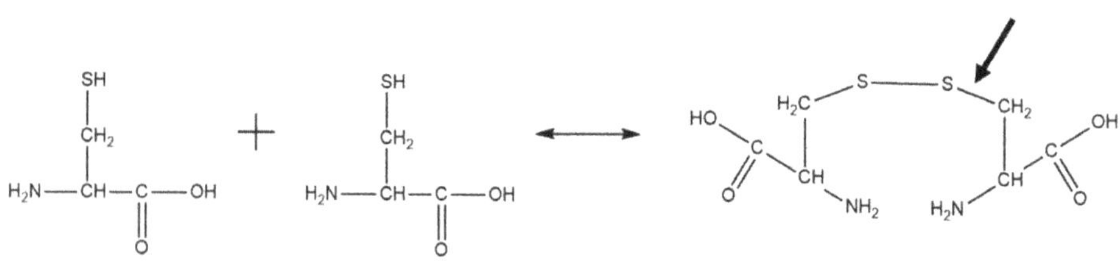

Cysteine　　　　Cystine

Figure 5.14. β-elimination of cysteine.

Disulfide interchange of cysteine can occur at neutral or alkaline pH and catalyze by thiols groups (Bobrowski et al. 2007).

Given the different types of chemical instabilities in proteins, it is necessary to analyze candidate proteins by a battery of biochemical assays to be able to identify likely degradation mechanisms.

Figure 5.15. Disulfide exchange in a protein.

Condensation reactions like glycation, the nonenzymatic glycosylation of proteins, can occur in the presence of reducing sugars. This is a covalent bonding between a protein and a reducing sugar such as glucose, fructose, lactose, maltose, and galactose. The carbonyl group of a sugar reacts with the amine group in proteins, a Maillard Reaction. Glycation may impair the biological activity of a protein. Several factors can affect the glycation of proteins, such as cell culture conditions, time, and temperature. The reaction happens with any amino group, and the amino acid most susceptible is lysine. pH will influence which end products are favored.

N-terminal cyclization. Glutamate and glutamine, when located at the N-terminus of recombinant proteins, can spontaneously cyclize to form pyroglutamate, although glutamate cyclization is slower than glutamine (Schilling et al. 2008). Pyroglutamate has been identified in a variety of proteins, including the light and heavy chains of antibodies (Chelius et al. 2006). Pyroglutamic acid formation does not appear to have a significant impact on the safety and efficacy of protein therapeutics.

Salt bridges and electrostatic interactions are important stabilizing factors at low salt concentrations and could be explained by the Debye-Hückel theory. Nevertheless, at high salt concentrations, these effects become less important (Robinson and Jenks 1965). Ionic strength is a critical magnitude affecting protein conformational stability; we should define it before moving further. In 1921, Lewis and Randall defined the ionic strength of a solution (I), as the semi sum of the products of the molality of each of the ions in the solution times the square of the charge of the ions.

$$I = \frac{1}{2}\sum_i m_i Z^2{}_i \tag{5.18}$$

m_i is the molality of the *i* ion, and Z_i is the charge of the ion. Thus, a buffer of any given composition has a different ionic strength at different pH values. Buffers of different compositions may have different ionic strengths at the same pH. Hence, when studying the pH dependence of solubility, ligand binding, or conformational stability of any given protein, this fact needs to be taken into consideration. The wise formulator in order to separate the contribution of pH and ionic strength could first estimate the buffer with the largest ionic strength and then adjust all the other buffers to a given ionic strength by adding an inert salt. Different buffer species at a given pH could be interacting with the protein of interest in such a way as to provide a large difference in the physical and chemical properties of proteins.

Some reactions will show a strong dependence on rate and equilibrium with ionic strength rather than the concentration of a particular salt. It has been shown for the sperm whale myoglobin that as the electrostatic interactions are lessened with increasing ionic strength, the native state becomes destabilized (Matthews and Gurd 1986). The electrostatic contribution to protein stability may be estimated from the change in summed electrostatic free energy terms for all charge sites, according to the modified Tanford-Kirkwood theory (Mathews et al.1981; Friend and Gurd 1979a, b).

CHAPTER 6

SOLUTION ENERGETICS

The "personality" of a protein or its "psychological profile"—if we can call it that—depends on the fine interplay between its intrinsic conformational stability dictated by its amino acids' constituents and the environment or the solution variables, where the protein is immersed. The fine interplay between intrinsic conformation and solution variables is critical for biological function. The right amount of stability and flexibility (protein dynamic) is required for optimal function, as we have discussed in the previous chapters. Evolution has allowed flexibility as a key feature of proteins. This in turn makes these macromolecules very labile, and they could easily lose their three-dimensional structure in response to environmental changes such as temperature, pH, light stresses, oxygen, agitation, excipients, detergents, metals, etc.

Honig and Yang propose that the major phenomenon opposing protein stability is the desolvation of polar groups upon protein folding (Honig and Yang 1995). Hydrogen bonding in proteins has been proposed to be somewhat destabilizing (Honig and Yang 1995) and indifferent or a minor stabilizing force (Dill 1990) and a principal contributor to the stability of the native state (Kauzmann 1959; Rose and Wolfenden 1993; Habermann and Murphy 1996). Although today we have accumulated a large amount of information about protein conformation and stabilization, we still do not fully understand their solution behavior or their *personality,* even though some proteins have similar conformations but different solution behavior. This is like if we attempted to understand people's personality just by looking at their appearance.

During most of the late nineteenth century and beginning of the twentieth century, protein chemists were involved in a heated debate about whether proteins had defined structure and amino acids were linked together. The debate was stimulated by *colloid chemistry,* which was the popular way, at the time, of thinking about large particles, considering the properties of inorganic colloids.

Thus, colloidal referred to items of a small size floating in a medium of one of three substances: a solid, a liquid, or a gas. Some of the common colloids include fog, hairspray, or mist, which are gas particles suspended in a liquid; whipped cream—a liquid suspended in a gas; hand cream—a liquid suspended in a liquid; pigmented ink or blood—liquid particles embodied in a liquid; Styrofoam—a solid in a gas; jelly—a solid in a liquid; and cranberry glass—a solid in a solid.

Some organic substances, including cellulose, polysaccharides, and proteins, behaved in some respect like *inorganic colloids,* and the idea was that proteins were just clusters of small molecules held together with a lack of definitive structure in contrast with another group that advanced the notion that proteins were indeed *macromolecules* with a defined structure. It was not until 1930 that the debate was settled in favor of the notion that proteins are indeed macromolecular structures and not colloids as we know them today.

Mixtures involving a fluid solvent can be classified as solutions colloids and suspensions. Nowadays most scientists working on formulation development for therapeutic proteins use the term *colloidal stability* to refer to the long-term integrity of a protein particle and its ability to resist phenomena such as sedimentation or particle aggregation, self-association/oligomerization. This is typically defined by the time that dispersed phase particles can remain suspended. Thus, a solution is a homogenous mixture that is formed when the particle size of the solute is less than 10^{-7} cm and is surrounded by the solvent molecules that in most biochemical reactions is water at a concentration of 55 M.

Because of the mixing, the substances in solution do not settle or separate as a function of time. Solutions are clear (transparent to light), but they may have a color. By contrast, colloids are mixtures of solute particles sized between 10^{-7} to 10^{-5} cm. Colloids remain well mixed over time and do not separate; they are cloudy and translucent to opaque. When a colloid is illuminated, the light scattering produced by the particles is the Tyndall effect (i.e., the scattering of a beam of light by a medium containing small, suspended particles). However, suspensions differ from colloids in that they will separate as a function of time (settle into layers) and have a solute particle size greater than 10^{-5} cm. The suspended substance can usually be separated by filtration (e.g., red cells in the blood).

Several methods have been employed to measure the protein concentration dependence on the diffusion interaction parameter K_D (Garidel et al 2015, Kingsbury et al. 2020) by either light scattering or analytical ultracentrifugation, where a positive value will indicate *good colloidal stability* or protein repulsion and a negative value will give *poor colloidal stability* or more protein attraction, leading to opalescence, aggregation, phase separation, or precipitation (figure 6.0).

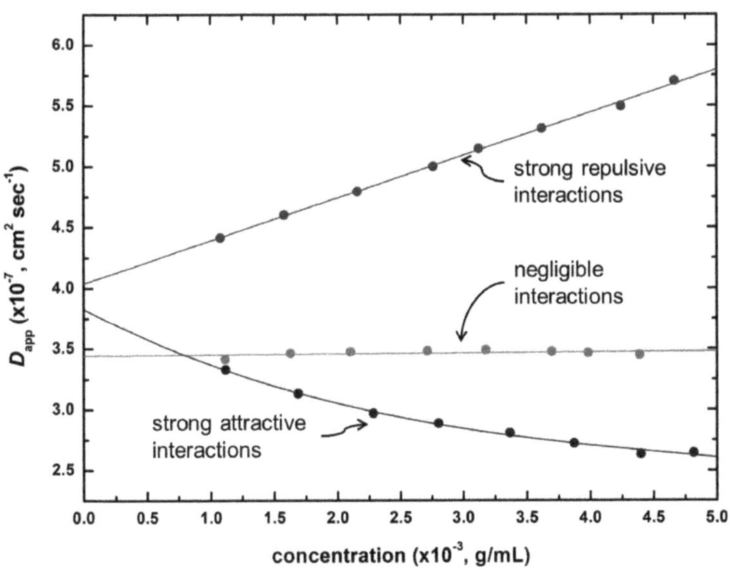

Figure 6.0. Diffusion interaction parameter as a function of protein concentration for three different monoclonal antibodies.

Unfortunately, the commonly used terminology of good or bad *colloidal stability* of proteins brings confusion, considering the old controversy of proteins being thought to be colloids rather than true macromolecules (see Chapter

2). In addition, proteins are not homogenous particles as colloidal chemistry claimed. Everything that happens in protein solutions is related to the *energy of the solution* in a crowded cytosol. Thus, the energy of the solution is given by: i) protein conformational stability as dictated by the Gibbs free energy of stabilization and depends on other parameters of the amino acid sequence of the protein and the solution variables; ii) and the relative intermolecular binding energy as can be calculated from an experimentally induced protein liquid-liquid phase separation or by measuring the diffusion interaction parameter (K_D) by either light scattering or analytical ultracentrifugation. We will examine in detail these important concepts that have a tremendous influence on the "social behavior of proteins."

A protein molecule in solution may undergo a host of interactions with solvent, small molecules and ions, itself, and other macromolecules. If we image the thermal motion in biological systems and the molecular storm where protein molecules are colliding with water molecules and other ions, the solution behavior of proteins will depend on the environment surrounding a protein. Below are shown two monoclonal antibodies: one with a charge mutation in the antigen-binding region and one that is a wild type. Their solution behavior has been measured by the diffusion interaction parameter, namely $K_{D.}$ As depicted in figure 6.1, both monoclonals show different diffusion interaction parameters (K_D) as a function of pH and buffer species.

Optimal behavior is observed by the wild type in histidine buffer at pH 6.0, while the mutant at the same pH shows slightly lower diffusion interaction parameter. Acetate buffer at pH 5.2 also appears beneficial for the monoclonals. By contrast, pH > 6.5 in phosphate buffer shows less optimal *solution energetics* for both the mutant and the wild type. This solution energetics is commonly referred to as good or bad colloidal stability.

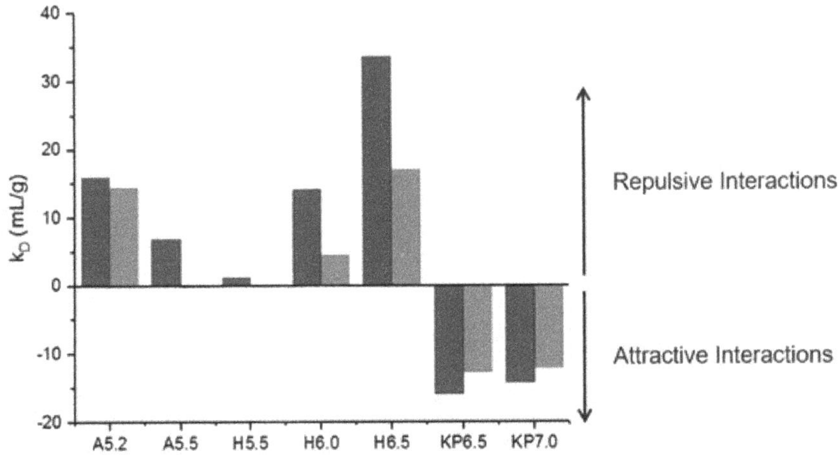

Figure 6.1. Two monoclonal antibodies. The light gray color is the monoclonal with a mutation in the antigen-binding region, where a negative charge has been changed to a polar residue. The darker color is the control monoclonal without the mutation. Diffusion interaction parameters have been measured by light scattering in different buffers at differ pH values. A is acetate, H is histidine, and KP is potassium phosphate, all at 20 mM. The numbers on the x axis represent the pH values.

Measuring Conformational Stability in Proteins

The common approach to measure the conformational stability of proteins is using DSC, where the unfolding of the protein as a function of temperature increments is recorded and the onset of melt and enthalpy can be directly measured as well as the melting temperature (T_m); the latter gives an estimate of the amount of the protein that exists—50 percent native versus 50 percent nonnative (see Chapter 4).

Another approach used to determine the conformational stability of proteins is the analysis of solvent denaturation curves. Two types of denaturant solvents are used for chemical denaturation studies: either guanidinium hydrochloride (GuHCl) or urea. Urea and GuHCl are not powerful chemical denaturants but produce their effects by a very large number of weak interactions (Schellman 2002). The use of urea to monitor protein unfolding started in the laboratory of Walter Kauzmann around 1948. Around 1950, Kauzmann suggested to John Schellman that the binding of urea to unfolded proteins was probably related to the interaction of urea with the peptide group, which has a similar constitution.

In 1951, John Schellman (1924–2014) completed his PhD in theoretical chemistry with Walter Kauzmann at Princeton in just three years. His first scientific paper was on the dielectric polarization of ice and appeared in *Physical Reviews* in 1951 (Schellman and Kauzmann 1951). He found time during his PhD to take part in Kauzmann's experimental program of determining the physical-chemical basis of protein denaturation. This program led to the modern understanding that protein denaturation is protein unfolding. During his career, he studied the question of how protein denaturants work. The standard explanation had been that denaturants bind at sites exposed in the unfolded protein and pull the unfolding reaction, and the binding reaction is of the standard site-binding type, like the binding of protons to glutamate side chains (Baldwin 2002).

In 1955, Schellman gave the equation for binding to multiple equivalent sites (Schellman 1955), an equation that afterward was generally used to interpret denaturation experiments. However, in 1974, Nick Pace found that the free energy of the unfolding reaction depends directly on the denaturant concentration, whereas the site-binding equation predicts a dependence on the logarithm of the denaturant concentration. In 1978, Schellman showed that the observed dependence is a natural consequence of the interaction between the denaturant and protein being of the weak interaction type, rather than the site-binding type (Schellman 1978). In the following years, both Schellman and Timasheff probed ever more deeply into the physical chemistry of the denaturation reaction and showed the importance of an exchange between the solvent (water) and the denaturant for understanding the measured properties (Baldwin 2002).

A method to quantitate protein unfolding was first employed by Charles Tanford (Tanford 1964) when he demonstrated that in the presence of urea, many nonpolar amino acids in the interior of the protein become exposed to the solvent in the unfolded protein. In principle, any physical technique that could distinguish the native and unfolded (nonnative) state of a protein could be used to monitor the unfolding transition, such as spectroscopic techniques (fluorescence, circular dichroism, ultraviolet absorption, biological activity, etc.).

Carlos Nick Pace was born in Salt Lake City, Utah, in 1940. He received his BS in chemistry from the University of Utah in 1962 and his PhD in biochemistry from Duke University in 1966, working with Charles Tanford, who was interested in protein denaturation and solubility in various solutions. As an independent researcher, Pace used the ideas and techniques he learned in Tanford's laboratory to study protein folding and denaturation (Kresge et al. 2008). In 1974, Nick Pace's laboratory demonstrated that plots of the free energy of unfolding (or lnK) against denaturant concentration were linear for both urea and guanidinium hydrochloride (Greene and Pace 1974). The method was called the linear extrapolation procedure. The approach is the most widely used method for measuring protein stability, and the paper has been cited more than three hundred and seventy times.

In the paper, Pace and Raymond F. Greene, Jr. used optical rotation measurements to follow the unfolding of ribonuclease, lysozyme, α-chymotrypsin, and goat β-lactoglobulin in urea and guanidine hydrochloride. By analyzing the proteins' denaturation curves, they were able to make linear plots of the free energy of unfolding (ΔG) versus denaturant concentration. Deviations have been found for some systems, mainly for guanidinium chloride at low concentrations, but most systems obey a linear relationship in the transition region within experimental error. The plot's intercept gave a measure of the protein's stability, whereas the slope gave an "m value" that measured how strongly ΔG depended on urea concentration. Pace's follow-up paper on m values (Myers et al. 1995) is the third most cited paper in the history of protein science, with about eight hundred citations.

By adding the denaturant, the equilibrium between the folded protein and unfolded/not native will shift toward the nonnative condition. The analysis is done at equilibrium to measure the fraction of folded molecules and unfolded/

nonnative denoted by f_N and f_u respectively. The analysis also assumes that there are only two states in the protein, namely native and nonnative/unfolded. It is known that there must be some intermediates in the pathway between natives and nonnatives. Still, these may be assumed to be insignificant at equilibrium, so the error introduced in the analysis is small.

Assuming the two-state mechanism (only folded/native [f_N] and not folded forms [f_u] of the protein are present), then:

$$f_N + f_u = 1 \qquad (6.0)$$

The observed value at any point on the transition curve (see figure 6.2) is given by:

$$y = y_F f_F + y_U f_U \qquad (6.1)$$

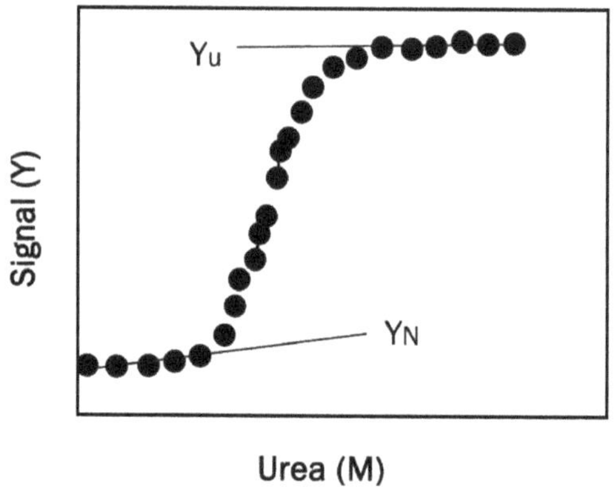

Figure 6.2. Unfolding of a model protein as a function of urea concentration at 25°C.

y is the parameter chosen to follow the unfolding/denaturation reaction, and y_F and y_U represent the values of y characteristic of the folded and unfolded (nonnative) protein. The values of y_F and y_U for any point in the transition region are obtained by extrapolation of the pre-and post-transition baselines, which is generally achieved by a least square analysis. Combining equations 6.0 and 6.1 yields:

$$f_U = (y_F - y)/(y_F - y_U) \tag{6.2}$$

The equilibrium constant K_U, and the free energy change ΔG_U for the folding/unfolding reaction can be calculated using:

$$K_U = f_U/(1 - f_U) = f_U/f_F = (y_F - y)/(y - yU) \quad \text{(6.3) and}$$

$$\Delta G_U = -RT \ln K_U = -RT \ln (y_F - y)/(y - y_U) \tag{6.4}$$

R is the gas constant (1.987 cal/deg/mol), and T is the absolute temperature (K). To obtain an estimate of $\Delta G(H_20)$ (or the ΔG in the diluted buffer), the values of the equilibrium constant K_U are obtained in the presence of the denaturants urea or guanidine hydrochloride. An extrapolation is done to zero denaturant concentration (figure 6.3). This model assumes that the linear dependence of ΔG_U on denaturant concentration observed in the transition region continues to zero denaturant concentration (Greene and Pace 1974; Santoro and Bolen 1988; Bolen and Santoro 1988). The linear extrapolation method was originally based on empirical observation of the linear dependence of $\Delta G°_{N\text{-}U}$ on denaturant concentration in the unfolding transition zone (Green and Pace 1974). For a number of proteins, the estimates of $\Delta G°_{N\text{-}U}$ from different denaturants, such as urea and guanidinium hydrochloride (GuHCl), are in good agreement, indicating that the $\Delta G°_{N\text{-}U}$ quantity is independent of denaturant and is a property of the protein alone.

In making a comparison between some chemical denaturation experiments— in some cases from those obtained by calorimetric unfolding after conversion to the same temperature—the $\Delta G°$ from thermal unfolding was found to agree with $\Delta G°_{N\text{-}U}$ determined from denaturant-induced unfolding (Schellman et al. 1980; Hawkes et al. 1984; Jackson et al. 1993). The denaturant binds to the protein backbone and to amino acid side chains. Binding can be electrostatic or hydrophobic. In addition, the denaturant can change the hydrophobicity of the hydration layer. The binding affinity will be different for different amino acid residues.

These are thermodynamic measurements; therefore, the unfolding reaction must reach equilibrium before the measurements are initiated. The time to reach the equilibrium will be protein-dependent and could be tested by measuring *y* with a proper technique (e.g., fluorescence, circular dichroism, or UV/visible spectroscopy, titration of enzymatic activity, etc.). The reversibility could be tested by measuring *y* and then diluting the denaturant; the solution should return to the pretransition region. If the unfolding curve shows more than one step, this indicates that the unfolding is more complex, and usually occurs with large proteins (> 12,000 Da) with multidomain.

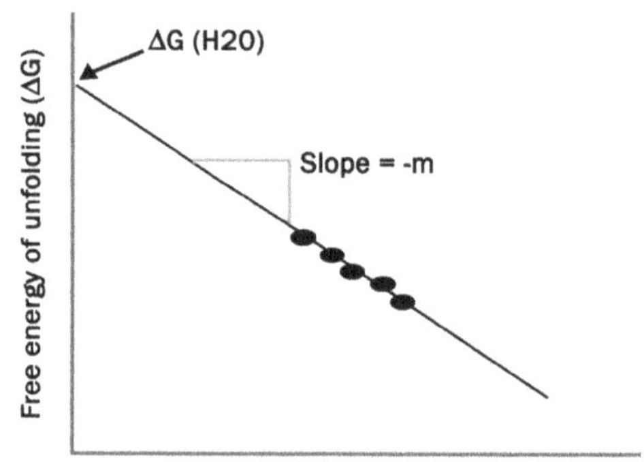

Figure 6.3. Linear extrapolation procedure to obtain an estimate of ΔG (H_20),

$$\Delta G_U = \Delta G(H_20) - m\ [denaturant] \qquad (6.5)$$

The value of m is a measure of the dependence of free energy on denaturant concentration and depends on the amount and composition of the polypeptide chain exposed to solvent during the unfolding reaction. Extrapolation to "physiological" conditions (zero concentration of chemical denaturant, 37°C, for instance) typically leads to a positive (although not too large) value of the unfolding free energy change (Quesada-Soriano et al. 2006; Duy and Fitter 2005; Ramani et al. 2005), indicating that K < 1 and that the unfolding equilibrium is shifted toward the folded state.

It has been shown that increasing sucrose concentration shifts the unfolding transitions to higher denaturant concentrations when phosphoglycerate kinase (PGK) is denatured by guanidine hydrochloride (GuHCl) in different sucrose concentrations (Taylor et al. 1995). Analysis of the data by the linear extrapolation procedure showed that, contrary to expectation, 10–30 percent (w/v) sucrose does not increase the free energy of unfolding the protein in buffer since all the curves converge to approximately the same ΔG_{H20} value of between -8000 and -9000 cal/mol (Taylor et al. 1995). Hence, sucrose does not stabilize the folded state of PGK over the unfolded state but does increase the concentration of denaturant required to bring about the transition (Taylor et al. 1995).

Other models have been proposed to explain the experimental results obtained; the most prevalent is the solvent exchange model (weak binding model) by Jon Schellman (Schellman 1994), which is more complex, and we will not describe it here. Nevertheless, the linear extrapolation model (see Green and Pace 1974; Bolen and Santoro 1988; Santoro and Bolen 1988), which we just described, is sufficient to obtain quantitative values of the conformational stability of proteins expressed by the ΔG_{H20}. Hence, analysis of proteins with mutated amino acids and/or under different solution variables (pH, ionic strength, temperature, presence of excipients, etc.) could permit the identification of conditions where the protein will have better conformational stability. This is highly desirable in therapeutic proteins usually developed as liquid drug products.

The methodology described does not permit an understanding of the natural propensity of a protein to socialize/interact with other proteins. We know that if a protein is in the nonnative/unfolded state, it will have hydrophobic regions exposed to solvent and therefore (thermodynamically unfavorable), to shield those regions from solvent, there will be a greater tendency to socialize/ aggregate, precipitate, and come out of the solution. There are proteins that have good conformational stability, but they will still tend to associate and precipitate; or as has been defined by others, they have *poor colloidal stability.* Or we can say that *the energy of the solution is suboptimal.*

Dissecting the Energy of Protein-Protein Interactions

The energy of the solution dictates if a protein will socialize with other proteins, leading to condensation, such as oligomerization, opalescence, viscosity, phase-separation, precipitation, or crystallization. Thus, a perfectly folded protein with good conformational stability could become insoluble at non-denaturing temperatures or solution conditions and high protein concentrations, as in the case of some therapeutic antibodies for subcutaneous administration. Weak interactions between protein molecules drive these protein condensation phenomena and are the physical manifestation of changes in the overall *energy of the solution.* Among the forces involved are steric repulsions, short-range interactions (London dispersion forces), and electrostatic interactions.

Those interactions, occurring at diluted protein concentrations, could be extrapolated to high protein concentrations by measuring the diffusion interaction parameter and osmotic second virial coefficient by light scattering or analytical ultracentrifugation. The fact that an osmotic coefficient could be measured in a system indicates that molecule A knows of the presence of molecule B, indicating that the system is not ideal. Table 6.0 shows a summary of the most used first principles methods to measure protein-protein interactions.

In previous chapters, we introduced analytical ultracentrifugation and its role in the understanding that proteins are indeed macromolecules with a defined structure. Usually, analytical ultracentrifugation is employed with light scattering as orthogonal tools to interrogate protein oligomerization/association. Since light scattering is a technique based on first principles as well as analytical ultracentrifugation, we will spend some time describing its principles and applications.

Table 6.0 First principles methods to measure protein-protein interactions.

Concentration dependence	Analytical Method
Molecular weight	Static light scattering (SLS) or Rayleigh scattering
	Sedimentation equilibrium
Diffusion coefficient	Dynamic light scattering (DLS)
Sedimentation coefficient	Sedimentation velocity

Light Scattering

Growing up, children wonder why the sky is blue. The answer lies in the scattering of sunlight. The atmosphere is composed of several gases, including around 78 percent nitrogen, 20 percent oxygen, 0.93 percent argon, and others, including the greenhouse gases carbon dioxide, methane, nitrous oxide, and ozone. Also, there are large amounts of small particles. We know that the sunlight that we call white light is in fact composed of seven colors (violet, indigo, blue, green, orange, and red), which we can see when the rainbow is visible. Each color represents a different wavelength and energy.

Blue light has shorter waves, with wavelengths between 450 and 495 nanometers. Typically, the human eye can detect wavelengths from 380 to 700 nanometers. Red light has longer waves, with wavelengths around 620 to 750 nm. Blue light has a higher frequency and carries more energy than red light. Blue light, due to its relatively smaller size, can impact the small dust particles in the atmosphere, resulting in the scattering of the blue light that we see as blue sky. When we approach the sunset or the sunrise, the sun is low and the sunlight must travel farther in the atmosphere. All the blue light is scattered in the sky, and we are only left with the red light we see. The phenomenon known as Rayleigh scattering was named in honor of John William Strutt, 3rd Baron of Rayleigh (1842–1919), who described the phenomenon in an article entitled: "On the Light From the Sky: its Polarization and Colour," which appeared in the London, Edinburgh, and Dublin *Philosophical Magazine* (vol. 41, 1871) while he was a Fellow at Trinity College in Cambridge, UK.

What Can Light Scattering (LS) Measure?

For a solute in solution, light scattering (LS) can determine:

- Molar mass (SLS).
- Root mean square (rms) radius, sometimes known as the *radius of gyration* (r_g), which provides information on size, compactness, and folding (SLS).
- Second virial coefficient, A_{22}, B_{22}.
- Translational diffusion coefficient, *Dt.* (DLS)
- Dt can be used to calculate the *hydrodynamic radius* (r_h). (DLS)

For additional reading on light-scattering principles, see Johnson and Gabriel 1981. For specialized applications, see Kirkwood and Goldberg 1950; Timasheff et al. 1957; Wyatt 1993; Murphy 1997; and Saluja et al. 2007.

Principle 1

The amount of light scattered is directly proportional to the product of the polymer molar mass and concentration.

$$I(\theta)_{\text{scattered}} \propto Mc\left(\frac{dn}{dc}\right)^2 \qquad (6.6)$$

dn/dc is the refractive index increment.

To characterize the scattering from a solution of macromolecules, it is first necessary to know their polarizability. This may be determined from a measurement of the change Δn of the solution's refractive index n with the molecular concentration change Δc by measuring the dn/dc (= $\Delta n/\Delta c$) value using a differential refractometer.

Principle 2

The angular variation of the scattered light is directly related to the size of the molecule. Angular variation depends only on rms radius of polymer (*radius of gyration).* The rms radius is a measure of its size weighted by the mass distribution about its center of mass. r_g is an indication of degree of compactness = folding.

Macromolecules much smaller than the wavelength of the incident light can be treated as though they were essentially point scatterers. For such very small molecules, the scattered light in the plane perpendicular to the polarization of the incident light is independent of scattering angle. It is the same at every scattering angle. For practical purposes, in globular proteins smaller than ~50 kDa, the intensity of the scattered light is uniform in all directions. For larger macromolecules, however, there are variations in the phase of the scattered light from different parts in the macromolecule. This can lead to the destructive or constructive interference of the scattered light.

Two Types of Light Scattering

1. "Classical" light scattering (also known as "static" or "Rayleigh" scattering or multi-angle laser light scattering, MALLS)

This is used to obtain a direct measure of molecular mass. It is very useful for determining whether the native state of a protein is a monomer or a higher oligomer, and for measuring the masses of aggregates or other nonnative species. It can be used for measuring the stoichiometry of complexes between different proteins (e.g., receptor-ligand complexes or antibody-antigen complexes).

In a typical Rayleigh scattering experiment (classical light scattering), a single frequency polarized light beam (e.g., from a laser) is used to illuminate a solution containing a suspension of the macromolecules of interest.

The electric field of the polarized light beam is generally produced perpendicular to the plane in which the intensity and angular dependence of the subsequently scattered light are to be measured. The intensity carries information about the molar mass, while the angular dependence carries information about the size of the macromolecule.

2. Dynamic light scattering (DLS), also known as "photon correlation spectroscopy" (PCS) or "quasi-elastic light scattering" (QELS).

This uses scattered light to measure the rate of diffusion of the protein particles. This motion data is conventionally processed to derive a size distribution for the sample, where the size is given by the *Stokes radius* or *hydrodynamic radius* of the protein particle. The hydrodynamic size depends on both mass and shape (conformation). Dynamic scattering is particularly good at sensing the presence of very small amounts of aggregated protein (<0.01 percent by weight) and studying samples containing a very large range of masses. It can be quite valuable for comparing stability of different formulations, including real-time monitoring of changes at elevated temperatures.

In a QELS (dynamic light scattering) measurement, the time-dependent fluctuations in the scattered light are measured by a fast photon counter. The fluctuations are directly related to the rate of diffusion of the molecule through the solvent. Therefore, the fluctuations can be analyzed to determine the hydrodynamic radius (rh) for the sample. As light scatters from the moving macromolecules, this motion imparts a randomness to the phase of the scattered light such that when the scattered light from two or more particles is added together, there will be a changing destructive or constructive interference. This leads to time-dependent fluctuations in the intensity of the scattered light.

The diameter that is measured in dynamic light scattering is called the hydrodynamic diameter and refers to how a particle diffuses within a fluid. The diameter obtained by this technique is that of a sphere that has the same translational diffusion coefficient as the particle being measured. The translational diffusion coefficient will depend not only on the size of the particle "core" but also on any surface structure, as well as the concentration and type of ions in the medium. This means that the size can be larger than measured by electron microscopy, for example, where the particle is removed from its native environment. The analysis can be done in batch mode, with no column separation, which could lead to dilution and aggregate disruption.

Figure 6.4. shows an example of a protein with a radius of ~4 nm (figure 6.4a) and a well-resolved aggregate of ~200 nm (figure 6.4b). The large aggregate represents approximately 20 percent of the scattered light that is proportional to mass and mass increase with radius. Then it can be determined that the mass of the aggregate in figure 6.4b represents ~0.35 percent of the total mass.

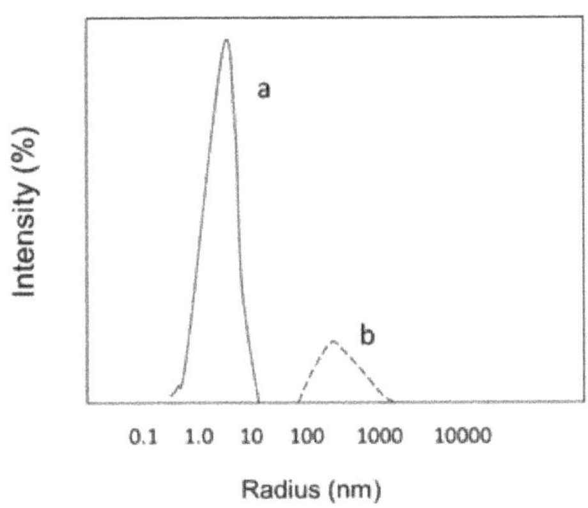

Figure 6.4. Example of a dynamic light scattering analysis of a protein with a small number of aggregate species.

DLS measures the scattered light of a sample in very short time periods and correlates the data. Since large molecules or particles move slower than small molecules, a defined correlation function is obtained. From the correlation function, the diffusion coefficient *(D)* of the molecules can be calculated by fitting the data. Finally, the hydrodynamic radius (r_h) of the particles and molecules can be calculated using the Einstein-Stokes equation:

$$D = \frac{k_B T}{6\pi \eta r_h} \quad (6.7)$$

k_B is the Boltzmann-constant. T is temperature. η is solvent viscosity, and *r*h is the hydrodynamic radius.

The second osmotic virial coefficient (B_{22} value) is a physical parameter used to measure the “colloidal stability” of a protein solution (Manning et al., 2018). In this book, we refer to this parameter as the protein solution’s energy or energy of the solution. The second osmotic virial coefficient (B_{22} value as measured by sedimentation equilibrium in the analytical ultracentrifuge) and called A_{22} when estimated by light scattering (Winzor et al., 2023), is a physical parameter that is used to directly measure the energy of a protein solution (Hofmann et al. 2016) and is a predictor of the protein aggregation propensity (Quigley and Williams 2015; Hoo et al. 2003; Blanco et al. 2014). The subscript 22 refers to the notation of component 2 in solution thermodynamic as protein.

Static light scattering (SLS) measurements require that multiple protein concentrations be evaluated, as the A_{22} value comes from the slope of a Zimm plot, which graphically represents the light scattered from the protein as a function of scattering angle and protein concentration (Nguyen et al. 2003). The simplest approach is to measure the scattering at different protein concentrations and only at one angle (90 degree) and extract the relevant values from the Debbye plot (see figure 6.5). The sign of the osmotic second virial coefficient can give information on the preferential interaction of the protein with the solvent or repulsion between the macromolecules in solution (positive value) or preferential attractions between proteins (negative value).

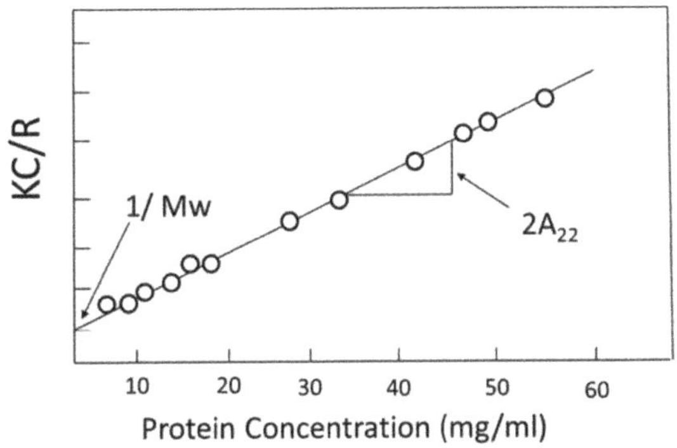

Figure 6.5. Debye plot for a monoclonal antibody by measuring light scattering at a 90-degree angle. The Y-intercept value of this plot is equal to the 1/Mw of the molecule in solution, and the slope is equal to 2x the virial coefficient A_{22}.

Liquid-Liquid Phase Separation

Of the different types of condensation reactions, several pharmaceutical proteins have shown the spontaneous and reversible formation of liquid-liquid phase separation (LLPS) that takes place upon changes in temperature or solution conditions. As described by Benedeck and coworkers (Wang et al. 2011, 2014, 2017, 2021; Wang and Latypov 2019), LLPS is the spontaneous segregation of a homogeneous protein solution below a specific temperature (usually below 4 °C) into a coexisting protein-dilute and protein-rich phase (figure 6.6).

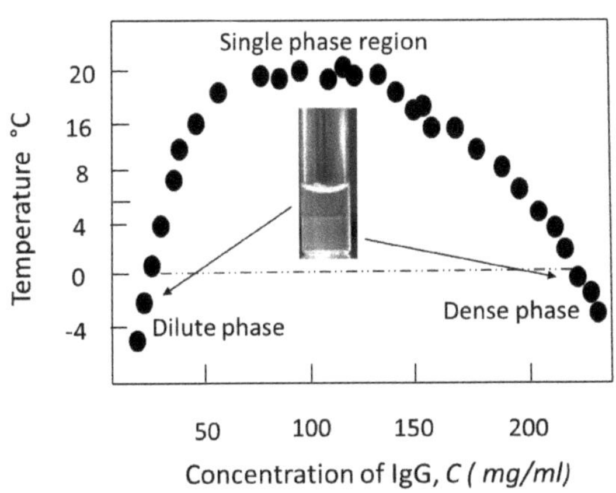

Figure 6.6. Coexistence curve for an IgG undergoing liquid-liquid phase separation.

The chemical potential of the proteins in the diluted and condensed phases is essentially equal, therefore coexisting in perfect thermodynamic equilibrium (in contrast to aggregation, for example). However, the condensed phase has a lower entropy due to the proximity of each protein molecule, limiting their solution mobility and larger binding energy due to the neighbor-to-neighbor attraction. Hence, the observation of LLPS is a manifestation of changes in the *energy of the solution* induced by protein-protein interactions.

LLPS is not easily observed in many proteins, but can be induced in solutions above 4°C by using preferentially excluded cosolvents. Of application to induce LLPS is PEG, which is preferentially excluded from the domains of proteins by steric effects (Arakawa and Timasheff 1985; Lee and Lee 1987; Bhat and Timasheff 1992). PEG is a hydrophilic polymer that affects the solubility of proteins by increasing inter-protein attractions without changing the nature of protein-protein interactions. When two protein molecules are close enough so that PEG molecules cannot fit in the space between them, the unbalanced local osmotic pressure adds an additional attraction between the two protein molecules. Native interactions are not affected. The dense phase (precipitate) mimics a highly concentrated protein.

As described by Wang et al. 2014, for most proteins, the specific attractive interactions between PEG and protein are negligible as compared to the depletion force. The Helmholtz free energy of a protein-PEG mixture solution of volume V is a sum of the protein free energy (ΔG_p) and the PEG free energy (ΔG_{PEG}):

$$\Delta G = \Delta G_p(N_p, V) + \Delta G_{PEG}(N_{PEG}, V_{PEG}) \qquad (6.8)$$

N_p and N_{PEG} are, respectively, the numbers of protein and PEG molecules, and V_{PEG} is the volume accessible to PEG molecules.

Thus, PEG-induced LLPs can be used as a tool to understand the energy of the solution of similar proteins that recognize the same receptor and the effect of various excipients on that solution energy. As described by Wang et al (2014), the solubility as a function of protein concentration is given by the following equation:

$$\varepsilon_{binding} = -(kT \ln (u_0 C'_1) + \pi_2 \Delta u) N_A/1000 \qquad (6.9)$$

C_1 is the protein concentration in the supernatant. N_A is Avogadro's number. k is the Boltzmann constant. T is the absolute temperature, and Δu is the difference between solvent volumes made inaccessible to PEG by a protein molecule in the diluted and condensed phase. The larger the Δu is, the stronger the PEG's effect in inducing LLPS.

The depletion interaction of PEG depends on the size (molecular weight) of the PEG. The parameter $\mathcal{E}_{binding}$ represents the intrinsic energy of the solution or binding energy and can be calculated from a simple PEG precipitation experiment. A protein sample is incubated for several hours in the presence of PEG, then centrifuged at a given temperature, and the equilibrium protein concentration in the supernatant is measured. The binding free energy of the native interaction between protein molecules can be calculated by extrapolating the protein supernatant concentration to zero PEG concentration (Wang et al. 2014; Thompson et al. 2016).

The binding energy calculated from the LLPS experiment is the effective potential energy of the protein in the condensed liquid phase. Thus, the binding energy measures the overall attractive protein interactions. This interaction is favored by the solution conditions (formulation) and the chemistry of the interacting groups at the surface of the protein. Large positive values of $\mathcal{E}_{binding}$ imply a strong attraction between protein molecules and therefore *poor solution energy*. Conversely, low or negative values of binding energy indicate *favorable solution energy* and weak protein-protein interactions (figure 6.7).

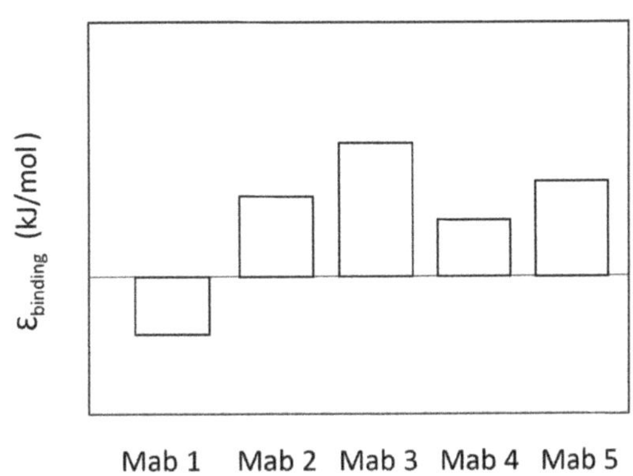

Figure 6.7. Shows the binding energy for different monoclonal antibodies. For mab1, inter-protein interactions are repulsive, whereas all other IgGs exhibit strongly attractive interactions.

The Energy of Activation, Collision Theory, and the Arrhenius Equation

For a chemical/biochemical reaction to proceed at the molecular level, in general, several conditions must be fulfilled: 1) molecules must interact/collide, 2) they must have correct orientation, 3) they must have sufficient energy to overcome the activation barrier to reaction. These conditions are expressed in the simplest rate equations A + B, forming P. Hence, the rate of reaction at any time is defined as the overall rate of formation of the products or loss of reactants. This is expressed in terms of the concentrations of reactants and products:

$$\text{Rate} = d[P]/dt = -d[A]/dt = -\,d[B]/dt \qquad (6.10)$$

The rate equation for this simple reaction would be:

$$\text{Rate} = k[A][B] \qquad (6.11)$$

k is the rate constant, and [A] and [B] are the molar concentrations of reactants.

Our intuition tells us that a reaction proceeds faster if we heat it up. However, many chemical reactions have a negative temperature coefficient. Biochemical processes that can withstand temperatures greater than 50°C for long periods are rare. In the field of proteins, irreversible changes (denaturation) occur that lead to aggregation and precipitation. Svante August Arrhenius (1859–1927), a Swedish scientist who received the Nobel Prize in Chemistry in 1903, showed in 1889 that for most chemical reactions there is a logarithmic relationship between the rate constant and the reciprocal of the temperature in degrees Kelvin:

$$k = Ae - Ea/RT \qquad (6.12)$$

Ea is the energy of activation (see Chapter 4). A is a pre-exponential factor or collision factor that depends on the frequency and orientation of the colliding molecules. The exponential energy term expresses the (Boltzman) probability that the colliding molecules will have sufficient energy to overcome the activation energy barrier at any given absolute temperature (T), and R is the gas constant (1.987 Cal mol^{-1} K^{-1}).

Or in a linear form:

$$\ln k = \ln A - \frac{E_a}{RT} \tag{6.13}$$

The integrated form of the Arrhenius equation is:

$$\ln \frac{k_2}{k_1} = \frac{E_a}{R}\left(\frac{1}{T_1} - \frac{1}{T_2}\right) \tag{6.14}$$

k_2 and k_1 are the specific reaction rate constants at T_2 and T_1, respectively. At constant concentrations of reactants, the rate constants can be replaced with velocities since v is proportional to k. The Arrhenius plot may not hold for proteins, even below the T_m, since other reactions could become rate-limiting as a function of temperature (figure 6.7). For example, oxidation of a critical residue can induce a conformational change, leading to aggregation and precipitation. Deamidation could also be linked to unfolding aggregation and precipitation. For a given temperature, the higher the activation energy barrier, the slower the reaction. For a given reaction, the higher the temperature, the more molecules with high energies and the faster the reaction.

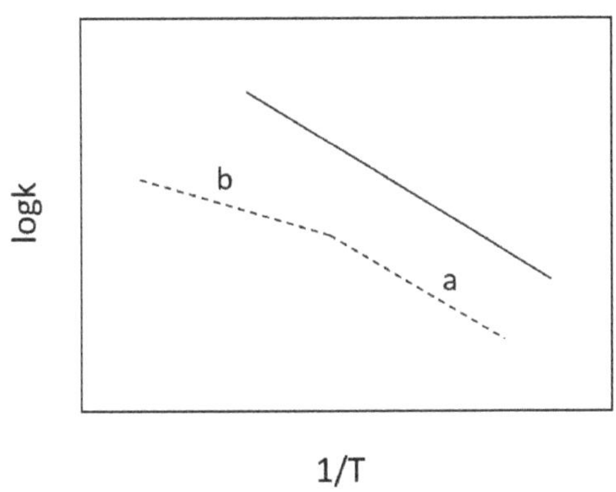

Figure 6.8. Arrhenius plot for two different proteins. The activation energy for a reaction (Ea) can be measured with the reaction rate constant (k) at different temperatures and plotting log k vs. 1/T. (---) Deamidation becomes rate limiting as a function of temperature where two slopes are obtained, indicating a different energy of activation. (a and b). (–) is a protein where no other reaction is rate limiting as a function of temperature.

The effect of temperature on the rate of reactions is usually expressed in terms of a temperature coefficient, $Q_{10,}$ which is the factor by which the rate increases by raising the temperature 10°C.

$$Ea = \frac{2.3RT_2T_1 \log Q_{10}}{10} \quad (6.15)$$

A Q10 of 2 equals an Ea of 12,600 cal/mole in the temperature range of 25 and 35°C. This observation reflects that the Ea of many reactions has roughly the same magnitude.

As we have seen, all chemical reactions require an effective collision between reactants to produce a reaction. Many collisions occur between water molecules, proteins, and other chemicals in a crowded environment in the cell. The process requires energy that we obtain from food, and many enzymes break down food into fundamental building blocks to repair tissues and continue the irreversible steps of life, where energy is dissipated to keep us alive. Plants absorb the energy from light in the process of photosynthesis (see Chapter 8).

One essential molecule that supplies all the energy for chemical and mechanical reactions is ATP (adenosine triphosphate), a molecule containing carbon, hydrogen, nitrogen, oxygen, and phosphorus. It has been called the *energy currency of the cell* since it is the preferred source of energy used by the cell. It donates its phosphoryl group to release large amounts of Gibbs free energy, available for work. ATP is created in the mitochondria, and many enzymes involved in metabolism (glycolysis, citric acid cycle, and oxidative phosphorylation) have ATP as modulators.

Enzymes Decrease the Activation Energy Without Affecting the Equilibrium

We have examined that the rate of a chemical reaction can be increased by temperature. The second way to accelerate a chemical reaction is to add a catalyst. Catalysts accelerate chemical reactions by combining transiently with the reactants and decreasing the *transition state* for the reaction (figure 6.9). Thus, a catalyst lowers the activation energy of chemical reactions, allowing a much more significant fraction of the molecules in each population to react per unit time than in the absence of the catalysts.

The importance of enzymes in biological systems is critical to reaching equilibrium conditions very fast. In the absence of enzymes, some of the most critical metabolic steps may take years. The most specialized proteins are enzymes that have unique catalytic powers. They have a high degree of specificity for their substrate; they function under diluted conditions at specific pH. The enzyme's active site may provide R groups of specific amino acid residues suitable for proton donors (e.g., -COOH, $-NH_3^+$, -SH) or proton acceptors (e.g., -COO-, $-NH_2$, -S-) in general acid-base catalysis.

While a discussion of the enzyme kinetics is beyond the present scope, for background on enzymology and catalysis, the reader is referred to specialized books (see Gutfreund 1972; Segel 1975; Jencks 1987; Price and Stevens 1999).

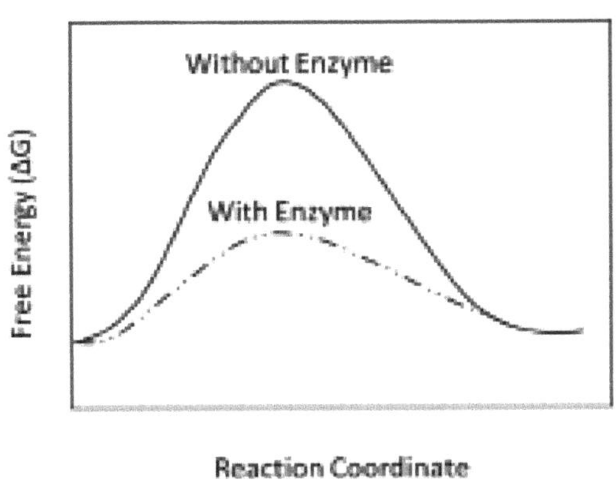

Figure 6.9. The effect of an enzyme on the free energy of activation of a chemical reaction.

Briefly, if we consider an enzyme that binds only one substrate:

$$E + S \underset{k_{-1}}{\overset{k_1}{\longleftrightarrow}} ES \longleftrightarrow \ldots\ldots \longleftrightarrow EP \underset{k_{-n}}{\overset{k_n}{\longleftrightarrow}} E + P \qquad (6.16)$$

If the substrate concentration [S] is large compared to enzyme [E] and the product [P] concentration is essentially zero, a steady state rate can be established, which is described by the Michaelis-Menten equation:

$$v = d[P]/dt = V_{max} [S][E]/ K_m + [S] \qquad (6.17)$$

V_{max} is the maximum velocity at saturating [S] (usually > 20 folds over K_m), and K_m is the Michaelis constant, which is the amount of substrate it takes for an enzyme to reach $V_{max}/2$. The relation between v and [S] is described as a rectangular hyperbole. The constants K_m and V_{max} are a function of several rate constants.

$$V_{max} = k_{cat} [E] \qquad (6.18)$$

kcat is the *turnover number,* the number of substrate molecules each enzyme site converts to product per unit time and in which the enzyme is working at maximum efficiency. It is a first-order rate constant or the catalytic constant for the enzyme.

k_{cat}/K_m is the catalytic efficiency and reflects both binding and catalytic events. It indicates how the velocity varies according to how often the enzyme and substrate combine. It is the best value to represent the enzyme's overall ability to convert substrate to product and compare different enzymes that catalyze the same reaction (e.g., wild type versus mutant).

The pKa of a weakly dissociable group changes in value with temperature (determined by its heat of ionization), depending on the environment. Hence, the extent of ionization at any pH will in part be determined by the temperature; it follows that the pH-activity curve of any enzyme is likely to be temperature dependent (figure 6.10).

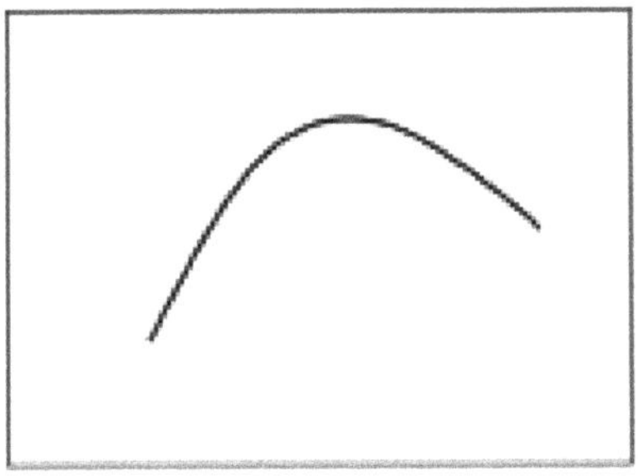

Figure 6.10. pH-activity curve for an enzyme.

A special class of enzymes, called regulatory enzymes, can sense various metabolic signals and change their catalytic rates accordingly. Some of those enzymes are inhibited by an excess of product, stimulated by an excess of substrate or other metabolic molecules that could bind to sites different from the substrate binding site. This type of regulation is called feedback inhibition, and those enzymes are called allosteric enzymes. Allosteric derives from the Greek *allo*, "other," and *stereos*, "space" or "site." Hence, allosteric enzymes are those that have other sites.

Allosteric enzymes usually have multi-polypeptide-chain subunits. They deviate from the Michaelis-Menten curve in substrate binding and reaction velocity as a function of substrate concentration. There appears to be communication between the binding site for the modulator and the catalytic site for the substrate, and they undergo conformational change upon binding of the modulator. The classic example is the binding of oxygen and CO_2 to hemoglobin (see Chapter 8), where oxygen binding to one of the subunits of hemoglobin produces a conformational change in the protein propagated to the remaining subunits, increasing the affinity for oxygen for the other subunits. In other words, substrate binding changes affinity for binding to other sites, and we call it *cooperativity.* The affinity can either increase (positive cooperativity) or decrease (negative cooperativity).

In a highly cooperative binding of a ligand at one site in a macromolecule (enzyme), the affinity of the other binding sites is altered so that each site is immediately filled if ligand molecules are available. The Hill equation (see eq. 6.19) is a modified form of the Michaelis-Menten equation and is commonly used to study the kinetics of reactions that exhibit a sigmoidal behavior.

$$V = \frac{V_{\max}[S]^n}{(K_{0.5})^n + [S]^n} \tag{6.19}$$

V is the rate of the reaction. V_{max} is the maximum velocity of the reaction. [S] is the substrate concentration. $K_{0.5}$ is the half-maximal concentration constant (a substrate concentration that gives rise to a reaction velocity that is 50 percent of V_{max}), and n is the Hill coefficient. The Hill coefficient is unitless. It provides a measure of the cooperativity of substrate binding to the enzyme, macromolecule, etc.

If $n > 1$, the reaction/process is thought to exhibit positive cooperativity with respect to substrate binding to the protein. When $n = 1$, there is no cooperativity with respect to substrate binding to the protein. If $n < 1$, the reaction exhibits negative cooperativity with respect to substrate binding. The equation is a statement of an idealized situation. When cooperativity is at its maximum, there is immediate filling of each site after the first site is filled. Experimentally, this situation is not obtained, i.e., a value of n equal to the number of binding sites is never observed (when $n>1$). Hemoglobin has four binding sites for oxygen; however, n is observed to be 2.8. Thus, the observed value of the slope of a Hill plot (Hill coefficient), n_h, is a qualitative indicator of the degree of cooperativity.

To undergo the allosteric transitions, the macromolecule requires an input of chemical energy that pays the "energetic toll" for switching to different conformations. Hemoglobin is one of those molecules that make those transitions from one functional form to another. The energy for these processes to take place comes from the intrinsic energy of binding to substrate, allosteric effectors ligands (Wyman 1964). However, the energy to trigger those processes also comes from the molecule itself. For example, conformational strain present in the ligated molecule could be relieved by dissociation of the ligand. This could release free energy available to other processes. In the case of hemoglobin, this occurs at the level of tertiary and quaternary structure changes upon ligand binding, such as the oxygenation of hemoglobin linked to subunit interactions (Edsall 1980).

To determine free energies of cooperativity, when the ligands are bound too strongly for direct affinity measurements, a thermodynamic linkage approach has been used (Smith and Ackers 1975; Ackers and Smith 1987; Daugherty et al. 1991). Thus, cooperativity is measured for each of the nine ligated species of hemoglobin by the difference in ΔG of its dimer to tetramer assembly reaction relative to the nonligated species D in equilibrium with T. If we consider the binding of one ligand X per tetramer, we obtain the equilibrium shown in figure 6.11.

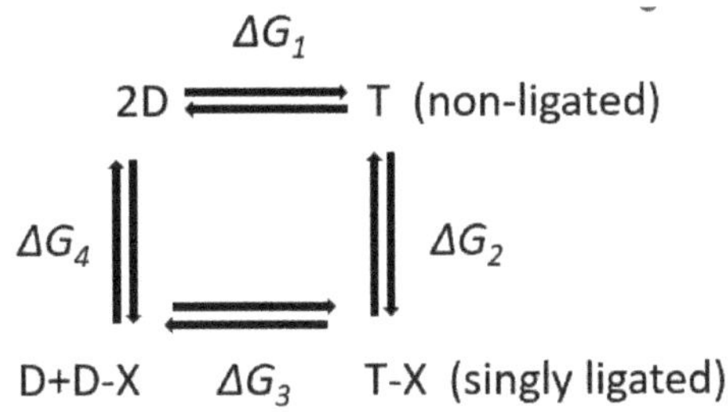

Figure 6.11. The thermodynamic box for the equilibrium between dimer-tetramer and the binding of ligands may alter the free energy of assembly.

The dimer-tetramer assembly alters the free energy of ligand binding.

Thus, ΔG_1 is different from $\Delta \boldsymbol{G}_3$, and $\Delta \boldsymbol{G}_2$ is different from ΔG_4. Since ΔG is a state function, the free energy change in going from 2D to T-X is independent of the reaction pathway and:

$$\Delta G_1 + \Delta \boldsymbol{G}_2 = \Delta G_3 + \Delta G_4$$

Then:

$$\Delta \boldsymbol{G}_3 - \Delta G_1 = \Delta G_2 + \Delta G_4$$

The linked thermodynamic box permits dissection of the cooperativity of free energy of ligation for a tightly bound ligand.

Proton Fluctuations and Protein Association

It has been shown that the charge fluctuations on a protein involving all ionizable groups may result in a nonspecific attraction between two identical molecules (Kirkwood and Shumaker 1952). A theory of site-site interactions due to proton/charge fluctuations has been proposed where a specific attraction may be established. One striking feature of the fluctuating charge interactions is that the pH dependence of the attraction passes through a maximum. This suggests that protein interactions that have a bell-shaped pH dependence are controlled at least in part by the fluctuating charges (Kirkwood and Shumaker 1952; Timasheff 1966). Thus, in the pH region or the pK for the residues involved, such a position of the groups would be reinforced by the fluctuating proton attraction, resulting in additional stabilization of a particular conformation of the molecule (Timasheff 1966).

In qualitative terms, the overall energy of a protein solution is given by the combination of:

- The delicate interplay between the conformational stability obtained by a DSC experiment plus the chemical denaturation approach to obtain $\Delta G(H_2 0)$.
- The tendency of the protein to form superstructures (oligomerization/self-association, particulation), illustrated by the binding energy given by the PEG-induced liquid-liquid phase separation or the diffusion interaction parameter (K_D) obtained by light scattering or analytical ultracentrifugation.

Therefore, the overall energy of a protein solution has the *intrastructural* component $\Delta G_{folding}$ and the *interstructural* components K_D and/or $\mathcal{E}_{binding}$.

CHAPTER 7

DANCING PROTEINS

We have learned that folded proteins show a high degree of order and structure linked to biological function. Folded proteins undergo a constrained collective motion related to their function. For a well-structured protein, there is an allowed conformational dynamic across a wide range of time scales and amplitude (Lermyte 2020). This includes the motion of side chains, loops, and collective motions sampling the conformational state around the stable native structure (Henzler-Widman et al. 2007). A stable three-dimensional structure of a protein is vital to understanding its function in the conventional structure function paradigm (Anfinsen 1973).

Two decades ago, the concept of a well-structured protein and its linkage to biological function began to be challenged. In the human genome, nearly 40 percent of protein-coding genes contain disordered regions of >30 amino acids in length (Ward, et al. 2004; Minezaki et al. 2005). There is recognition now of many proteins without well-folded structures, namely intrinsically disordered proteins (IDPs) (Uversky 2002; Tompa 2002), also called *dancing proteins*. Multiple alternative names have been used for these proteins, such as intrinsically unstructured, intrinsically disordered, intrinsically unfolded, intrinsically denatured, flexible natively denatured, natively unfolded, natively disordered protein, and partially folded proteins.

IDPs are polypeptide chains lacking a defined conformational structure (Wright and Dyson 1999). Comparative analysis of the amino acid sequence of IDPs for those of ordered proteins demonstrates the noticeable enrichment in the content of disorder-promoting amino acids, such as Ala, Arg, Gly, Gln, Ser, Glu, Lys, and Pro (see Appendix for abbreviations), paralleled by the significant depletion in the content of order-promoting amino acids, such as Ile, Leu, Val, Trp, Tyr, Phe, Cys, and Asn. Analysis of the number of highly charged polypeptides revealed that the intrinsic preference of a polypeptide backbone for forming structure depends on charge content (Mao et al. 2010).

Accurate prediction of a protein's predisposition to be intrinsically disordered is a prerequisite for the further understanding of principles and mechanisms of protein folding and function and is key for elaborating a new structural and functional hierarchy of proteins. Therefore, IDP prediction has attracted the attention of many researchers, and several prediction tools have been developed. The compositional differences between ordered proteins and *dancing proteins* (IDPs) facilitated the development of various disorder predictors. These predictors were initially elaborated based on amino acid composition. Later, the predictors were developed based on some basic physical principles and machine learning algorithms, which use the characteristic features of IDPs and intrinsically disordered protein regions (IDPRs), such as net charge, hydrophobicity, and other sequence features (He et al. 2009).

It is now evident that a significant fraction of the human proteome comprises proteins that, under physiological conditions, lack fixed, ordered three-dimensional structure as a whole, or have segments that are IDPs. In general, they exhibit more extensive plasticity and are less constrained in their motion than folded proteins. Since these proteins have different IDPRs, those regions might have well-defined and specific functions. Due to their conformational flexibility, they could be multifunctional and interact with different molecules or other structurally unrelated proteins. Thus, IDPs are not homogenous but have a mosaic architecture reflecting their highly heterogeneous organization. Some IDPs bind to their target site with low affinity, while others, despite their flexible conformation, can interact with ultra-high (picomolar) affinities, retaining their structural disorder and dynamic character (Borgia et al. 2018; Fuxreiter 2018).

Folding Funnel and the Free Energy Landscape for Ordered Proteins. How Does This Concept Apply to Intrinsically Disordered Proteins?

The conformational properties of free IDPs are influenced by a variety of factors, including (a lack of) sequence complexity (Ruff et al. 2019), posttranslational modifications (Bah, et al. 2015; Tsytlonok et al. 2019), proline isomerization (Zosel et al. 2018), and distribution of charged residues (Das and Pappu 2013).

The free energy landscape is the cornerstone in the study of protein folding. Its most fundamental aspect is that it is globally funneled, such that the folding is energetically biased (Bryngelson, et al. 1995; Dill and Chan 1997; Levinthal 1969). Indeed, this notion resolves the well-known paradox of Levinthal (Levinthal 1969) and accounts for why proteins fold in milliseconds to seconds instead of requiring astronomical timescales (Zwanzig et al.1995; Karplus 2011). The folding of numerous intrinsically disordered proteins, which is central to their functions, occurs only through the binding with their partners (Tompa 2005; Wright and Dyson 2009; Uversky 2013).

Uncovering the molecular details of the interaction underlying the folding upon binding of intrinsically disordered proteins is of fundamental importance in molecular biology and is of practical value in protein engineering. Site-directed mutagenesis is a powerful technique to probe effects on protein-protein interaction arising from specific amino acids in the sequence (Fersht 1999; Morrison and Weiss 2001). Related computational methods have also been developed, such as computational alanine scanning of protein-protein interfaces (Massova and Kollman 1999). These mutation-based approaches, however, necessarily invoke perturbations to the underlying protein structures, which sometimes exert disruptive effects unexpectedly and intricately (Xiao et al. 2013). Figure 7.0 shows the hypothetical energy landscaping for an IDP following four unfolded states in solution, separated by shallow energy barriers, suggesting high structural flexibility.

Posttranslational modifications like glycosylation can stabilize one stage. Further stabilization can be attained by binding to a partner, decreasing the folding energy. As we saw in canonical proteins, there is a hindrance in forming folded IDP (see Chapter 3, figure 3.13). In the case of IDP folding, multiple states in the folding funnel are separated by a shallow energy barrier and a lack of cooperativity to attain the final three-dimensional structure, as we see in canonical proteins. This unique behavior may be due to the lack of hydrophobic amino acids in IDPs compared to adequately folded proteins (Uversky et al. 2000; Oldfield et al. 2014).

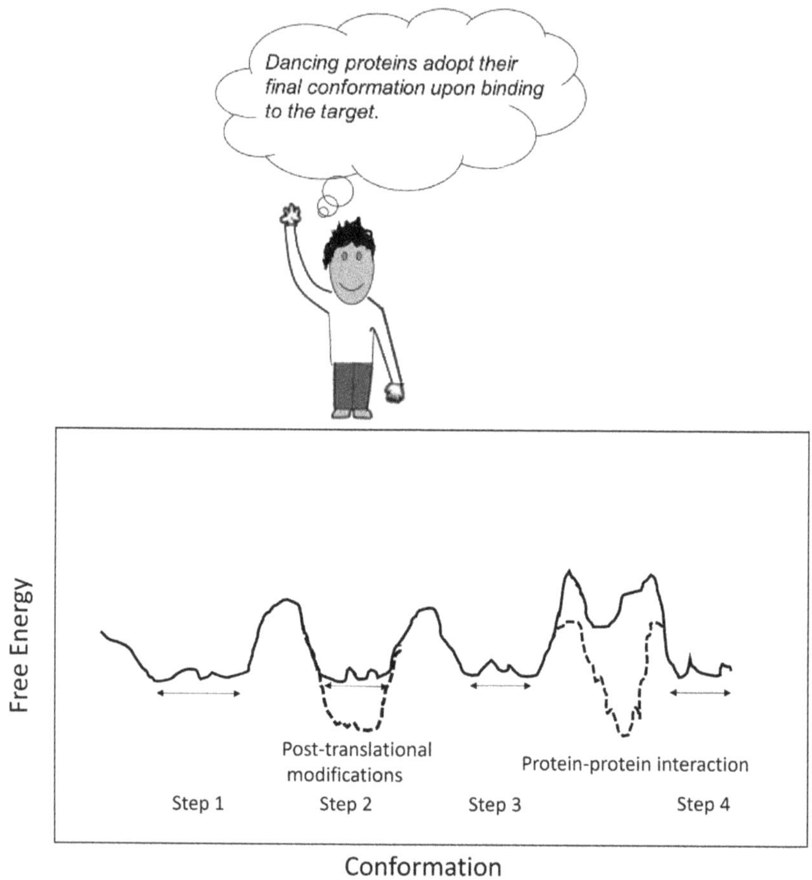

Figure 7.0. Hypothetical energy landscaping for an IDP following four unfolded states in solution.

Fuzzy Complexes Formed by Dancing Proteins

IDPs and IDPRs are promiscuous when binding to partners. This involves interaction with multiple partners. The illustrative examples of disordered proteins that bind to around 10–100 binding partners are p21, p27, p53, BRCA1, XPA, β-synuclein, and estrogen receptors (Uversky 2011).

Fuzzy complexes refer to IDPs or IDPRs bound to a target protein while retaining their plasticity. Several fuzzy complexes have been characterized, and they usually include the interactions of a well-structured protein with an IDP or IDPRs. Also, there is evidence of IDP-IDP, IDPR-IDPR, and IDP-IDPR interactions, creating the so-called super fuzzy complexes (Wang and Wang 2019; Hadži et al. 2021). Using a statistical thermodynamic model, it has been shown that the target-bound ensembles are determined by a simple code that combines the IDP sequence and the distribution of IDP-target interaction hotspots. These two parameters define the conformational space of target-bound IDPs and rationalize the observed structural heterogeneity of fuzzy complexes (Hadži, et al. 2021). Despite the observed IDP dynamics and conformational heterogeneity, such complexes are functionally specific and play crucial roles in several physiological processes.

The first characterized extremely fuzzy complex is 4.1G-CTD/NuMA (Wu et al. 2017). Protein 4.1 is a ubiquitously expressed adaptor protein, which serves as a hub that organizes signaling complexes involving many membrane proteins. Target-bound IDP ensembles optimize allosteric coupling (Hilser and Thompson 2007; Li and Hilser 2018). Identification of crucial parameters shaping the fuzzy ensembles suggests the extent to which the bound-state IDP conformations are templated by the target or encoded by its sequence. This question has been addressed in several studies, reaching different conclusions (Toto et al. 2020; Crabtree et al. 2018). For example, IDP regions with the helical preference of the p53 protein adopt different conformations on different targets (Uversky 2016), suggesting that the IDP conformation is context-dependent and templated by the target.

In contrast, some IDPs may adopt practically identical conformations when bound to different targets (Crabtree et al. 2018; Drobnak et al. 2013), suggesting that IDP conformations are encoded by their sequence alone. Some IDPs can fold into a specific three-dimensional structure after binding to natural partners. Most of the functional advantages of IDPs/IDPRs over their ordered counterparts are directly or indirectly related to some specific features originating from the unique peculiarities of disorder-based interactions, enabling them to participate in one-to-many and many-to-one interactions.

Many other IDPs/IDPRs are known to preserve significant levels of disorder even in their bound state. In the usual consideration of a protein structure function relationship, a protein molecule is often considered a semi-homogeneous entity, entirely ordered or disordered, or consisting of homogeneously ordered and/or disordered regions/domains. However, this conception is another oversimplification, because proteins typically comprise differently folded regions with different levels of conformational stability. In the ordered protein realm, this heterogeneity not only gives rise to multidomain proteins but also defines submolecular *foldon* units, i.e., a part of a protein that folds and unfolds as a unit.

Units, even in small single-domain proteins, are involved in transport and catalytic processes. A very important recent addition to this spectrum of *foldon*-related entities is *unfoldons* (Bertz and Rief 2008), which are regions of ordered protein with a dormant disorder that needs to be transiently awakened to make the protein functional.

Intrinsically Disordered Proteins in Disease

Proteins, as the engines of life, are deeply involved in maintaining life processes; their dysfunction can result in various pathological conditions. Indeed, a broad range of human diseases known as protein conformation or protein-misfolding diseases arise from the failure of a specific peptide or protein to adopt its three-dimensional conformation state (protein misfolding) to yield a functional state. Misfolding consequently causes protein aggregation (and/or fibril formation), opalescence, precipitation out of the solution, and toxic function (Stefani and Rigacci 2013). Some proteins have an intrinsic propensity to assume a pathologic conformation, which becomes evident with aging or at persistently high concentrations (Uversky 2008).

Interactions with some endogenous factors (e.g., chaperones, intracellular or extracellular matricWes, other proteins, or small molecules) can change the conformation of a pathogenic protein and increase its propensity to misfold (Ashraf et al. 2014). Misfolding and misfunction can originate from a point mutation(s) or result from exposure to internal or external toxins, impaired post-translational modifications (PTMs) (e.g., phosphorylation, advanced glycation, deamidation, racemization), and an increased probability of degradation (Uversky 2014). The formation of various aggregates represents the most visible consequence of protein misfolding, and these aggregates form the basis for the development of various protein deposition diseases.

IDPs/IDPDs are tightly controlled inside the cell, and their dysregulation and misbehavior are frequently associated with the development of various pathological conditions. The emerging trend of defining several human neurodegenerative disorders as syndromes of protein folding and oligomerization has led to beta-amyloid protein (Aβ), one of the intrinsically disordered proteins associated with neurodegenerative diseases like Parkinson's, prion disease, and Alzheimer's disease (AD). There is still a significant gap between the site-specific structural information and the complex structural diversity of Aβ amyloids.

The description of the structural polymorphisms of Aβ amyloids can provide valuable information on the molecular basis of AD onset-progress and is essential for comprehending the Aβ aggregation pathways (Korsak and Kosyreva 2015). Aβ molecules can spontaneously self-aggregate in vitro into different species (Jan et al. 2010) under distinct conditions. Aβ peptides can form soluble oligomers and protofibrils, which could be intermediates of a fibrillation process. Studies have revealed the high neurotoxicity of these species and their close links with Alzheimer's disease. Bioinformatics tools have revealed that although many viral membrane proteins are canonical proteins, universally ordered, the intrinsic disorder is still present in these proteins, indicating IDRs might have crucial functions. For instance, influenza A virus, virulent strains 1918 H1N1 and H5N1, differs from less virulent or nonvirulent strains H3N2 and 1930 H1N1 in their disordered content of the hemagglutinin protein (Goh et al. 2009; Xue et al. 2014).

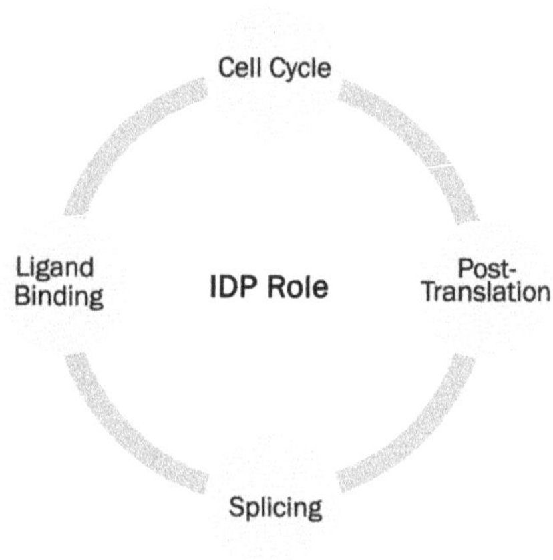

Fig. 7.1 IDP involvement in various cellular processes.

Prostate-associated gene 4 (PAGE4) is an archetypal IDP implicated in human prostate cancer (Salgia et al. 2018). Detailed structure function studies on PAGE4 and its interaction with AP-1 may yield additional insights that can potentially be used to identify small molecule compounds that disrupt the PAGE4/AP-1 complex.

Another example of the involvement of IDPs in human disease is their participation in cell signaling and cancer-associated proteins. Breast cancer type I susceptibility protein (BRCA1) is one of the most typical IDPs, with a long central region of one thousand four hundred and eighty amino acids, shown to be disordered by nuclear magnetic resonance and circular dichroism spectroscopy (Mark et al. 2005). In-silico studies using molecular dynamic simulations and computer-assisted drug design could further support such attempts and help screen for small molecule inhibitors (Ekins et al. 2007). An FDA-approved compound, trifluoperazine dihydrochloride, was found to bind to a disordered region of multifunctional protein nuclear protein 1 (NUPR1) and arrest pancreatic ductal adenocarcinoma (PDAC) development (Neira et al. 2017). This tells us that the organization of nature and biological systems is more complex than we think.

CHAPTER 8

THE WORKHORSES OF THE CELL

The macromolecules named proteins are the essential components of life. Life as we know it today is not possible in the absence of proteins. Proteins are indeed the workhorses of the cell, and the success of living organisms depends on the work performed by the proteins. Proteins are synthesized and regulated depending upon the functional need in the cell. The blueprints for proteins are stored in DNA, which programs what a living organism could become and is used as a template by highly regulated transcriptional processes to produce messenger RNA (mRNA). The message coded by mRNA is then translated into defined sequences of amino acids that form a protein. Transcription is the transfer of information from DNA to mRNA, and translation is the synthesis of protein based on an amino acid sequence specified by mRNA (figure 8.0).

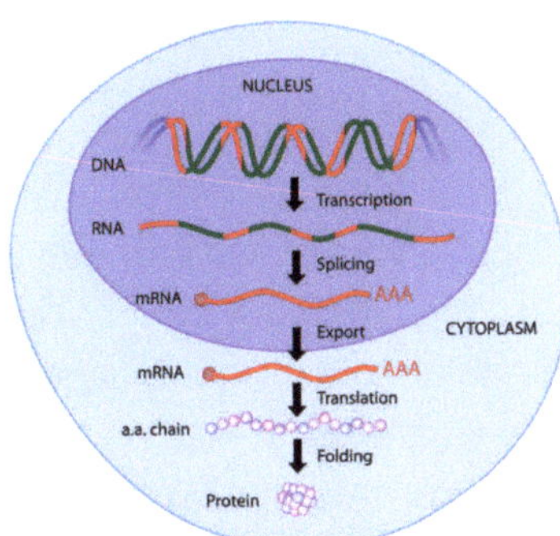

Figure 8.0. The transfer of information from the nucleus of the cell to form proteins.
Alamy: Image ID: CT2772 order reference: OY72206947, August 27, 2022.

The biological activity of a protein depends on its three-dimensional structure, dictated by its sequence of amino acids. In other words, the order of the building blocks of amino acids in the polypeptide chain are as we reviewed in the preceding chapters, but there are exceptions to this rule as we saw in Chapter 7.

We have learned that there are twenty different amino acids that form the proteins in living systems. The properties of the lateral chains of amino acids have a profound impact on the folding of the proteins. While the polypeptide chain is being formed in the ribosome, the folding of the protein (in a three-dimensional conformation) takes place, minimizing the free energy (Chapter 4). In other words, the "most comfortable" conformation is adopted. In principle, if we know the sequence of amino acids in the protein, the forces that act on i) the atoms of the protein, and ii) the molecules present in the dissolvent, then it could be possible to predict the three-dimensional structure of a protein. However, the situation is more complex and not possible to do for large proteins.

One of the principal objectives of the study of proteins has been to decipher their structure and, with that, to uncover their mechanism of action in health and diseased states. Based on this approach, several general principles have emerged, like the existence of substructures shared by different proteins, which probably fulfill similar functions. It is also important to understand how proteins have evolved and diversified to play different functions in living organisms (Chapter 1).

In all the protein functional diversity, we can see a common denominator. In the great majority of cases, proteins work by specific interaction with different molecules or interact among themselves. In the case of structural proteins, the interaction is between identical molecules, leading to a larger structure, such as a fiber or a tubule. Other proteins have shown an affinity for a variety of different molecules. The antibodies, for example, bind to different antigens with high affinity; hemoglobin captures oxygen in the lungs and then releases it into the peripheral tissues. The proteins that form receptors in the cell membrane recognize messenger molecules (like hormones and neurotransmitters) that could also be proteins with high specificity for the receptors.

A dynamic equilibrium governs the interaction between the protein and the receptor. In other words, the interactions are reversible and not covalent; they are bound and released constantly. The percentage of bound molecules depends on the relative quantities of both substances and the association constant (K_b) between them (see Chapter 5 for details). The strength of the interactions depends on several factors, such as the tight match that the two molecules could have, and that is dictated by the electrostatic interactions or repulsions among electrically charged regions.

The proteins that are enzymes have a high degree of flexibility to recognize a specific molecule (substrate) that is bound in dynamic equilibrium; however, an enzyme is capable of producing a change in the substrate that it is bound to. The change includes, in general, the breaking of a chemical bond in the substrate, the addition of a chemical group, or the alteration of some of the covalent bonds.

In the mechanism of the enzymatic reactions, we can distinguish three steps. First, the substrate binds to the enzyme, then the chemical reaction takes place, and at the end, there is a release of the transformed substrate into a product. The three steps are reversible. For example, if molecule A is bound to the enzyme and is converted into molecule B, it is possible that B could be bound to the enzyme and converted to A.

The direction of the enzymatic reaction, either the formation of B from A or the formation of A from B, is dictated by the thermodynamics of the system that we examined in Chapter 4. If the free energy is decreased in one direction, then the reaction will take place in that direction. The enzyme will not affect the equilibrium, but it will accelerate the obtaining of the equilibrium by decreasing the *energetic barrier*. The enzymes decrease the *energy of activation* by having a higher affinity to the transition state than to the substrates. By the laws of thermodynamics, the free energy of the products should be lower than the free energy of the substrate. Therefore, in the absence of enzymes, most of the biochemical reactions are extremely slow, and many processes of life will not take place.

One important aspect of proteins is their ability to change conformations within certain limits in order to exert their biological function. The pioneering studies of Max Perutz, elucidating the three-dimensional structure of the hemoglobin by X-ray diffraction, illustrate the changes that hemoglobin undergoes upon binding of oxygen or $C0_2$; these changes are transmitted to neighboring hemoglobin molecules, which form a complex that explains the concept of cooperativity.

Proteins that play key functions in nature are highly conserved among species. One example is the acetylcholine neuro-receptor responsible for propagating the nerve impulse. This receptor is highly conserved where the homology of the human protein sequence is remarkably like the one found in sea rays. Other fascinating protein molecules are the ones involved in vision.

Toward the second half of the twentieth century, William A. H. Rushton from Cambridge University wrote: "The molecules respond to light like people to music. Some do not absorb anything; others limit themselves to move a finger or a foot. But there are others that get up, dance, and change partners." Now, after the elucidation of the molecules involved in the cascade of reactions of the visual excitation, the words of William Rushton resonate after the demonstration of the cells involved and the complex formed by cis retinal and the protein opsin, forming the rhodopsin. The remainder of the chapter presents a brief summary of seven different proteins that play critical roles as engines of life.

The Engines of Life in Some Unique Functions

Capturing the Energy from the Sun

Early in the "molecular revolution," the Harvard University laboratory of American botanist Lawrence Bogorad (1921–2003), a pioneer of photosynthesis research, constructed the first restriction map of a plastid DNA (Bedbrook and Bogorad 1976). They were the first to map the eighteen genes for chloroplast rRNAs (Bedbrook et al. 1977) and the first to map, clone, and sequence a gene for a plant protein, rbcL, the plastid gene for the large subunit of ribulose 1.5 bisphosphate carboxylase oxygenase, known as RuBisCO (Coen et al. 1977; reviewed in Bogorad 2003).

I met Dr. Bogorad when I attended an international course in plant molecular biology that he was leading in Valdivia Chile (see figure 8.1). This was my first introduction to the isolation of mRNA, a very unstable molecule. RuBisCO is a key enzyme of life, present in the chloroplasts of plant cells (Ellis 1979). The enzyme permits the conversion of sun energy. Atmospheric carbon dioxide is converted by plants and other photosynthetic organisms to energy-rich molecules, such as glucose. Most life is ultimately sustained by photosynthesis and its rate-limiting carbon-fixing enzyme, RuBisCO. As the most important enzyme on earth, RuBisCO catalyzes the reaction that draws inorganic carbon into the biosphere during the Calvin-Benson cycle.

In higher plants, algae, and cyanobacteria, RuBisCO occurs as a hexadecamer consisting of eight large (50 kDa) subunits and eight small (13 to 15 kDa) subunits. Whereas the large subunits are encoded in the chloroplast genome in plants, the small subunits are nuclear gene products that travel as preproteins to the chloroplast, where they lose their signal peptides before combining with large subunits to form a stable holoenzyme.

RuBisCO is an enzyme involved in the first major step of carbon fixation, a process by which atmospheric carbon dioxide is converted by plants and other photosynthetic organisms to energy-rich molecules, such as glucose. RuBisCO is part of the Calvin cycle of plants, wherein a first step catalyzes the fixation of atmospheric CO_2, forming two molecules of 3-phosphoglycerate from CO_2 and ATP and NADH as the source of energy. One molecule of 3-phosphoglycerate will continue in the cycle to become ribulose-1,5-bisphosphate again, and the other will follow a series of transformations until it reaches glucose, which is the source of energy (see figure figure 8.2 for the net reaction).

This process is vital for plants, but the small flaw is that in addition to fixing CO_2, it also fixes O_2, a process known as photorespiration, which is less efficient than the carbon fixation process. Carboxylation (CO_2 fixation) is preferable to oxidation (O_2 fixation) in a 3:1 ratio, so photorespiration represents a 33 percent inefficiency. This photorespiration process is due to the low affinity of RuBisCO since it can act with both CO_2 and O_2 as substrates.

These facts make the efficiency of the RuBisCO 70 percent, while if it were specific for CO_2, the efficiency would be greater. One possible justification for this small flaw would be that the mechanism evolved into a world without free O_2 in the atmosphere, so the oxygen problem was found much later.

Figure 8.1. International course on molecular biology of plants. Organized by PNU-UNESCO.

Group of students and professors, including Dr. Luis Meza (standing, first from left). Professor Lawrence Bogorad (standing, second from left). Dr.Michele Delseny, (standing fourth from left), member of the French Academy of Science; Dr. Michel Castroviejo (front row left) and Dr. Simon Litvak from the University of Bordeaux (standing, second from right). The author is in the front row (second from right). Campus of the Universidad Austral de Chile, November 1983, Valdivia, Chile.

$$6CO_2 + 6H_2O \xrightarrow{\text{Sunlight}} C_6H_{12}O_6 + O_2$$

Carbon Dioxide, Water, Chlorophyll, Glucose, Oxygen

Figure 8.2 Photosynthesis equation.

The Transport of Oxygen

Hemoglobin is the protein that transports oxygen from the lungs to the tissues and takes the carbon dioxide and hydrogen ions from the tissues to the lungs. One molecule of hemoglobin is composed of four polypeptide chains, two α-chains of one hundred and forty-one amino acids, and two β-chains of one hundred and forty-six amino acids each. The α and β chains differ in their amino acid sequence, but they form similar spatial or three-dimensional structures.

The oxygen-binding pigment in hemoglobin is called heme, which gives blood its red color. The heme group is composed of a ring of carbon atoms, nitrogen, and hydrogen called porphyrin, with one atom of iron in the center. One polypeptide chain of hemoglobin plus the heme group constitutes what is called a subunit. In the hemoglobin molecule, the four subunits are bound together, forming a tetramer (figure 8.3).

In the skeletal muscle and the heart, there is another protein related to hemoglobin, called myoglobin, made of a hundred and fifty-three amino acids (like the β-structure of hemoglobin). Myoglobin is a compact globular protein with a complex and intricate three-dimensional structure, in contrast to other proteins, such as keratin, belonging to the class of fibrous proteins, which usually are less soluble. Myoglobin facilitates the transport of oxygen from the blood to the mitochondria—the organelle that constitutes the primary site of consumption of oxygen in the cell.

In 1965, Francois Monod and Jean Pierre Changeux of the Pasteur Institute of Paris, together with Jeffries Wyman of the University of Rome, recognized that enzymes could activate themselves and had certain characteristics like hemoglobin (Monod et al. 1965), something they called *allosteric proteins.* All allosteric proteins are composed of several subunits, so each molecule contains several sites with the same catalytic activity, just as hemoglobin contains several heme groups that bind the oxygen, and all of them show similar cooperativity effects (Kanaori et al. 2011).

Monod and his colleagues knew that deoxyhemoglobin and oxyhemoglobin had different structures, which made them suspect that enzymes could also be found in at least two structures. They postulated that such structures should be distinguished by the arrangement of the subunits and by the number and strength of the links between them. If there were only two alternative structures, the one that showed fewer and weaker links between the subunits could fully develop their catalytic activity (or affinity toward oxygen); this structure was called R, from "relaxed." The activity would be dampened in the structure with more and stronger bonds between the subunits; this form was called T, from "tense."

In any of these structures, the catalytic activity (or affinity toward oxygen) of all subunits in a molecule must always remain the same. This postulate of symmetry allowed the properties of allosteric enzymes to be described by means of an elegant mathematical theory with only three independent variables: K_R and K_T, which in hemoglobin represent the oxygen equilibrium constants of the structures R and T, respectively; and L, which means the number of molecules in the structure T divided by the number in the structure R, measuring this quotient in the absence of oxygen.

The term allostery was coined because the regulatory molecule that activates or deactivates the enzyme has a different structure from the structure of the molecule whose chemical transformation catalyzes the enzyme (see Ackers and Halvorson 1974; Mills and Ackers 1979; Ackers 1979; Ackers and Smith 1987). Thus, in the case of hemoglobin, the binding of oxygen makes a change from T to R more favorable, and this change is cumulative, making the binding of oxygen to the other subunits more favorable. This alters the oxygen-binding curve from the normal saturation binding (noncooperative hyperbolic curve) seen for myoglobin to a sigmoidal curve seen in hemoglobin.

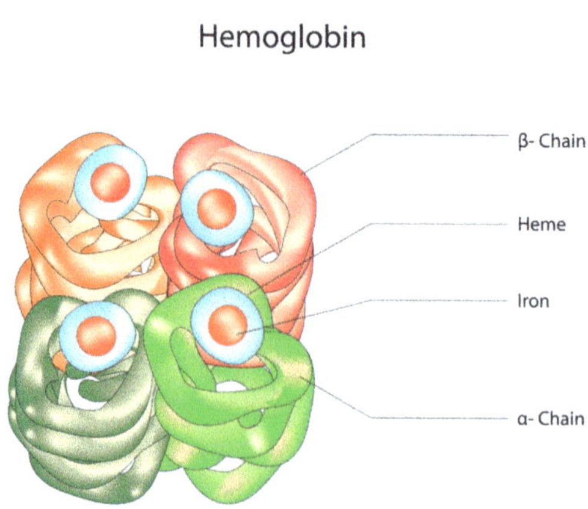

Figure 8.3. Hemoglobin structure.

Image ID: 2JKPF4J. KKT Madhusanka / Alamy Stock Photo.

Reference: OY73309826 September 22, 2022.

The Transmission of the Nerve Impulse

Nicotinic acetylcholine receptors (nAChRs) constitute a family of ligand-gated channels, originally classified based on their activation by the alkaloid nicotine, with acetylcholine (ACh) being the endogenous ligand. nAChRs are pentameric transmembrane proteins that form cationic channels permeable to Na^+, K^+, and Ca^{2+} (Changeux, 2012). The acetylcholine receptor (AChR) is by far the best-characterized neurotransmitter receptor and one of the best-characterized integral membrane proteins (Barrantes 1983, 1988). Found at the motor endplate of voluntary muscles, it is a highly regulated, intrinsic membrane protein responsible for communicating electrical stimuli across the neuromuscular synapse. The neurotransmitter serving as the chemical vehicle for this transfer is acetylcholine, thus classifying the AChR as a cholinergic receptor.

More specifically, the AChR is pharmacologically classified as nicotinic, meaning it is responsive to nicotine and blocked by curare-like compounds. The second class of cholinergic receptors, known as muscarinic, is stimulated by muscarine and blocked by atropine.

A major problem hindering research on the AChR for many years was its low concentration in mammalian sources. Rapid biochemical and biophysical advances toward the characterization of the AChR became possible when it was found that the electric organ of certain fish (*Torpedo, Narcine, and electrophorus*) contains a high concentration of a functionally similar protein (McCarthy et al. 1986; Barrantes 1983; Lentz and Wilson 1988a, b). Instead of containing contractile machinery, the cells of the electric organ are organized into stacks of large cells known as electroplaques (Aronstam 1982). This tissue is ontogenically derived from skeletal muscle in which the original contractile capacity of generating electric potentials has been developed.

Studies comparing structural homology in terms of cross-reactivity to various antibodies suggest that the structure of the AChR has been remarkably conserved through evolution (Lindstrom et al. 1983). The AChR is an integral transmembrane membrane protein that projects from both sides of the bilayer and appears to occur as monomeric and dimeric forms exhibiting sedimentation coefficients of 9 S and 13 S and Stokes radii of 73 and 95 Å, respectively (Meunier et al.1971;

Raftery et al. 1972). Using membrane osmometry, the overall molecular weight has been determined to be 270,000 Da for the monomer (Martinez-Carrion et al. 1975) and it exhibits an isoelectric point of 4.9 (Brackes et al. 1975).

The AChR is an oligomeric protein composed of four nonidentical tightly associated subunits designated α, β, ϒ, б. The subunits exhibit apparent molecular weight in SDS gels of forty thousand (α); fifty thousand (β); sixty thousand (ϒ); and sixty-five thousand (б), with the stoichiometry of 2:1:1:1 (Raftery et al. 1980; Conti-Tronconi et al. 1982).

The dimeric form of the AChR is predominant in *Torpedo* and is formed by a disulfide bond between the б subunits of adjacent monomers (McCrea et al. 1987). Dimerization may not be required for function (Anholt et al. 1980). Each of the four subunits contains carbohydrate moieties since each of the appropriate bands on SDS gel stain with periodic acid-Schiff's base (Vandeln et al. 1979). The primary sequences of each subunit, although not identical, display considerable sequence homology; of the aligned sequences, 19 percent are occupied by three identical residues, with the fourth being chemically similar (Claudio et al. 1983; Noda et al 1983).

Hydrophilicity profiles as well as the secondary structure predicted from the amino acid sequence are similar for each of the subunits, suggesting that they are oriented in a pseudo-symmetric fashion with respect to the membrane. In addition, each subunit contains four strongly hydrophobic regions, which may correspond to transmembrane sequences. Reconstitution of the nAChR in a lipid environment has shown that cholesterol is essential to maintain the nAChR functionality (Criado et al. 1977; Pérez-Ramírez 1994).

Cholesterol is necessary for homeostasis of AChR levels at the plasmalemma and for ion translocation (Borroni and Barrantes 2011). The functional role of AChR subunits is not well understood; however, reconstitution of purified AChR into model membranes with different lipids composition indicates that the α_2,β,ϒ, б complex constitutes the basic functional unit (Gonzalez-Ros et al. 1980; Criado and Barrantes 1982; Pérez-Ramírez 1994).

The structure and action of the acetylcholine receptor has been explored by electron microscopy (Unwin 2003). The analyses show that the protein subunits have a predominantly β-sheet fold in the ligand-binding domain and an α-helical fold in the pore domain. The binding of ACh initiates rotations of the inner β-sheets of both α subunits. These rotations are communicated to the inner, pore-lining helices and cause them to switch cooperatively into an alternative configuration (Unwin 2003). The structure of the muscle-type (from Torpedo electric organs) nAChR has been resolved to 4 Å by electron microscopy. The best diffracting crystals were obtained by co-crystallization with nicotine and a cholesterol analog and allowed for the collection of a complete dataset to 3.9 Å resolution (Morales-Perez et al. 2016).

The production of high-resolution crystals of the AChR has been hampered by the hydrophobicity of the receptor and the tight interaction with lipids. Understanding the membrane topology and the three-dimensional structure of the receptor is key to begin to understand its mechanism of action (Pérez-Ramírez and Martinez-Carrion 1989). Fluorescent probes sensitive to the physical state of the lipid environment have been used as reporter groups for the lipid matrix of the AChR (Clarke et al. 1987), and pyridoxal phosphate has been used as a chemical probe to identify the external (Dwyer 1988) and cytoplasmic surface of the receptor (Pérez-Ramírez and Martinez-Carrion 1987, 1989).

Analysis of the primary structure from AChR subunits from *Torpedo californica* has led to several models for the transmembrane organization of the polypeptide chain in all the subunits and the localization of domains in this complex macromolecule (Noda et al. 1983; Finer-Moore and Stroud 1984; Guy 1984). See figure 8.4 for a proposed structure for the receptor.

AChRs are associated with a growing list of diseases. The disease first and best characterized by the direct involvement of AChRs is autoimmune myasthenia gravis, in which an antibody-mediated autoimmune response to muscle AChRs impairs neuromuscular transmission (Lindstrom et al. 1988). A mutation in a neuronal AChR has recently accounted for the first example of a mutation that causes a specific form of epilepsy (Steinlein et al. 1995). Substantial loss of neuronal AChRs is found in both major neurodegenerative disorders Alzheimer's disease and Parkinson's disease (Whitehouse et al. 1988; Perry et al. 1995).

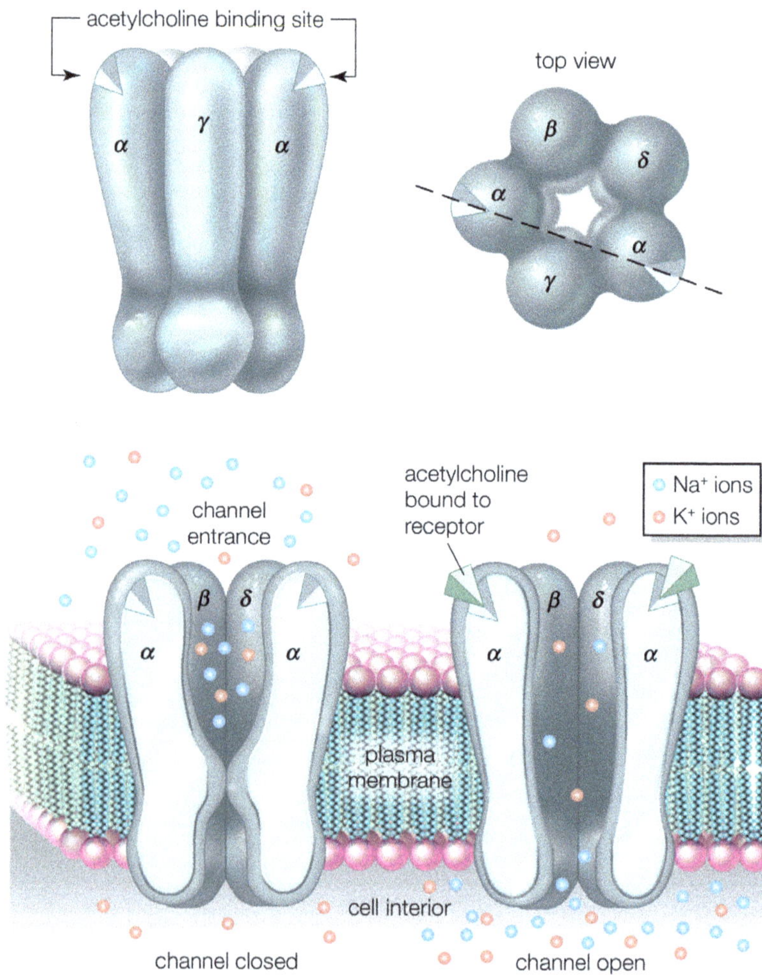

Figure 8.4 Structure of the nicotinic acetylcholine receptor.
Image ID: BCE62F. Alamy: order reference: OY73299804, September 22, 2022

Bone Formation

Bone morphogenetic protein-2 (rhBMP-2) is a human protein categorized as part of the TGFβ superfamily. It induces bone (Wang et al. 1990) by signaling for the differentiation of progenitor cells into osteoblast cells (Uludag et al. 1999) and by upregulation of mRNA for myriad proteins, including α1(I) collagen (Lecanda et al. 1997). Several studies have shown rhBMP type molecules are involved in morphogenesis, axial growth, hard tissue development, and repair, including organs and tissues as different as bone, cartilage, kidney, the periodontal ligament, dentine, and the central nervous system (Reddi 1994;1998; Ripamonti et al. 1997; Thomadakis et al. 1999).

Translation of the gene for rhBMP-2 produces a three hundred and ninety-six amino acid pro-peptide. Subsequent N-terminal digestion by a co-expressed enzyme cleaves the first two hundred and eighty-two amino acids, resulting in a mature rhBMP-2 monomer, one hundred and fourteen amino acids in length, and an N-terminus of glutamine (Q). The enzyme cleaves the pro-peptide in a second location, yielding an extended form of rhBMP-2, one hundred and thirty-one amino acids in length with threonine (T) on the N-terminus.

In solution, glutamine in the N-terminus of mature rhBMP-2 cyclizes to pyroglutamic acid (<Q) (Pérez-Ramírez and Stecker 2005; Morin et al. 2006) until equilibrium is reached. The result is three possible rhBMP-2 monomer isoforms: the mature (Q283), the extended isoform (T266), and the matured cyclized isoform (<Q283). Hence, purified rhBMP-2 is predominantly found as six different dimers (see figure 8.5) denoted as: <Q283/<Q283, Q283/Q283, <Q283/T266, Q283/T266, and T266/T266 (Pérez-Ramírez and Stecker 2005; Morin et al. 2006). A single cysteine bond forms the dimer, while the remaining seven cysteines of each monomer comprise an intra-chain cysteine knot, a structure typically found in TGFβ proteins (Burt and Law 1994).

One unique aspect that sets rhBMP-2 (and rhBMP-4, rhBMP-7 (OP1) apart from other TGFβ proteins, such as activin, is its discrete and limited local action (Rahman et al 2015). This short-range effect is attributed to a conserved, multi-basic triplet in the N-terminus that binds to localized components, such as heparan sulfate proteoglycan (Ohkawara et al. 2002) and heparin (Ruppert et al. 1996). The N-terminus of mature rhBMP-2 is reminiscent of known or postulated heparin-binding sites in other growth factors (Goetz et al. 1994; Roghani et al. 1994; Spivak-Kroizman et al. 1994).

Solubility studies have shown that rhBMP-2 falls out of solution when ionic strength reaches 100 mM at a pH approaching physiological (Abbatiello and Porter 1997), consistent with an affinity precipitation mechanism that requires the anion-dependent formation of protein complexes. Interestingly, heparin has shown improved solubility of rhBMP-2 at pH >6 (Ruppert et al. 1996). The unusual solubility behavior of rhBMP-2 at physiological pH is a physical manifestation of protein oligomerization related to its biological function. Indeed, it has been shown that rhBMP-2 undergoes reversible self-association behavior (Pérez-Ramírez and Steckert 2005). The self-association behavior of rhBMP-2 is modulated by the N-terminus of the dimeric protein (Pérez-Ramírez and Steckert 2005; Morin et al. 2006).

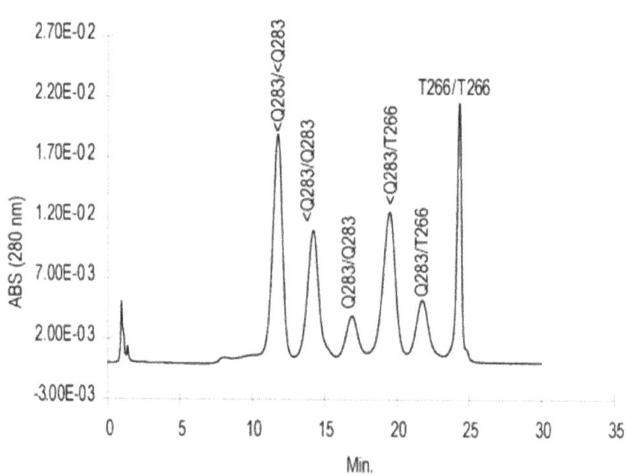

Figure 8.5. Separation of rhBMP-2 isoforms by cation-exchange chromatography.
Reprinted with permission from Biomacromolecules, 2006, 7, 131–138.

The nature of the interaction between the rhBMP-2 and the collagen delivery matrix (to implant the combination product) in terms of surface layers and interactions has been investigated (figure 8.6). Kinetic and thermodynamic analysis of the interactions between rhBMP-2 and the collagen matrix suggest that the association is specific and not a consequence of adsorption or aggregation caused by the matrix or solution conditions. The rhBMP-2/collagen association reaches an equilibrium that changes based on the rhBMP-2 isoform content, on pH, and, to a lesser extent, on ionic strength (see tables 8.0 and 8.1).

It is unknown whether a true monomer–multimer equilibrium exists on the surface of the collagen matrix or a continuous multilayer formation. Results indicate that rhBMP-2 binds to the collagen matrix with affinities on the order of 10^3 to 10^4 M^{-1}. Maximum binding, ν, is primarily a function of pH for heterogeneous rhBMP-2 and the extended (T266/T266) isoform. However, binding saturation of the <Q283/<Q283 isoform is unaffected by pH. Overall, binding saturation is higher than the calculated saturation of a rhBMP-2 monolayer, suggesting both hydrophobic and ionic interactions in a multilayer formation (Morin et al. 2006).

The study performed by Morin et al. 2006 does not distinguish the sequence of events leading to multilayer formation by rhBMP-2 on the surface of the collagen delivery matrix (figure 8.6). However, sedimentation velocity analyses have shown the formation of discrete and reversible rhBMP-2 oligomers (10–11S self-associated species) in solution, depending on isoform composition and pH (Pérez-Ramírez and Stecker 2005). These findings suggest that increasing mainly the pH in the formulation could optimize the amount of rhBMP-2 delivered by the collagen matrix to the site of interest. Hence, solution variables that could modulate the free energy of interaction between rhBMP-2 and the delivery system could help in identifying the desired pharmacokinetic profile.

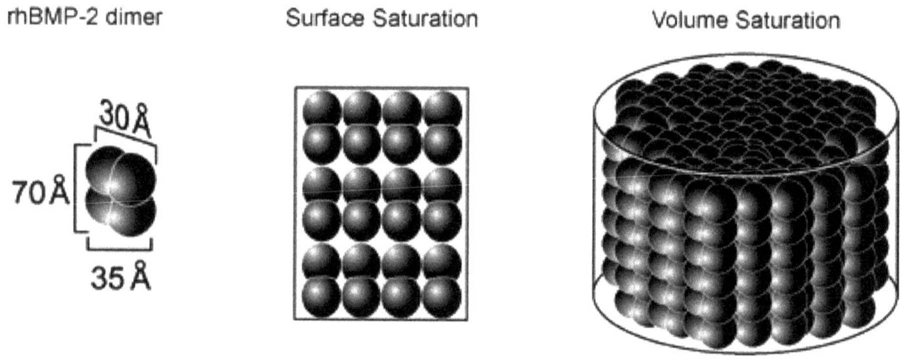

Figure 8.6 Schematic of rhBMP-2 and collagen interactions as surface saturation and volume saturation.

Reprinted with permission from Biomacromolecules, 2006, 7, 131–138.

pH	NaCl mM	ν or ν_T[a] µg/mg	K_b or K_{b1}[b] (M^{-1})	app $\Delta G_1^{\circ c}$ kcal mol^{-1}	K_{b2}[d] (M^{-1})	app $\Delta G_2^{\circ c}$ kcal mol^{-1}
4.4	none added	80	6.6×10^3	–5.19	N.D.[e]	f
4.4	12	170	5.4×10^3	–5.08	N.D.	
4.4	25	140	6.3×10^3	–5.18	N.D.	
4.9	none added	110	2.4×10^4	–5.94	N.D.	
4.9	12	260	2.6×10^4	–6.02	8.4×10^3	–5.35
4.9	25	260	2.3×10^4	–5.95	1.0×10^4	–5.45
5.4	none added	150	6.3×10^4	–6.54	1.2×10^4	–5.56
5.4	12	260	4.3×10^4	–6.32	8.1×10^3	–5.33
5.4	25	260	7.8×10^4	–6.67	9.0×10^3	–5.39

Table 8.0 Binding Strength of Unfractionated rhBMP-2 to Collagen

[a]ν_T is the total number of binding sites per unit of substrate (collagen) for biphasic binding (where a nonlinear Scatchard plot observes two or more modes of binding), comparable to ν for simple binding (determined from the x-intercept of Scatchard plots where one mode of binding is observed as a linear Scatchard plot). [b]K_b or K_{b1} is the equilibrium-binding constant of rhBMP-2 to collagen matrix determined from the slope of the linear trend line of Scatchard plots. [c] They are the apparent standard free energies for observed monophasic and biphasic interaction between rhBMP-2 and collagen matrix at 25°C, where $\Delta G^\circ = -RT. \ln K_b$. [d]$K_{b2}$ = equilibrium binding constant is determined from the trend line of the final portion of data in a nonlinear Scatchard plot. [e] N.D. represents those Scatchard plots that were linear, and therefore, biphasic interaction was not detected. [f]Data points in respective Scatchard plots were deemed too scattered to generate a slope and determine a binding constant.

As shown in the table, the apparent equilibrium binding constant, K_b, tended to increase with increasing pH, but K_b did not tend to increase with increasing NaCl. Examination of the apparent standard free energy of interactions showed the same trend. Binding was more favorable at pH 5.4 than at pH 4.4. The K_b values were obtained as the slope of the visual best fit of a single line or two tangent lines to the data points. An approximate K_b error can be assessed as the variability in the visual best fit of tangent lines to the data points. Adapted with permission from Biomacromolecules, 2006, 7, 131–138. Copyright 2006 American Chemical Society.

Isoform	pH	NaCl mM	ν or ν_T[a] μg/mg	K_b or Kb_1[b] (M^{-1})	app ΔG_1°[c] kcal mol^{-1}	Kb_2[d] (M^{-1})	app ΔG_2°[c] kcal mol^{-1}
<Q283/<283	4.4	none added	40	e			
<Q283/<283	4.4	25	140	2.5×10^4	−5.00	4.8×10^3	−5.02
<Q283/<283	5.4	none added	60				
<Q283/<283	5.4	25	120	9.9×10^4	−6.81	9.0×10^3	−5.39
T266/T266	4.4	none added					
T266/T266	4.4	25	30				
T266/T266	5.4	none added	60				
T266/T266	5.4	25	210	7.2×10^4	−6.52	9.0×10^3	−5.43

Table 8.1 rhBMP-2 Isoforms Binding to Collagen Matrix

[a]ν_T is the total number of binding sites per unit of substrate (collagen) for biphasic binding (where two or more modes of binding are observed by a nonlinear Scatchard and is comparable to ν for simple binding (determined from the x-intercept of Scatchard plots where one mode of binding is observed as a linear Scatchard plot). [b]K_b or Kb_1 is the equilibrium binding constant for rhBMP-2 interaction with the collagen matrix, as determined from the slope of linear trend line of Scatchard plots. [c]apparent standard free energy of interaction for rhBMP-2 and collagen matrix at 25 °C, where $\Delta G^{\circ} = -RT \ln Kb$. and are the apparent standard free energies for observed monophasic and biphasic interaction between rhBMP-2 and collagen matrix at 25 °C. [d]Kb_2 = equilibrium binding constant determined from trend line of final portion of data in a nonlinear Scatchard plot. [e]For entries with no data recorded, data points in respective Scatchard plots were deemed too scattered to generate a slope and determine a binding constant. Reprinted with permission from Biomacromolecules, 2006, 7, 131–138. Copyright 2006 American Chemical Society.

Silk Proteins

Silk fibroin is a natural polymer consisting of disulfide-bound heavy and light chains of approximately 390 KDa and 25 KDa respectively (Inoue et al. 2000). The primary structure consists of twelve hydrophobic protein domains rich in alanine and glycine residues, with a high propensity toward forming antiparallel β-sheets (physical cross-links) structure. Eleven small hydrophilic amorphous domains separate these blocks. Silk fibroin possesses valuable features as a biomaterial, such as good biocompatibility in all aqueous manufacturing processes, controllable degradation rates, and excellent mechanical properties (Panilaitis et al. 2003; Horan et al. 2005; Wang et al. 2008; Yucel et al. 2009; Kojic et al. 2012).

The ability of silk to serve as a viable sustained release matrix for drugs is well established. Sustained local delivery of drugs can have many advantages over systemic delivery in applications such as tumor regression or osteoarthritis. The most significant advantage is the potential improvement in efficacy by maintaining drug levels within the therapeutic window for longer periods of time (Langer 1980; Uhrich et al.1999). Also, delivering drugs directly to the disease site eliminates the dependence on physiological targeting mechanisms, making higher levels of therapeutic available at the disease site.

Thus, silk fibroin poses many features that are potentially attractive in a sustained local delivery matrix for proteins: aqueous processing into a variety of formats for implantable or injectable drug depots, many control points for manipulating release, and preserved stability during encapsulation. These features may allow silk to overcome the limitations that have prevented other polymer-sustained delivery systems from gaining widespread use with protein therapeutics.

The release of various small molecules from silk films, silk-coated powder reservoirs, and microspheres has been characterized (Hines and Kaplan 2011; Pritchard and Kaplan 2011; Tsioris et al. 2012). The release of proteins from silk-based biomaterials has also been evaluated (Hofer et al. 2012; Guziewicz et al. 2011).

A series of release studies have been conducted to improve the understanding of the silk-mab matrix interaction for sustained local delivery of antibody therapeutics (Guziewicz et al. 2013). It is hypothesized that the hydrophobicity of the silk matrix governs antibody release. In the silk hydrogel matrix, hydrophobic interactions determine the amount of antibody trapped irreversibly in the matrix and the rate of untrapped antibody release. Secondarily, in the lyophilized state (lyogel), hydrophobic β-sheet density changes hydration properties, leading to alterations of release profiles (Guziewicz et al. 2011, 2013).

Silk lyogels can be prepared in low-density β-sheet networks (LDBN) and high-density β-sheet networks (HDBN). The transition from LDBN to HDBN occurs between 20 mg·$(cm^3)^{-1}$ and 42 mg·$(cm^3)^{-1}$ β-sheet structure in silk lyogels (Guziewicz et al. 2011). In the LDBN form, near-complete rehydration is observed. In the water-rich environment, any enhanced silk-antibody interactions are rapidly reversed, and release behavior comparable to that of the hydrogel is observed.

In the HDBN form, significant hydration resistance is observed. Lower water content within the HDBN lyogels limits water competition with the silk-antibody interactions. The decreased kinetics of water-silk interactions, replacing antibody-silk interactions, results in the sustained release of untrapped antibodies. These findings indicate that release rates could be controlled by modifying the hydration properties of silk lyogels (Guziewicz et al. 2011, 2013).

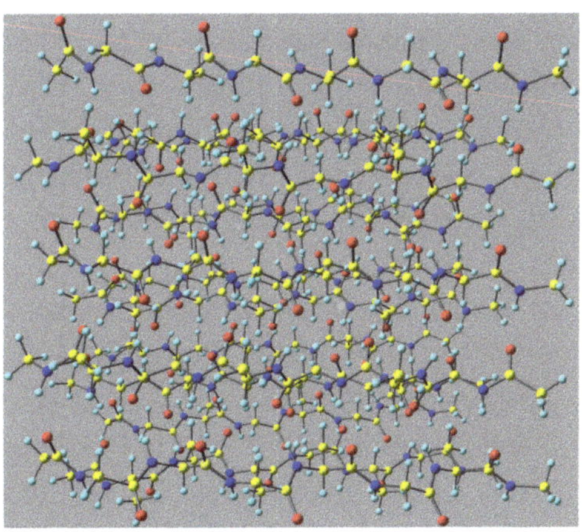

Figure 8.7. The protein fibroin is an insoluble protein present in silk created by spiders.

Source: Under license by Alamy. April 10, 2023. Image ID: 2BWA86B. order reference: OY80203410.

Tubulin and Microtubules

Microtubules are core components of the eukaryotic cytoskeleton with critical roles in cell division, shaping, motility, and intracellular transport. Tubulin is the subunit protein of microtubules, which is assembled into protofilaments aligned longitudinally along the axis of the cylinder (figure 8.8). Native tubulin, when isolated in solution, exists as a dimer ($\alpha\beta$) of two subunits, α and β, which have molecular weights close to fifty thousand and similar primary structures. The in vitro assembly of microtubules requires that the exchangeable (E-site) nucleotide binding site be occupied by GTP or triphosphate analog. The incorporation of tubulin into a microtubule is rapidly followed by hydrolysis of the E-GTP to GDP + Pi.

Microtubules are not formed in the presence of GDP. They are destabilized by Ca^{++} ions and low temperatures (< 20°C). Since the core of microtubules is unstable, being formed of GDP-tubulin (Goodson and Jonasson 2018), the stabilization of the assembled organelles has been attributed to GTP-caps (Goodson and Jonasson 2018), the presence of which has been proposed as the control of the dynamic instability of microtubules (Tran and Walker 1997). Carlier et al. (1989) have shown that this stabilization is due to a transient tubulin-GDP-Pi complex. Tubulin interacts with several drugs, such as colchicine, podophyllotoxin, and vinblastine, which inhibit microtubule formation, this being the source of antimitotic and other intracellular actions (Andreu and Timasheff 1982). It has been identified in several features in the interactions of these drugs with tubulin. Vinblastine forms complexes with tubulin in a rapidly reversible exchange with the free drug in solution (Lee et al. 1975; Na and Timasheff 1986); colchicine binding is essentially irreversible (Engelborghs 1998; Andreu et al. 1991).

Both colchicine and vinblastine induce tubulin to form polymers with geometries other than microtubules. The binding of colchicine engenders in the protein a slow GTPase activity in the E-site (Andreu and Timasheff 1981; Pérez-Ramírez et al. 1994; Pérez-Ramírez and Timasheff 1994). Reversible binding analogs of colchicine also induce the same enzymatic activity (Andreu et al. 1991) and inhibit microtubule formation, the exact mode (whether stoichiometric or sub stoichiometric) being defined by the fine structure of the drug (Pérez-Ramírez and Timasheff 1995; Pérez-Ramírez et al. 1996; Rossi et al. 1996).

The binding of drugs affects tubulin stability differently: colchicine stabilizes it (Andreu et al.1984), while vinca alkaloids accelerate its aging (Prakash and Timasheff 1992).

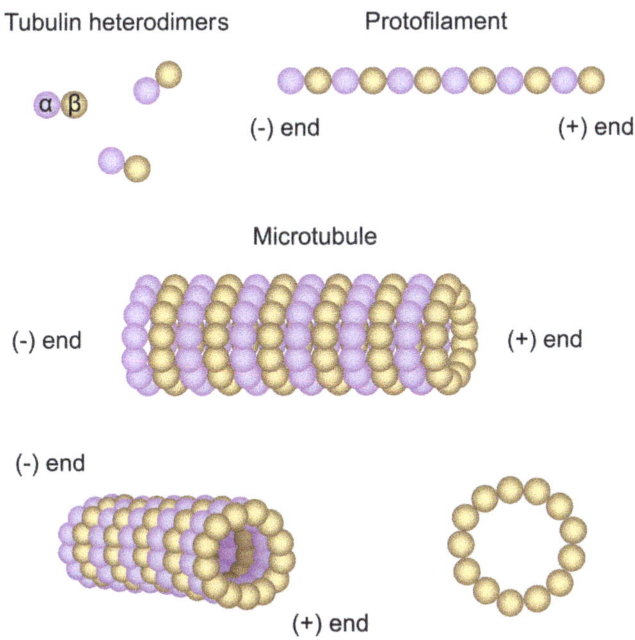

Figure 8.8. Tubulin and microtubules.

Source: under license by Alamy. Order reference: OY80252643. invoice number: IY03433845. April 12, 2023.

It has been shown by thorough solution thermodynamic studies that the binding of colchicine (a three-ring molecule) to tubulin conforms to a bidentate mechanism (Medrano et al. 1991; Engelborghs 1998). In the case of colchicine, the steric hindrance caused by ring B imposes the order of binding of the drug moieties (see figure 8.9 for colchicine and analogs structure) to subsites in the protein with the tropolone methyl ether (ring C) of colchicine binding first (Medrano et al. 1989,1991) by stacking and hydrogen bond interactions (Menendez et al. 1989), followed by a conformational change and the binding of the trimethoxy phenyl moiety (ring A) by hydrophobic interactions (Andreu and Timasheff 1982).

The question of the locus of colchicine binding to tubulin has been the subject of several studies over the past decade, mainly by affinity and photoaffinity techniques. The results have been inconclusive, as both the α and β subunits have been identified as the site (McLoughlin and O'Boyle 2020). On the other hand, studies of mutants and tubulin isoforms (Huzil et al. 2007) suggest the β-subunit as containing the binding site, although allosteric interactions are not excluded. Sedimentation equilibrium studies in the analytical ultracentrifuge using reversible binding analogs of colchicine have been able to identify the colchicine binding site as being located on the β-subunit, most probably in the proximity of the α-β interface.

In order to gain insight into the requirements for binding to tubulin and inducing substoichiometric inhibition of microtubule assembly, the ring C subsite interactions has been probed with biphenyl analogs that are derived from allocolchicine (ALLO). ALLO contains a carbomethoxyl ($COOCH_3$) in position 4' (see figure 8.9). A scrutiny of this function using biphenyls with ring A of ALLO and position 4' occupied by CO (TKB) or OCH_3 (TMB) led to the discovery that, although all bind to tubulin with similar affinity (TKB somehow more strongly), only the ketone (TKB) inhibits microtubule assembly sub-stoichiometrically (Pérez-Ramírez et al. 1996). That substoichiometric inhibition is modulated by a ketone group in the proper position in ring C is supported by the finding that tropolone methyl ether (ring C of colchicine), which binds very weakly to tubulin ($Kb \sim 10^3\ M^{-1}$), is a substoichiometric inhibitor of microtubules.

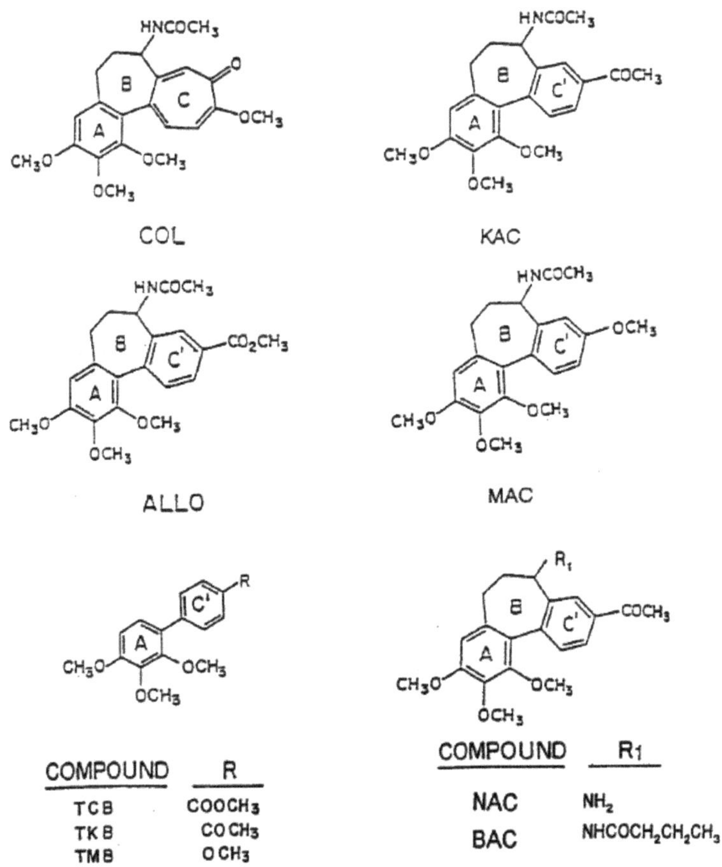

Figure 8.9. Colchicine and analogs.

Reprinted with permission from Pérez-Ramírez et al., Biochemistry 1998, 37, 1646–1661.

The binding of the drugs to the colchicine binding site of tubulin was demonstrated by the inhibition of the binding of COL by either KAC or MAC, as shown in figure 8.10. The interaction of KAC with tubulin quenched the intrinsic fluorescence of a tryp group in the protein (figure 8.10b). When liganded to tubulin, KAC developed fluorescence, with excitation and emission maxima at 315 nm and 430 nm (figure 8.10b). Substoichiometric amounts of KAC, BAC, NAC, and MAC were found to inhibit the in vitro microtubule assembly, and KAC is a stronger microtubule assembly inhibitor than COL (Pérez-Ramírez et al. 1998).

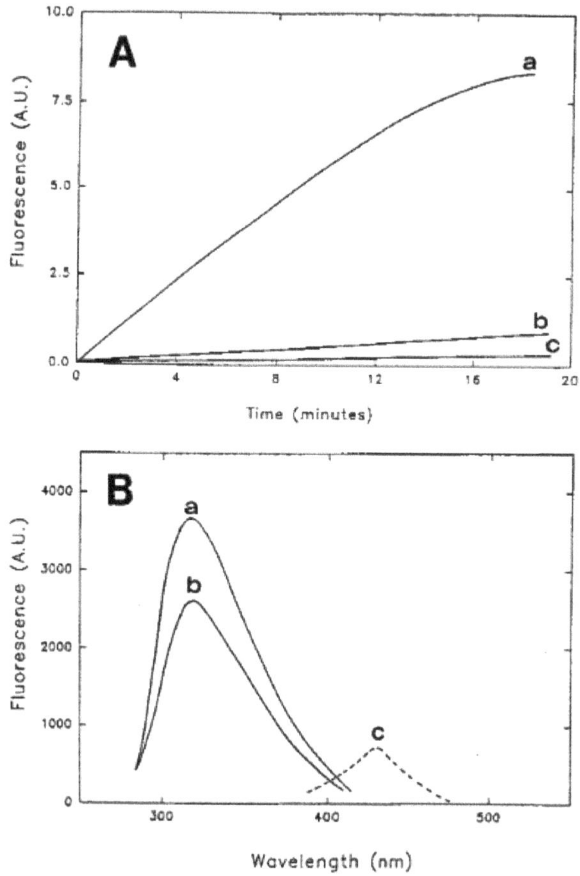

Figure 8.10. (A) Inhibition of COL binding to tubulin by KAC and MAC at 25°C in 10 mM sodium phosphate buffer, pH 7.0, and0.1 mM GTP. Fluorescence time courses are shown of the binding of 110-5M COL to 7.310-6M tubulin in the absence of analogues (a), or in the presence of 110-4M MAC (b) and KAC (c), respectively. COL then MAC or KAC were added consecutively to the protein solution in the fluorometer cuvette at 25°C as small volumes of concentrated solution. They were mixed simultaneously to start the reaction. The excitation and emission band widths were 10 nm using the photomultiplier gain in low. The excitation was at 357 nm and emission was monitored at 435 nm (4). (B) Fluorescence changes produced by the interaction of KAC with tubulin. Fluorescence emission spectra: (a) 7.310-6M tubulin (excitation at 280 nm); (b) 7.310-6M tubulin with 110-5M KAC (excitation at 280 nm); (c) 1.010-5M KAC with 7.310-6M tubulin (excitation at 315, nm spectrum amplified 55 times). The fluorescence intensity is presented in arbitrary units (AU).

Reprinted with permission from Pérez-Ramírez et al., Biochemistry 1998, 37, 1646–1661.

The binding of COL to tubulin also has, consequently, the induction of a GTPase activity in tubulin (Andreu and Timasheff 1981; Pérez-Ramírez et al. 1994; Pérez-Ramírez and Timasheff 1994). To test whether the GDP-tubulin-COL complex formed as the product of the GTP hydrolysis was conformationally the same as that formed on replacement of the E-site GTP by GDP, the GTPase reaction was followed by sedimentation velocity.

Figure 8.11 shows the Schlieren profiles obtained in the analytical ultracentrifuge by monitoring the GTPase activity of the tubulin-COL complex. Sedimentation velocity examination of the product of the reaction showed that GDP-tubulin-colchicine, generated by hydrolysis of the E-site GTP, was indistinguishable from that produced by nucleotide exchange at the site, the protein assuming the "curved" conformation in both cases.

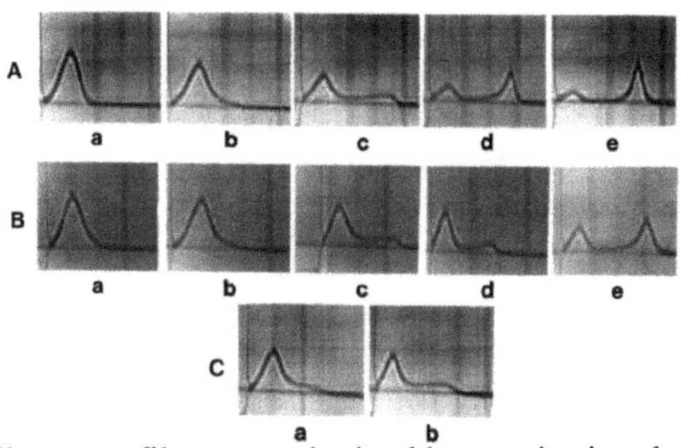

Figure 8.11. Schlieren profiles were obtained by monitoring the GTPase activity of the tubulin-colchicine complex by sedimentation velocity. The times at which the photographs were taken after the centrifuge came up to speed are indicated in parentheses. (A) Protein was incubated at 37 °C, and samples were run (a) 0 (5 min), (b) 2 h (5 min), (c) 5 h (5 min), (d) 8.25 h (2 min), and (e) 11.8 h (2 min) after the start of incubation. (B) Incubation was at 30 °C, and samples were run (a) 0 (5 min) (b) 2.9 h (5 min) (c) 6.5 h (2 min), (d) 22.1 h (2 min), and (e) 45.9 h (2 min) after the start of incubation. (C) Schlieren patterns obtained when tubulin-colchicine-GTP and tubulin-colchicine-GDP were prepared by nucleotide exchange and mixed in the given proportions. In panel C, profile a (5 min) has approximately the same GTP/GDP ratio as profile b in panel A, while profile b in panel C (5 min) has the same ratio as profile c in panel A, as determined by HPLC. In all cases, photographs were taken at a bar angle of 65°. Sedimentation is from left to right.

Reprinted with permission from Perez-Ramirez et al., Biochemistry 1994,33, 6253–626.

Aminoacyl-tRNA Synthetase

The aminoacyl-tRNA synthetases are an essential and universally distributed family of enzymes that play a critical role in protein synthesis, pairing tRNAs with their cognate amino acids for decoding mRNAs according to the genetic code (Rubio Gomez and Ibba 2020). The fidelity of the genetic information translation depends on the correct assignment to each amino acid a tRNA specific.

$$\text{amino acid} + \text{ATP} \longrightarrow \text{amino acid-AMP} + \text{PPi} + \text{tRNA} \longrightarrow \text{aminoacyl-tRNA}^{\text{amino acid}} + \text{AMP}$$

Where in the scheme above, amino acid is one of the twenty amino acids that conform proteins, tRNA represents a transfer RNA specific for a given amino acid. Therefore, the role of the synthetases is to assign amino acids to the triplet during translation. The aminoacyl tRNA synthetase plays a critical role in decoding the genetic message, transforming with high fidelity the language of nucleotides in the language of amino acids. A specific adjustment between the enzyme and the RNA substrate does this.

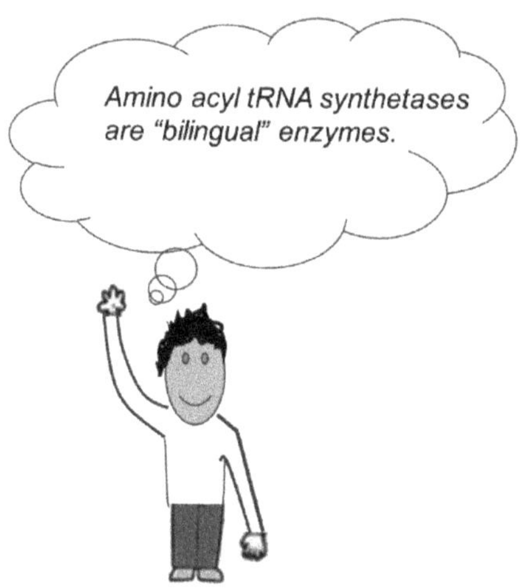

All the aminoacyl tRNA synthetases recognize the common substrate ATP, but they have restricted specificity for the other two substrates, amino acid and tRNA. The synthetases differ in size, structure, and amino acid sequence. The quaternary structure can be of the type $\alpha_2\alpha_4$ or $\alpha_2\beta_2$. The size varies between three hundred and thirty-four (Tryp-tRNA synthetase) to one thousand one hundred and twelve residues (Phe-tRNA synthetase of *E. coli*). The molecular masses vary between fifty-one thousand (Cys-tRNA synthetase) and three hundred and eighty-four thousand (Ala-tRNA synthetase).

Although their sequences differ, nine enzymes share a group of nine amino acids found only in tRNA synthetases (Webster et al. 1984). A subgroup of this sequence is the terminal tetrapeptide His X Gly His, where X is usually a hydrophobic residue like Ile. The enzymes with this motif share a second tetrapeptide, Lys Met Ser Lys. The folding pattern of Tyr, Met, and Gln-tRNA synthetases show amino-terminal domains like a Rossmann folding of alternating β-sheets and α-helix to form the site for ATP binding and the amino-acyl-adenylate. The fold occupies the same site in the three enzymes and begins with a carboxy-terminal of a β-sheet, forms a loop, and finishes in the end-terminal of an α-helix.

The two His of the tetrapeptide in Tyr and Glu-tRNA synthetase interact with specific phosphates of the bound ATP. The sequence Lys Met Ser Lys participates in the binding of ATP and perhaps in the binding of the extreme 3' of the tRNA in the fold (Ureta 2010).

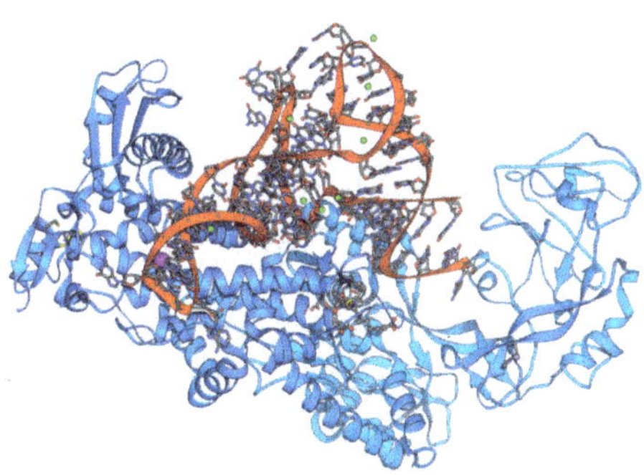

Figure 8.12. Molecular model showing bacterial isoleucyl-tRNA synthetase complexed with aspartyl tRNA (transfer ribonucleic acid). Isoleucyl-tRNA synthetase attaches the correct amino acid, in this case isoleucyl, onto the corresponding tRNA molecule during protein translation.

Source: Alamy Stock Photo ID:DP27WC. Reproduced with authorization of copyright holder.

The structure of modern tRNA is formed by two domains with different functions. It has been proposed that the primitive systems of aminoacylation used only one domain, and the domain of the anticodon would be a later addition. To compensate for the increase in size of the tRNA, new domains would have been added to the primitive synthetases. This allowed the anticodon to position itself far away from the activated amino acid and, in such a way, to diversify the tRNA specificity as discussed by Tito Ureta (Ureta 2010). These enzymes originated more than 3.5 million years ago, probably together with the development and early evolution of the genetic code.

In 1982, the first catalytic RNA or ribozyme was reported: the self-splicing intron of the Tetrahymena pre-rRNA (Cech 2002). Additional examples of natural ribozymes were soon found, and research in the field focused on their enzymic mechanism and secondary and tertiary structure (Cech 2002). In Chapter 1, we briefly mentioned the possibility that RNA evolved before coded proteins and DNA genomes, such that the biosphere began with an RNA core where much of the translation apparatus and related RNA architecture arose before RNA transcription and DNA replication (Fine and Pearlman 2023). Based on that hypothesis, it is possible to think that the aminoacylation could have originated in ribozymes that catalyzed the binding of specific amino acids to tRNA precursors. Then, the ribozymes could have been replaced by ribonucleoprotein complexes that could be the precursors of the current synthetases. The codified synthesis of proteins could have emerged during that evolutive process.

Only a limited number of amino acids could have participated in the generation of proteins. An indirect support for this hypothesis is the synthesis of a ribozyme able to recognize an activated amino acid and transfer it to the 3' extreme of a tRNA; in other words, working as a trans, like a proper aminoacyl-tRNA synthetase (Lee et al. 2000).

CHAPTER 9

PROTEINS AS MEDICINES

What is there that is not poison? All things are poison and nothing is without poison.Solely the dose determines that a thing is not a poison.

—PARACELCUS (1493–1541)

A drug is any chemical compound either synthesized in a laboratory or obtained from plant, animal, or marine origins, intended to change the body's normal physiological functions. All medicines are drugs, but all drugs may not be medicines. For example, when morphine is administered to a patient to reduce pain, it is a medicine, but when the same morphine is taken without any clinical issues as such, it is called a drug, but not a medicine. In the United States, according to the federal Food, Drug, and Cosmetic Act (FD&C Act), a law enacted by Congress, medicine is defined as a substance intended for use in the diagnosis, cure, mitigation, treatment, or prevention of disease and recognized by an official pharmacopeia. It is a substance (other than food) intended to affect the structure or any function in the body. It could be a substance intended for use as a component of medicine but not as a device or a component, part, or accessory of a device.

Biological products are included in this definition and are covered by the same laws and regulations. However, there are differences in their manufacturing processes (chemical processes versus biological processes). The European directive (2001/83EC) defines medicine as any substance or combination of substances that may be used in or administered to human beings either with a view to restoring, correcting, or modifying physiological functions by exerting a pharmacological, immunological, or metabolic action or to make a medical diagnosis. Medicines include antibiotics, vaccines, anticoagulants, contraceptive pills, enzyme replacement therapies, RNA, gene and cell therapies, and monoclonal antibodies. Each of these substances has either preventive, curative, or replacement properties, or modifying physiological functions with regard to human or animal diseases.

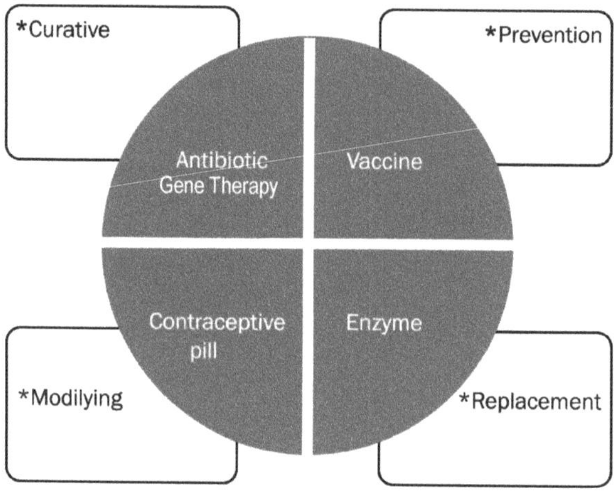

Figure 9.0. Schematic of the different types of medicines and their action.

There are different categories of medicines that can be classified according to whether they:

- are made at either industrial scale or pharmacy compounding.
- are available through a prescription or not.
- are available over the counter or pharmacy only.
- have a particular composition or nature.
- are new chemical entities (NCEs).
- are biological drugs.
- are generic drugs.
- are biosimilar drugs.
- are homeopathic drugs.
- are herbal drugs.

New Chemical Entities (NCEs)

NCEs can have origins in microorganisms, such as antibiotics. Penicillin, discovered in 1928 at St. Mary's Hospital, London, by Alexander Fleming, belongs to this category. This discovery led to the introduction of antibiotics, which greatly reduced the number of deaths from infection. Other NCE active substances are derived from plants, like Taxol and Taxotere; classified as a "plant alkaloid," a "taxane," and an antimicrotubule agent derived from the bark of the

Pacific yew tree. Taxol is used to treat breast, ovarian, lung, bladder, prostate, melanoma, esophageal, and other types of solid tumor cancers. It has also been used in Kaposi's sarcoma.

Flavonoids are a group of natural polyphenolic compounds, secondary metabolites of plants that may play an important role in human health protection due to their antioxidant properties. Other active compounds derived from plants are morphine, codeine, atropine, and quinine. Very potent neurotoxins such as cobratoxin or α-bungarotoxin are derived from snake venoms and bind to muscle-type and neuronal-type acetylcholine receptors. This bond will block the receptor's ability to bind acetylcholine and thereby inhibits the ion flow through the postsynaptic membrane, which will lead to paralysis. Cone snails are marine predators that use venom to immobilize their prey. The venom of these mollusks contains a cocktail of peptides that mainly targets different voltage- and ligand-gated ion channels.

Biological Medicines

A biological medicine is manufactured in a living system, such as a microorganism, plant, or animal cell, or contains components of a living organism. Types of biological drugs include vaccines, blood and its components, allergens, genes (DNA), gene-coding sequences (mRNA), cells, tissues, and recombinant proteins. Due to the complexity of these products, extensive characterization is required, including an understanding of their physical-chemical properties (discussed in previous chapters) and biological behavior (toxicity, safety), as well as knowledge of the manufacturing process and its control, since the final product is highly dependent on the process employed.

Small molecules are typically taken as a pill and absorbed into the body via the gut; by contrast, biological large proteins cannot be administered by the oral route since they get easily degraded. However, both small molecules and large biologicals interact with a variety of biological targets to treat disease, with the advantage of a small molecule drug being that it can enter cells easily because of its size.

A biological target is a molecule within a living organism to which some other entity (like an endogenous ligand or a drug or medicine) is directed and/ or binds. For example, several monoclonal antibodies target unique cell membrane receptors that are only expressed in tumor cells, thereby inhibiting the function of that receptor. Simvastatin, a small molecule drug, targets enzymes that the body uses to make cholesterol.

There is a big effort in discovering and validating new targets that may provide the basis for developing new medicines. In the case of using small molecule drugs to act on the target, thousands of compounds are screened using high throughput technologies in order to select the ones that have good affinity to the target and biological activity against the target. In the case of biologics against a given target, they are made by genetic engineering and given to the patient. One example of this approach is enzyme replacement therapy as in Pompe disease, where the inability of the body to produce enough of the enzyme β-glucosidase can be repaired by injection of the enzyme made by genetic engineering.

A common approach is to block a receptor protein with a monoclonal antibody. In this case, transgenic mice (genetically modified mice that have foreign DNA introduced to them to produce human-like antibodies) are exposed to the target to induce the formation of antibodies against that target. The cells that produce the antibodies are extracted to amplify them and create new cell lines to produce the antibodies against the target.

One advantage of biologics-based medicines is their low clearance and longer half-life in the body when compared to small molecules; therefore, less frequent dosing is usually required when using a biologic as a medicine, which provides convenience to the patients, a differentiation factor when a patient selects a given treatment (Pérez-Ramírez 2020). There are other health products, such as medical devices, cosmetics, and dietary supplements, that are not the focus of this book.

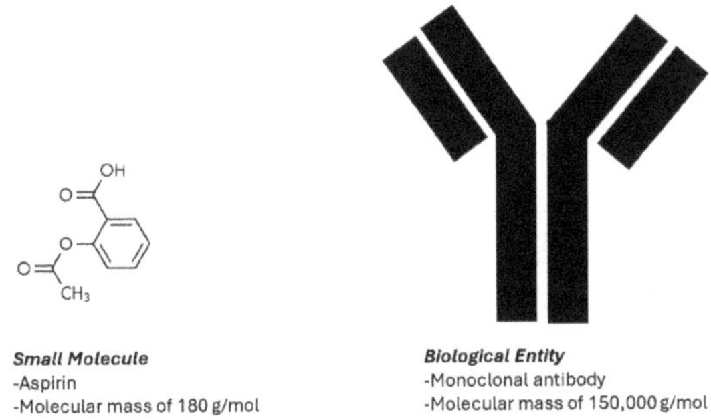

Figure 9.1. Comparison of a small molecule, like aspirin (acetylsalicylic acid), with a biological macromolecule (monoclonal antibody).

Table 9.0. Difference between a new chemical and new biological entity

Parameter	New Chemical Entity	New Biological Entity
Size	Small	Large
Structure	Unique	Complex
Stability	Stable	Complex
Manufacturing	Predictable chemical process	The process makes the product
Toxicity	Usually high	Usually low
Administration	Oral	Parenteral (Injectable)
Dossier	Detailed process controls and justifications	Detailed process controls and justifications

One of the main goals of medicines is to have safe and efficacious therapies that could limit the health risk to patients. Before the birth of biotechnology, active compounds were purified from human plasma or other fluids with the risk of increasing infections. Due to the advances in biotechnology, different therapies are now available.

Gene Therapies

A specific gene is delivered using a transport system (vector), an attenuated virus. The gene can express in the cell producing the protein that is lacking or at fault. A variation of gene therapy is the delivery of the messenger RNA to cells using a viral vector or complexed to lipid nanoparticles.

Cell Therapies

Cell therapy starts with collecting cells from a donor, expanded in vitro, and kept preserved by cryo-protection. These intact living cells are transferred into a patient to help lessen or cure a disease. The cells may originate from the patient (autologous cells) or a donor (allogeneic cells). The cells used in cell therapy can be classified by their potential to transform into different cell types. The cell therapy approach has been used in treating cancers, autoimmune disease, infectious disease, and repairing spinal cord injuries, among other indications.

Tissue Therapies

Tissue repair is an essential mechanism to maintain the integrity and function of the body in response to a variety of both acute and chronic injuries and disease states. Tissue therapies deal with the restoration of tissue architecture and function following an injury. A biological substitute, culture implantation reaches the site of action and provides effective treatments for acute and chronic wound healing applications, including chronic skin ulcers and burns. One of the major challenges to achieving clinically meaningful tissue repair is a better understanding of the mechanisms involved in both native and endogenous, as well as exogenous stem cell homing of transplanted cells.

Protein Therapies

This is the main target for the development of biological medicines. Roughly 60 percent of the portfolio of pharmaceutical companies is focused on developing protein therapeutics, with a large portion dedicated to monoclonal antibodies, recombinant enzymes, growth factors, and fusion proteins. Proteins typically act as separate independent units, but those units can be fused together (fusion proteins) to create proteins with new biological functions to treat a variety of diseases. Many proteins in the cell have dual and often opposing functions, owing to interaction with different receptors.

Vaccine Therapies

Vaccines are another major area of biopharmaceutical research and development. Vaccines are weakened or killed forms of disease-causing microorganisms or one of their surface proteins, sugars, toxins/antigens-inactivated proteins, and RNA. Recently, RNA as a therapeutic agent has gained significant attention due to the development of the first vaccines against COVID-19. The so-called RNA messenger is a molecule that carries a copy of the genes to be expressed as a protein. Thus, this molecule is transported from the cell's nucleus to the cytosol, where the cell machinery will decode the information and produce the protein.

RNA technology has been in development for many years. It is not a coincidence that the first vaccines for COVID-19 are all RNA-based, because RNA is a molecule that can be produced in vitro in a few days to a week. However, the production of a protein vaccine is more complex due to the intrinsic instability of proteins. I have included vaccines in a different category from protein therapies since the approaches to develop a protein-based vaccine are different from the ones required to develop a therapeutic protein. A vaccine is expected to induce an immune reaction to the injected foreign protein and later develop "memory," cellular immunity. However, a protein therapeutic must affect a biological function, without inducing an immune response to the protein injected (no induction of neutralizing antibodies). Patient convenience and quality by design are key factors in drug product development and differentiation (Pérez-Ramírez 2020).

Developing biological therapeutic medicines is a complex process that requires many years of extensive research and development. It is not uncommon that from the moment that a lead for a potential medicine is discovered in the laboratory, it may take six to eight years to be available in the market, with a total cost of research and development of approximately one hundred million to one billion, depending on the disease profile. Generally, there is one medicine with one chance in a thousand of success.

Global teams of hundreds of scientists/engineers are involved in research, development, and manufacturing. Once a potential therapeutic drug or biologic has been discovered, the process of developing the therapeutic for a particular disease begins with preclinical development in animal models and continues through increasingly complex and demanding phases of clinical testing in humans to support approval for marketing. Much of what is done throughout the process of drug development is driven by necessary regulations that require the sponsor of a new drug to demonstrate its safety and efficacy during each step of the process. Along the development path, a therapeutic protein is exposed to shear stress due to mixing, pumping, and filtration and to different contact surfaces and leachables that could result in aggregation, precipitation, and overall poor solution energetics (figure 9.2).

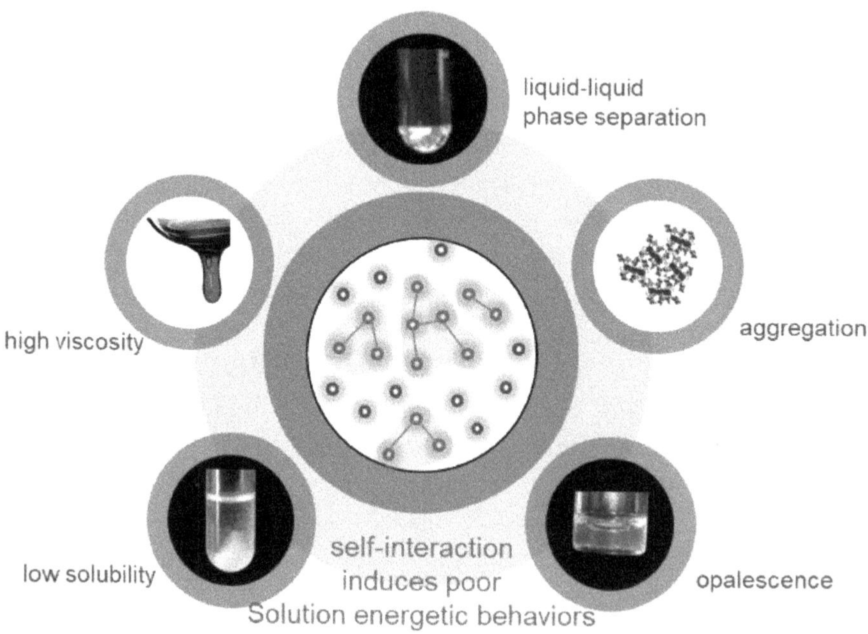

Figure 9.2 Poor solution behavior limits dosage form options and complicates manufacturing (Courtesy of Dr. J.Kingsbury).

Some vaccines may take longer to develop; it has been almost forty years since the AIDS epidemic began, and no vaccine is available yet. Medicines are unique in responding to a public health need and must comply with regulatory requirements and special market access. Medicines are subject to specific requirements throughout their life, from discovery to their manufacture and availability to patients. According to the World Health Organization (WHO), many diseases are still without a satisfactory treatment. Many challenges are due to the aging of the population and emerging diseases, such as COVID-19, avian flu, SARS, and those linked to challenging lifestyles (cardiovascular, obesity, immune-inflammatory, etc.).

Why are certain medicines tablets while others are liquids to be injected and kept refrigerated? Why are others liquid and kept frozen?

Medicines that must be kept frozen are very rare, but the different vaccines authorized for COVID-19 are all kept frozen to preserve their stability.

The Birth of Biopharmaceutical Products

The term biotechnology is considered the set of techniques that use living organisms or substances proceeding from them to elaborate or modify a product, improve plants or animals, or develop microorganisms for specific uses like creating medicines. Biotechnology was born in rudimentary form with the establishment of human societies and their need to develop organisms that would allow them to maintain food security and achieve territorial expansion.

The birth of modern biotechnology occurred with the appearance of molecular biology, a discipline that allowed us to decipher the structure of DNA in the twentieth century, as well as the mechanisms for translating genetic information located in DNA in proteins. All knowledge generated by modern biotechnology is based on the efforts of molecular biology, biochemistry, biochemical engineering, cell biology, microbiology, immunology, genetics, etc., allowing the comprehensive study and genetic manipulation of biological systems. This explosion also has its roots in the advances of chemistry and physics during the beginning of the twentieth century that led to the birth of biochemistry as a fundamental science to understand and explain the principles of the biological world using chemical-physical laws. This led to the discovery of the four building-block molecules that compose all living organisms, namely i) nucleic acids, ii) proteins, iii) carbohydrates, and iv) lipids.

Biotechnology gave rise to biopharmaceutical products, often referred to as biotech medicines. They are made in living factories and are microbes or cell lines using genetic engineering techniques. Baking is a process that can fall into the biotechnology category since it uses a living organism like yeast to produce a product, in this case bread.

Biopharmaceutical products emerged as a direct consequence of the explosive development of life sciences. Advances in genetic engineering allowed researchers to identify the genes that produced the protein of interest and cut the DNA sequence from a sample DNA with restriction enzymes. Those isolated genes are then introduced into a vehicle (vector) that can be engineered to carry DNA, such as a plasmid (circular units of DNA) or phages (viruses that infect bacteria). Once the gene of interest is inserted into the vector, the vector is used to carry the gene into the DNA of the host cells. Typical host cells used are Escherichia coli (E Coli) or mammalian cells grown in cultures, such as Chinese hamster ovary cells (CHO) and yeast.

Biotherapeutics created with biotechnology tools represent a broad range of molecules: hormones, growth factors, peptides, small proteins, cytokines, therapeutic replacement enzymes, blood factors, anticoagulants, monoclonal antibodies, their fragments, and fusion proteins, and there are now over a hundred and seventy launched protein products. As of December 2019, seventy-nine therapeutic mAbs have been approved by the US FDA, but there is still significant growth potential.

The first recombinant FDA-approved pharmaceutical was the human insulin (Humulin-US/Humuline-EU) produced by Eli Lilly in 1982 (Sandow et al. 2015). The insulin was obtained from genetically engineered Escherichia coli to treat diabetes.

Genzyme pioneered the concept of enzyme replacement therapy by introducing to the market in 1991 the product Ceredase to treat Gaucher's disease. Contemporary interest in treatments for rare diseases by major pharmaceutical companies stems largely from an exception among rarities: the astonishing commercial success of Cerezyme (Hoppe 2000). In 1994, Genzyme obtained approval for the recombinant product imiglucerase (Cerezyme), obtained from genetically engineered Chinese hamster ovary cells.

In 1997, the first full-length chimeric monoclonal antibody, rituximab, and the first full-length humanized monoclonal antibody, daclizumab, were approved for commercialization (Goswami et al. 2013; Liu 2014). This started the era of monoclonal antibody biotherapeutics, currently the fastest growing field of pharmaceuticals, with projected antibodies drug market revenue to surpass around $380 billion by 2030 (Lu et al. 2020; Wang et al. 2023). The annual sales of Humira (a monoclonal antibody that inhibits TNF to generate an anti-inflammatory effect) are above $12 billion (Ecker et al. 2015). Humira (the biggest blockbuster biotherapeutic of today) is a monoclonal antibody drug product that is administered at a high concentration in a liquid formulation in a prefilled syringe (PFS) for subcutaneous administration. It has set the bar high and paved the way for increasing convenience for patients.

Figure 9.3. Bioreactors/fermenters are used in the biopharmaceutical industry to encourage microorganisms to thrive, allowing sufficient fermentation to make pharmaceutical products such as proteins and antibiotics.

Biologics are challenging molecules to deliver due to their complex and labile structure. They are large hydrophilic molecules with very limited permeability across biological barriers. They have rapid clearance from the system and usually require high therapeutic doses. The combination of rapid clearance and high therapeutic dose and the trend toward lower dosing frequency and convenience to patients (Pérez-Ramírez 2020) has generated the need to create highly concentrated products. This has created other challenges, such as concentration-driven instability and viscosity, sensitivity to stress and interfacial interactions, and compatibility with medical devices.

The Development of Drug Products Is Complex

Development of a safe and efficacious medicine requires initial testing in animals, but it is impossible to predict the behavior of a given drug only from animal studies; no new drugs could ever be developed without thorough human testing to demonstrate their effectiveness. This process of testing is known as a clinical trial. Clinical trials are only one step in the lengthy and expensive process of drug discovery. Only one of about every five thousand screened substances originally synthesized or identified from natural products makes it through the clinical trial process and is approved as a new medicine (Pritchard 2003).

First, after basic research and screening, promising substances are moved into the development stage (figure 9.4). While in development, the drugs are tested in vitro. A formulation is developed to stabilize the molecule as a drug product, and in parallel, the production of the active component at sufficient scale is initiated, together with the development of analytical assays. The next step involves preclinical trials, in which a drug's effectiveness and toxicology are established in animal models, like mice and then nonhuman primates.

Following successful animal studies, the substance is moved through three phases of clinical testing that involve human subjects (figure 9.4). The first phase establishes the metabolism and the side effects (safety) of the drug treatment (phase I); the second phase gauges the efficacy of the treatment (phase II); and the final stage evaluates the overall risk-benefit ratio in patients (pivotal clinical trials/phase III), where the drug product is tested at a commercial scale before submitting the dossier to regulatory agencies.

The entire process of drug discovery must meet federally mandated standards of scientific practice, and all drugs that yield successful trial results must receive approval from regulatory agencies before being released in the market. Moreover, human and animal testing are also subject to scrutiny by, respectively, institutional review boards (IRBs) and institutional animal care and use committees (IACUCs), both before and during drug trials. Finally, long-term efficacy trials after a drug becomes available (phase IV) help document any additional effects and are an integral part of producing a safe product, a phase called pharmacovigilance.

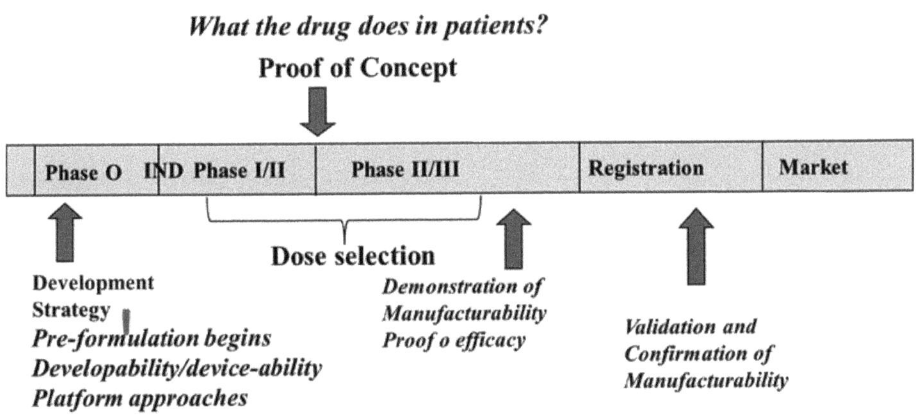

Figure 9.4. Project flow and critical gates to develop a biopharmaceutical product.

The nature of biological products, including the inherent variations that can result from the manufacturing process, can present challenges in characterizing and manufacturing products that often do not exist in the development of small molecule drugs (Saluja et al. 2018). Slight differences between manufactured lots of the same biological product (i.e., acceptable within-product variations) are normal and expected within the manufacturing process. As part of its review, regulatory agencies (e.g., FDA, EMEA) assess the manufacturing process and the manufacturer's strategy to control within-product variations. These control strategies are put in place to help ensure that manufacturers produce biological products with consistent clinical performance. Once the product is on the market, it has an exclusivity. In the US, biologic exclusivity is set at twelve years, and fifteen-plus years without competition has become the norm rather than the exception.

According to the FDA,

> Patents and exclusivity work in a similar fashion but are distinct from one another and governed by different statutes. Patents are a property right granted by the United States Patent and Trademark Office anytime during the development of a drug and can encompass a wide range of claims. Exclusivity refers to certain delays and prohibitions on approval of competitor drugs available under the statute that attach upon approval of a drug or of certain supplements. A new drug application (NDA) or abbreviated (ANDA) holder is eligible for exclusivity if statutory requirements are met. See 21 C.F.R. 314.108, 316.31, 316.34 and sections 505A, 505E, and 505(j)(5)(B)(iv) of the FD&C Act. Periods of exclusivity and patent terms may or may not run concurrently. Exclusivity was designed to promote a balance between new drug innovation and greater public access to drugs that result from generic drug competition (Frequently asked 2020).

Biosimilars

Biosimilar antibodies are "generic" versions of "innovator" (or "originator") antibodies with the same amino acid sequence but are produced from different clones and manufacturing processes. Consequently, biosimilar mAbs may include possible differences in glycosylation and other microvariations such as charge variants that may affect quality, safety, and potency (Vlasak et al. 2009).

Biosimilar mAbs have no clinically meaningful differences from an existing reference product approved by a regulatory agency. Minor differences between the reference product and the proposed biosimilar product in clinically inactive components are acceptable. For example, these could include minor differences in the stabilizer or buffer compared to what is used in the reference product. Any differences between the proposed biosimilar product and the reference product are carefully evaluated by regulatory agencies to ensure the biosimilar meets high approval standards. As mentioned above, slight differences (i.e., acceptable within-product variations) are expected during the manufacturing process for biological products, regardless of whether the product is a biosimilar or a reference. Bio-better antibodies target the same validated epitope as a marketed antibody but have been engineered to have improved properties, e.g., optimized glycosylation profiles to enhance effector functions or an engineered Fc domain to increase serum half-life (Correia 2010). In contrast to the low-cost generic versions of small molecules that are off patent, it is currently not possible to produce exact copies of large proteins and glycoproteins, such as antibodies, owing to their structural complexity.

Translational research is trying to translate (move) basic science discoveries more quickly and efficiently into practice. However, there is no linear correlation between translational research and clinical applications. There are several milestones that are identified in the development of a product.

- Fundamental research
    a. Analytical tools
    b. Broad expertise
- Preclinical research
    c. Relevant disease models
- Translation to humans
    d. Input from first in-human studies using preclinical models
- Translation to patients
    e. Observation/analysis of studies from the different phases of clinical development (e.g., phase I, clinical to pivotal clinical trials)
- Translation to population
    f. Health evaluations on societal application and impact

Proteins are complex macromolecules and play different roles in health and disease. Proteins come in a wide variety of shapes and personality, and they are generally between 5 to 220,000 KDa or around sixty to two thousand amino acids long. The structure and functional chemistry of proteins have been perfected over billions of years of evolutionary history and they indeed constitute the true engines of life.

GLOSSARY

Acetylsalicylic acid: See aspirin.

Achiral: Having no asymmetric atom; among the natural amino acids, only glycine is achiral.

Acid dissociation constant (K_a): The ratio of the concentration of the dissociated form of an acid to the undissociated form.

Activation energy (Ea): The energy required to overcome a barrier to reaction.

Active site: The region on an enzyme where the substrate binds and catalysis occur.

Active ingredient(s)/active substance: An active ingredient is any component that provides a pharmacological activity or other direct effect in the diagnosis, cure, mitigation, treatment, or prevention of disease, or affects the structure or any function of the body of humans or animals.

Activity coefficient (single ion): The effective concentration (i.e., activity) of an ion in a solution divided by the actual concentration of that ion.

Affinity: The measurement of the tenacity with which two or more molecules stick together.

Agonist: A substance that stimulates an activity of a target molecule.

Algorithm: A computational procedure for analysis or calculation.

Aliphatic: Having the properties of hydrocarbons.

Allosteric regulation: Allosteric regulation is the regulation of an enzyme by binding an effector molecule at a site other than the protein's active site. The change of one part of the protein affects the activity at another.

Allosteric effector: A ligand that controls the state and activation of an allosteric protein.

Alzheimer's disease: A neurodegenerative disease associated with the deposition of protein aggregates in the brain.

Amino acid: Organic acids where one carbon has been replaced by an amine group. The α-amino acid is the one in which the amine group is attached to the same carbon as the carboxylic acid.

Amyloidosis: A disease characterized by extracellular deposits of protein aggregates.

Amorphous: Having no apparent shape and order; noncrystalline.

Amphipathic: The property of having both hydrophobic and hydrophilic parts.

Anisotropic: With properties that differ depending on the orientation. See polarization of fluorescence.

Antibody: A protein that forms part of the vertebrate immune system. Its function is to recognize foreign molecules (antigens) and bind to them.

ANS or 8-anilino-1-naphthalene-sulfonate: A dye for specific residues that emits fluorescence upon binding to hydrophobic regions.

Anticodon: A triplet of bases within tRNA that binds to a complementary triplet based in messenger RNA.

Beer-Lambert Law: The absorbance of a solution A is given by $A= \mathcal{E}ct$, where c is the concentration of the solute, t is the thickness, and $\mathcal{E}$ is the molar extinction coefficient. Usually, c is measured in moles liter^{-1} and t is in cm (1 cm); therefore, $\mathcal{E}$ is liter mole^{-1} cm^{-1}.

Antigen: A foreign molecule that elicits the production of antibodies that bind it specifically.

Apoptosis: Programmed cell death.

Archaea: One of the basic divisions of life; contains, but is not limited to, numerous simple organisms adapted to unusual environments, such as high temperatures.

Aromatic: A molecule containing a set of conjugated double bonds, for instance the amino acids tryptophan, phenylalanine, and tyrosine.

Arrhenius equation: The formula for the dependence of reaction rate on temperature.

Aspirin: Also known as acetylsalicylic acid, a nonsteroidal anti-inflammatory drug used to reduce pain, fever, and/or inflammation.

Atomic resolution: A very detailed X-ray structure determination in which the experimental data unambiguously show the position of individual atoms.

Backbone: The repetitive part of protein structures; a linear change linked by peptide bonds.

Basic solution: A solution that has a pH greater than 7.0. Also known as an alkaline solution.

β-barrel: A barrel-shaped protein structure formed by a series of β strands.

β -bulge: One or more residues that loop out of a strand of β-sheet, interrupting the regular pattern of hydrogen bonding.

β-strand: A section of polypeptide chain that is in the extended or βregion of the Ramachandran plot, and which hydrogen bonds to one or more neighboring strands to form a βsheet.

β-hairpin: Two antiparallel strands of β-sheet connected by a short loop.

Blue shift: A change in absorbance that leads to a shorter wavelength. If the ground state interacts more strongly with the solvent than the excited state, the absorption band maximum shifts to a shorter wavelength.

Bohr effect: The phenomenon observed when exposure to carbon dioxide, which lowers the pH inside the cells, causes the oxygen affinity of hemoglobin in red blood cells to decrease.

Bragg's Law: The relationship between scattering angle θ and interplanar spacing within a crystal, d: $n\lambda = 2d \sin\theta$.

Buried surface area: The surface area of atoms in a protein inaccessible to a probe sphere the size of a water molecule.

Canonical structure: One of a roster of discrete conformations of certain antigen-binding loops of antibodies.

C terminus: The end of a polypeptide chain that has a free carboxylate group; that is, the right-hand end as normally written.

Carbamylation: A posttranslational modification resulting from the nonenzymatic reaction between isocyanic acid and free functional groups of proteins, with the free amino groups. Modification of lysine groups of proteins with cyanate as probe of structure and function.

Carbanion: An organic anion in which the negative charge is located on a carbon atom.

Catalysis: Speeding up the rate of a reaction.

Catalytic constant (kcat): A kinetic constant that is a measure of how rapidly an enzyme can catalyze a reaction when saturated with a substrate(s). The catalytic constant is equal to the maximum velocity (Vmax) divided by the total concentration of enzyme $[E]_{total}$ or the number of moles of substrate converted to product per mole of enzyme active sites per second under saturating conditions. Also known as the turnover number.

Catalytic triad: The three amino acids that make up the key active site residues of serine proteases: Ser, His, and Asp.

CD: See circular dichroism.

Channel: A gap in a membrane, allowing passage of ions or molecules.

Chaos: A dynamic state in which small changes in initial conditions may make large changes in subsequent behavior.

Chaotropic agent: A strong hydrogen-bonding substance that enhances the solubility of nonpolar compounds in water by disrupting regularities in hydrogen bonding among water molecules. Concentrated solutions of chaotropic agents, such as urea and guanidine salts, decrease the hydrophobic effect and are thus effective protein denaturants.

Chaperone: A protein that shields an unfolded protein and prevents it from aggregating. Some chaperones (chaperonins) help unfold proteins to refold.

Chimeric: A protein containing modules originating from different sources.

Chiral: A molecule whose enantiomers are nonsuperimposable. The structure and its mirror image are not superimposable. Chiral compounds are typically optically active.

Chloroplast: A membrane-enclosed organelle found in plants, where photosynthesis occurs.

Chlorophyll: A green pigment in photosynthetic membranes that is the principal light-harvesting component in phototrophic organisms.

Chromatography: A technique used to separate components of a mixture based on their partitioning between a mobile phase, which can be gas or liquid, and a stationary phase, which is a liquid or a solid.

Chromophore: A molecule that absorbs ultraviolet radiation. These are typically aromatic or conjugated (containing alternating double and single bonds).

Circular dichroism (CD): Absorption spectroscopy method based on the differential absorption of left and right circularly polarized light. Optically active chiral molecules will preferentially absorb one direction of the circularly polarized light.

Clipping: Refers to specifically clip C- or N-terminal amino acid residues in proteins when they may come in contact with proteases.

Cloning: The generation of many identical copies of a molecule, cell, or organism.

Codon: A three-nucleotide (triplet) messenger RNA sequence that codes for an amino acid.

Coenzyme: A nonprotein compound that is necessary for the functioning of an enzyme.

Cofactor: A non-amino acid molecule that forms part of a protein structure and assists function. There are two types of cofactors: essential ions and coenzymes.

Coiled coil: A simple structure formed by two α-helices winding around each other to form a super coil.

Collagen: A fibrous protein that forms structural material in the body.

Collision induced dissociation (CID): The fragmentation of proteins for mass spectrometry by acceleration to induce high-energy collisions with neutral gases.

Complementary-determining region (CDR): The region in an immunoglobulin. It is hypervariable in sequence, which contributes to the antigen-binding site.

Condensation: A reaction involving the joining of two or more molecules accompanied by the elimination of water, alcohol, or other simple substance.

Configuration: A spatial arrangement of atoms, which cannot be altered without breaking and reforming covalent bonds.

Conformation: The spatial disposition of atoms in a molecule. Any three-dimensional structure, or spatial arrangement of a molecule that results from rotation of functional groups around single bonds. Because there is free rotation around single bonds, a molecule can potentially assume many conformations.

Conformational disease: A disease arising from the misfolding of a protein.

Conjugate acid: The product resulting from the gain of a proton by a base.

Conjugate base: The product resulting from the loss of a proton by an acid.

Constant domain: A domain in an immunoglobulin contributing to its overall structure and response to antigen but not containing the combining site itself.

Convergent evolution: An occurrence in which a similar structural or functional feature in a protein is evolved independently more than once.

Converter enzyme: An enzyme that catalyzes the covalent modification of another enzyme, thereby changing its catalytic activity.

Coomassie blue: A stain used to visualize the positions of proteins on an electrophoresis gel.

Cooperativity: The phenomenon whereby the binding of one ligand or substrate molecule to a protein influences the affinity of the protein for additional molecules of the same substance. Also, the phenomenon whereby formation of structure in one part of a macromolecule promotes the formation of structure in the rest of the molecule.

$\Delta \tilde{C}p$: See specific heat capacity change.

C-terminus: The amino acid residue bearing a free α-carboxyl at one end of a peptide chain. Also known as the carboxyl terminus.

Crystal structure: A description of the ordered arrangement of atoms, ions, or molecules in a crystalline material. Determination of atomic positions from the diffraction of X-rays by a crystal permits to identify the structure of the molecule/macromolecule.

Cryptic genes: A gene that was once presumably functional but is no longer.

Curare: A nondepolarizing muscle relaxant that blocks the nicotinic acetylcholine receptor (nAChR).

Cytoplasm: The part of the cell enclosed by the plasma membrane, excluding the nucleus.

Cytoskeleton: Structural elements within the cell, giving shape and rigidity to the cell and providing tracks for intracellular transport proteins.

Cytosol: The aqueous portion of the cytoplasm minus the subcellular structures.

Dalton: A unit of mass equal to one atomic mass unit.

Denaturant: A chemical that destabilizes protein and (at sufficient concentration) causes it to unfold. Examples of denaturants are urea and guanidinium chloride.

Denaturation: The loss of native protein structure.

Denatured state: The unfolded (nonnative) state of a protein.

Desolvation: The process of dissociating or releasing water molecules electrostatically bound to a particle in aqueous solution. During ligand-protein or protein-protein interactions, the hydration shell of both molecules must be partially shed and rearranged to allow the formation of direct interactions between protein and ligand or protein-protein interactions.

Detergent: An amphipathic molecule consisting of a hydrophobic portion and a hydrophilic end that may be ionic or polar. Detergent molecules can aggregate in aqueous media to form micelles. Also known as surfactant.

Differential scanning calorimetry (DSC): A technique to evaluate phase transitions in molecules (proteins, membranes, lipids) by measuring the rate of heat absorbed by the sample when it is heated simultaneously with a sample of water. When the sample undergoes phase transitions, it absorbs heat energy at a faster rate than the water.

Diffusion-controlled reaction: A reaction that occurs with every collision between reactants molecules. In enzyme-catalyzed reactions, the Kcat/Km ratio approaches a value of 10^8 -10^9 M^{-1} s^{-1}.

Dielectric constant: The proportion by which a medium shields charges, compared with the vacuum. A measure of how far away electrostatic effects can be felt.

Diffusion limited: An enzyme that is so efficient that the rate of encounters between enzyme and substrate limits the rate of reaction.

Dihedral angles: A quantitate description of the conformation of a molecule in terms of rotations around bonds.

Directed evolution: The selection of protein molecules with desired properties by laboratory rather than natural methods.

Dipole: Something with two poles, like a magnet or a molecule that has positive and negative charges at opposite ends (electric dipole).

Dipole moment: A measure of the strength of the dipole, which for an electrical dipole is equal to *qr*, where *q* is the charge and *r* is the vector connecting the charges. A dipole moment therefore has both a magnitude and a direction.

Dissociation constant: A measure of the affinity of two molecules.

Disulfide bridge: Posttranslational modification in which two cysteine residues, which may be far apart in the sequence, come together in the structure and form an S-S bond.

Divergent evolution: A common mechanism in which the gene for an initial protein is duplicated, and the original copy then diverges over the course of evolutionary time. Structure is retained more than function, and much more than sequence.

DLS: See dynamic light scattering.

Domain: Polypeptide chain (or part of one) that can independently fold into a stable compact tertiary structure or fold.

Dosage form: A physical form in which a drug is produced and dispensed, such as a tablet, a capsule, or an injectable.

Drug product configuration: The physical form of the product, including the primary container, fill volume, amount of activity in the vial, extractable volume, and formulation.

Drug product formulation: The chemical composition of the dosage form.

Drug product/finished product: The finished dosage form that contains a drug substance, generally, but not necessarily in association with other active or inactive ingredients.

D_2O: See heavy water.

Dynamic light scattering (DLS): The light scattering technique measures how fast scattered light intensity changes over time, providing information on particle size and diffusion rate.

Electron density map: The three-dimensional plot that shows where in space electrons are concentrated within the repeating unit of the crystal.

Electronegative: An atom or group that is good at attracting bonding electrons toward it. Examples include N and O.

Electrophile: A positively charged atom, which is stacked by, and forms a bond to, a nucleophile.

Electrophoresis: A technique used to separate molecules by their migration in an electric field, primarily based on the net charge.

Electrospray ionization (ESI): A method for vaporizing proteins for mass spectrometry.

Electrostatic interactions: Interactions between charged or polar atoms.

Ellipticity: A measure of the degree of polarization of light.

ELISA: Enzyme-linked immunosorbent assay.

Enantiomers: Stereoisomers that are nonsuperimposable mirror images.

Endergonic reaction: A chemical reaction that is characterized by a positive free energy change. Such a reaction cannot occur spontaneously and requires the input of energy from outside the system to proceed.

Energy transduction: The interconversion of different types of energy.

Energy landscape: A diagram indicating the variation in energy in a protein with conformation.

Enthalpy (H): A thermodynamic state function that describes the heat content of a system.

Enthalpy/entropy compensation: The commonly observed tendency in a series of related measurements (e.g., binding interactions, dissolution, effects of mutations on binding or stability) for the free energy to remain roughly constant, but with large compensatory changes in entropy and enthalpy.

Entropy (S): A thermodynamic state function that describes the randomness or disorder of a system.

Enzyme: A biological catalyst, a protein, or RNA. Some enzymes may require additional cofactors for activity. Enzymes accelerate chemical reactions.

Enzyme-substrate complex (ES): A complex formed when substrate molecules bind noncovalently within the active site of an enzyme.

Equilibrium: The state of a system in which the rate of conversion of reactants to products is equal to the rate of conversion of products to reactants. The free energy change for a system at equilibrium is zero.

Equilibrium constant (K_{eq}): The ratio of the concentrations of products to the concentration of reactants at equilibrium.

Epimers: Isomers that differ in configuration at only one of several chiral centers.

ES: See enzyme-substrate complex.

Essential amino acid: An amino acid that cannot be synthesized by an animal and must be obtained in the diet.

Excipient: Inert substance that may facilitate the application, stability, delivery, handling, or marketing of the drug product.

Extensive property: Volume and energy.

Intensive property: Density and viscosity

Fab: A fragment of an antibody, including the antigen-binding site.

FAD: See flavin adenine mononucleotide.

Far-ultraviolet: Electromagnetic radiation much more energetic than visible light, wavelength range 100–200 nm.

FDNB: See Sanger reagent.

Ferredoxin: Small, mostly acidic, soluble proteins found ubiquitously in biological organisms. Ferredoxin serves as an electron transfer agent in biological redox processes and facilitates electron transfer events like photosynthesis.

Fibroin: A fibrous protein present in silk.

Fibrous protein: Proteins that form fibers that are usually long and cross-linked, like keratin, collagen, and silk. They are water-insoluble proteins.

First order reaction: A reaction whose rate is directly proportional to the concentration of only one reactant.

Flavin adenine mononucleotide (FAD): A noncovalently bonded coenzyme that acts as a redox carrier in catalysis.

Flavin mononucleotide (FMN): Riboflavin-5′-phosphate is a biomolecule produced from riboflavin (vitamin B2).

Fluorescence: Fluorescence is prompt photoluminescence that occurs very shortly after the photoexcitation of a substance.

Fluorophore: A chemical that is capable of fluorescence.

FMN: See flavin mononucleotide.

Folding funnel: A pictorial concept describing the folding of a protein along convergent trajectories from a denatured state with many conformations, to a native state with only one.

Folding pattern: The topology of the three-dimensional structure of a protein.

Folding rate: The rate of formation of the native state of a protein from the denatured state.

Free energy change (ΔG): A thermodynamic quantity that defines the equilibrium condition in terms of the change s in enthalpy (H) and entropy (S) of a system at constant pressure $\Delta G = \Delta H - T\Delta S$, where T is absolute temperature. Free energy is a measure of the energy available within a system to do work.

Fuzzy complexes: Refer to IDPs or IDPRs bound to a target protein, retaining their plasticity.

ΔG: See free energy change.

$\Delta G°$: See standard free energy change.

($\gamma\pm$): See mean ionic activity coefficient

Gel filtration: A chromatographic separation based on size. The resin (stationary phase) through which the solution is passed contains channels or pores into which small molecules can fit, but large ones cannot. Therefore, small molecules migrate more slowly through the resin and emerge from the column later.

Gene: A segment of DNA that is transcribed. Also referred to as a segment of DNA that encodes a functional protein or RNA.

General acid: An amino acid residue that donates a proton during a reaction and thereby catalyzes it.

General base: An amino acid residue that remotes a proton during a reaction and thereby catalyzes it.

Genome: The total amount of genetic information in an organism. It is equivalent to a single complete chromosome or to a set of chromosomes (haploid).

Genotype: The sequence of DNA in an organism.

Genomics: The determination, analysis, and comparison of the complete genetic material of organisms of different species.

Gibbs free energy: A measure of spontaneity and equilibrium under conditions of constant temperature and pressure.

Globular proteins: A major class of proteins, many of which are water soluble. Globular proteins are compact and roughly spherical, containing tightly folded polypeptide chains.

Glycation: See protein glycation.

H: See enthalpy.

Heat capacity: A measure of the heat energy required to increase the temperature of an object. It usually has the symbol Cp, meaning the heat capacity at constant pressure. It can be determined by measuring the enthalpy at different temperatures.

Heat of vaporization: The amount of heat required to evaporate 1 gram of a liquid.

Heat-shock gene: A gene whose transcription is increased in response to stresses, such as high temperature.

Heavy water: Deuterium oxide or D_2O. A form of water whose hydrogen atoms are all deuterium.

Helical wheel: A pictorial representation of the positions of the side chains in an α-helix, usually used to identify hydrophobic or hydrophilic protein surfaces.

Helix-coil transition: The transition between a helical conformation and a random coil conformation. This is an example of cooperativity.

Hemoglobin: A tetrameric heme containing globular protein in red blood cells that carries oxygen (0_2) to other cells and tissues.

Henderson-Hasselbach equation: The relationship between pH, pK, and the protonated and unprotonated forms in a buffer, where pH = pK + log [base]/[acid].

Helix-turn helix domain: A structural motif common in DNA-binding proteins.

Heterodimer: Association of two different proteins.

Heterologous binding: A protein-protein complex in which different monomers contribute different sets of residues to the site of interaction.

Heterotropic: The modulator and the substrate are different molecules.

HPLC: See high pressure liquid chromatography.

High-performance liquid chromatography (HPLC): A sensitive method for separating chemicals according to retention time on a column.

Hill plot: An analysis of binding data that reveals cooperativity.

Histone: The main protein component of chromatin. There are six classes of histones in humans, which act as a core around which DNA is wrapped (forming nucleosomes) and clamps to hold the nucleosomal DNA in place.

Homologous: Two proteins are homologous if they are related by evolution; that is, they share a common ancestor or at some point have diverged from each other. This necessarily implies that they have some sequence similarity, but in distantly homologous proteins this is less defined.

Homotropic: The modulator is also the substrate for the enzyme.

Housekeeping enzyme: An enzyme that is required in every cell and performs an essential task, such as basic metabolism.

Hydrodynamic radius (r_h): Also called the Stokes radius (Rs), this is the radius of a uniform sphere that tumbles and diffuses at the same rate as the protein and is larger than the radius of gyration roughly by the size of a single layer of water molecules.

Hydrogen bond: A weak electrostatic interaction when a hydrogen atom bonded covalently to a strongly electronegative atom also bonds to the unshared electron pair of another electronegative atom.

Hydrolysis: Cleavage of a bond within a molecule by the addition of elements of water.

Hydrolase: An enzyme that catalyzes the hydrolytic cleavage of its substrate.

Hydropathy plot: A graph of amino acid hydrophobicity against sequence, used to determine the location of transmembrane sequence.

Hydropathy index: A measure of the overall hydrophobicity of a stretch of amino acid residues in a protein.

Hydrophilic: Water loving. Chemicals that have good affinity for water. These typically contain polar or hydrogen-bonding groups.

Hydrophobic: Water hating. Chemicals that have no affinity for water and therefore tend to aggregate together away from water in aqueous solution. These usually have side chains that are hydrocarbons or aromatic rings.

Hydrophobic effect: The force that drives hydrophobic groups or molecules to associate with each other rather than with water. The hydrophobic effect appears to depend on the increased entropy of solvent water molecules that are released from an ordered arrangement around the hydrophobic group.

Hydrophobic interaction: A weak noncovalent interaction between nonpolar molecules or substituents that results from the strong association of water molecules with one another. Such association leads to the shielding or exclusion of nonpolar molecules from an aqueous environment.

Hydrophobicity: The degree to which a compound or functional group that is soluble in nonpolar solvents is insoluble or sparingly soluble in water.

IDP: See intrinsically disordered proteins.

IDPR: See intrinsically disordered protein regions.

Induced fit: Theory proposed by Koshland in 1958 that an enzyme structure changes on binding to substrate.

Initial velocity (v): The initial linear portion of the enzyme reaction when less than 10 percent of the substrate has been depleted or less than 10 percent of the product has formed.

Inner-filter effect: The absorption of the excitation/emission light of fluorophores by appropriate absorbers.

Intrinsically disordered protein (IDP): A protein that even under physiological conditions does not adopt a fixed compact stable tertiary structure.

Intrinsically disordered protein regions (IDPRs): Polypeptide segments that do not contain sufficient hydrophobic amino acids to mediate cooperative folding.

Ion pair: An electrostatic interaction between ionic groups of opposite charge within the interior of a macromolecule, such as a globular protein.

Ion pair of water (K_w): The product of the concentration of hydronium ions and hydroxide ions in an aqueous solution, equal to 1.0×10^{-14} M.

Isoelectric focusing: A technique for purifying and characterizing proteins according to the isoelectric point (pI). Proteins are mixed with suitable buffers and run on a gel with the use of a voltage applied across the gel. They migrate to a position determined by their pI.

Isoelectric point (pI): The pI or isoelectric point is the pH at which a protein has no net migration in an electric field that has a net zero charge (isoionic point). It is usually the pH at which proteins are less soluble. The value can be calculated from the amino acid sequence, but it is better to measure it experimentally.

Isomolal: The constant composition of the solvent (not dialyzed).

Isopeptide bond: A peptide bond between the α-carboxylate group of an amino acid and the ε-amine of a lysine group.

Isopotential: A region in space where every point in it is at the same potential. In the case of protein solutions in the presence of excipients/ligands, the isopotential is obtained by extensive dialysis equilibrium of protein and ligand/excipient.

K_a: See acid dissociation constant.

k_{cat}: See catalytic constant.

K_{cat}/K_m: See specificity constant.

K_i: See inhibition constant.

Kinetic order: The sum of the exponents in a rate equation, which reflects how many molecules are reacting in the slowest step of the reaction. Also known as reaction order.

Kinetics: The study of the rates of change, such as the rate of the chemical reactions.

KDa: Protein sizes are usually measured in kilodaltons (KDa). The mass of ^{1}H is 1 Da, ^{12}C is 12 Da, etc. The average mass of an amino acid, when linked together by peptide bonds to make a protein, is 110 Da. Therefore, a protein of two hundred residues has an approximate mass of 200 x 110 = 22,000 Da or 22 KDa.

K_m: See Michaelis-Menten constant.

K_w: See ion product of water.

Le Chatelier's principle: "If a chemical system at equilibrium experiences a change in concentration, temperature, or total pressure, the equilibrium will shift in order to minimize that change."

Light scattering: See dynamic and static light scattering.

Liquid-liquid phase separation: conversion of a homogeneous liquid phase into inhomogeneous phases with a diluted and concentrated phase in thermodynamic equilibrium.

LLPS: See liquid-liquid phase separation.

Lock and key model: A model suggested by Emil Fisher in 1894; it proposes that enzymes catalyze reactions by acting as a lock into which the substrate fits neatly.

Loop: A nonrepetitive polypeptide region that connects secondary structures within a protein molecule and provides directional changes necessary for a globular protein to attain its compact shape. Loops contain from twenty-two to sixteen residues. Short loops of five residues are usually called turns.

MALDI: Matrix-assisted laser desorption/ionization is a soft ionization technique used in mass spectrometry (MS) that produces rapid and efficient ionization of a wide variety of molecules.

Maximum velocity (Vmax): The initial velocity of a reaction when the enzyme is saturated with substrate, that is, when all the enzyme is in the form of an enzyme-substrate complex.

Mean ionic activity coefficient **($\gamma\pm$):** The average rational activity coefficient of the ions of an electrolyte that dissociates in solution into cations of charge z+ and anions of charge z−, given, according to the limiting law of the Debye-Hückel theory.

Melting temperature (T_m): The temperature at which 50 percent of the protein is unfolded.

Messenger ribonucleic acid (mRNA): A class of RNA molecules that serve as templates for protein synthesis.

Micelle: An aggregation of amphipathic molecules in which the hydrophilic portions of the molecules project into the aqueous environment, and the hydrophobic portions associate with one another in the interior of the structure to minimize contact with water molecules. Detergents form micelles.

Michelson interferometer: The Michelson interferometer produces interference fringes by splitting a beam of light so that one beam strikes a fixed mirror and the other a movable mirror. When the reflected beams are brought back together, an interference pattern results.

Microtubule: A protein filament composed of α and β tubulin heterodimers assembled around a hollow core in a helical arrangement, creating a strong, rigid fiber. Microtubules are components of the cytoskeleton network and can form structures capable of directed movement.

Metastable: A protein conformation that is stable under some circumstances but can be changed to a different conformation by perturbation of the conditions.

Mitochondrion: A membrane-enclosed eukaryotic organelle that produces most of the energy in the cell.

Molar extinction coefficient: The Beer-Lambert Law states that the absorbance of a solution A is given by $A = eCt$, where C is the concentration of the solute, t is the thickness, and *e* is the molar extinction coefficient. It is a characteristic of the molecule in question; it can be predicted from the amino acid sequence.

Molar mass: The weight in grams of one molar of a compound.

Molecular weight: See relative molecular mass.

Molten globule: Refers to a protein in mildly denaturing solution conditions. Such proteins have native-like secondary structure but not well-formed tertiary structure, including a larger radius than folded proteins. Similar properties have been observed in protein intermediates; these are often called molten globules.

Monomer: A small compound that becomes a residue when polymerized with other monomers.

M_r: See relative molecular mass.

Motif: Refers to 1) sequence motif or 2) structural motifs. A sequence motif is a sequence of amino acids that is characteristic of a protein fold. A structural motif is a group of elements of regular secondary structure. .i.e., coiled-coil, β-sheet, helix-turn-helix, metal-binding regions.

Mutation: A heritable change in the sequence of nucleotides in DNA that causes a permanent alteration of genetic information.

MS/MS: Tandem MS (or MS/MS, MSn) is a technique to break down selected ions (precursor ions) into fragments (product ions).

N: See plate number.

nAChRs: See Nicotinic acetylcholine receptor.

NAD: See nicotinamide adenine dinucleotide.

N-terminus: The end of the polypeptide chain that has a free amino group, namely the start or left-hand end as normally written.

Native state: The structure of a protein in its native (that is, physiologically active) state.

Near-ultraviolet: Radiation of higher energy than visible light, but less energetic than far-ultraviolet radiation; the wavelength ranges from 300–400 nm.

Natural selection: The term used by Charles Darwin to describe the mechanism by which evolution occurs, sometimes called survival of the fittest.

Neutral solution: An aqueous solution that has a pH value of 7.0.

Nicotinamide adenine dinucleotide (NAD): Is a coenzyme found in all living cells. NAD is called a dinucleotide because it consists of two nucleotides joined through their phosphate groups.

Nicotinic Acetylcholine receptor (nAchR): Ligand-gated ion channels that allow the passage of sodium, potassium, and calcium ions across the membrane when activated by acetylcholine or nicotine.

Ninhydrin: 2,-dihydroxyindane-1,3-dione chemical, mainly used to detect ammonia or primary and secondary amines.

Nitrogen fixation: The reduction of atmospheric nitrogen to ammonia. Biological nitrogen fixation occurs in only a few species of bacteria and algae.

N-linked glycoprotein: A glycoprotein in which one or more oligosaccharide chains are attached to the protein through covalent bonds to the amide nitrogen atom of the side chain of asparagine residues. The oligosaccharide chains of *N*-linked glycoproteins contain a core pentasaccharide of two *N*-acetyl-glucosamine residues and three mannose residues.

NMR spectroscopy: See nuclear magnetic resonance spectroscopy.

Noncompetitive inhibition: Inhibition of an enzyme-catalyzed reaction by an inhibitor that binds to either the enzyme or the enzyme-substrate complex.

Nonessential amino acid: An amino acid that an animal can produce in sufficient quantity to meet metabolic needs.

N-terminus: The amino acid residue bearing a free α-amino group at one end of a peptide chain. In some proteins, the N-terminus is blocked by acylation. The N-terminus residue is usually assigned the residue number 1, also known as the amino terminus.

Nuclear magnetic resonance spectroscopy (NMR spectroscopy): A technique used to study the structures of molecules. In NMR, the absorption of electromagnetic radiation by molecules in magnetic fields of varying frequencies is used to determine the spin states of certain atomic nuclei.

Nucleophile: An electron-rich group, negatively charged or containing unshared electron pairs, which is attracted to chemical species that are positively charged (electrophile).

Oligomer: A multi-subunit molecule whose arrangement of subunits always has a defined stoichiometry and almost always displays symmetry.

O-linked glycoprotein: A glycoprotein in which one or more oligosaccharide chains are attached to the protein through covalent bonds, usually to the hydroxyl oxygen atom of serine or threonine residues.

Oncogene: A gene whose expression causes cancer.

Open reading frame (ORF): A DNA sequence that is presumed to code for a protein.

Operon: A bacterial transcription unit consisting of several different genes co-transcribed from one promoter.

Optical rotatory dispersion (ORD): A variation in the optical rotation of a substance with a change in the wavelength of light.

ORD: See optical rotatory dispersion.

Ordered kinetic mechanism: A reaction mechanism in which both the binding of substrates to an enzyme and the release of products from the enzyme follow an obligatory order.

Oxidase: An enzyme that catalyzes an oxidation-reduction reaction in which 0_2 is the electron acceptor.

Oxidation: The loss of electrons from a substance to transfer to another substance (the oxidizing agent). Oxidation can take several forms, including the addition of oxygen to a compound. The removal of hydrogen from a compound to create a double bond or an increase in the valence of a metal ion. Common mechanism of protein degradation.

Oxidizing agent: A substance that accepts electrons in an oxidation-reduction and thereby becomes reduced.

PAGE: See polyacrylamide gel electrophoresis.

Partial specific volume: The volume taken up upon mixing into a solution. It can be taken as the inverse of the density. The partial specific volume of a protein (v_2^-) is an experimental quantity containing information about solute-solvent interactions and protein hydration. The average v2- for a protein is ~ 0.73 cm^3 /g.

PCR: See polymerase chain reaction.

Peptide: Two or more amino acids covalently joined in a linear sequence by peptide bonds. Meanwhile, proteins are long molecules made up of multiple peptide subunits.

Peptide bond: The bond between the C=0 of one amino acid and NH of the next. It is flat, with a nature of double bond character in a trans configuration; that is with the O and H facing in opposite sites.

Peptide group: The nitrogen and carbon atoms involved in a peptide bond and their four substituents: the carboxyl oxygen atom, the amide hydrogen, and the two adjacent α-carbon atoms.

Perrin fluorescence theory: Relates fluorescence anisotropy to the rotational correlation time (τc) of the labeled molecule and the fluorescence lifetime of the fluorophore (τ).

pH: A logarithmic quantity that indicates the acidity of a solution; that is, the concentration of hydronium ions in solution. pH is defined as the negative logarithm of the hydronium ion concentration.

Phase separation: The creation of two distinct phases from a single homogeneous mixture.

ϕ-value: The relation between interactions of a residue in the native state of a protein and the interactions in the transition state for folding.

ϕ (phi): The angle of rotation around the bond between the α-carbon and the nitrogen of a peptide group.

pH optimum: In an enzyme-catalyzed reaction, this is the pH at the point of maximum catalytic activity. In a biopharmaceutical protein, it is the pH at which the drug product exhibits the best biological activity and physical-chemical properties.

Phenotype: An observable characteristic of an organism: appearance, behavior, development, etc. It arises because of the genome, and it therefore is contrasted with genotype.

Phosphatase: An enzyme that removes a covalently attached phosphate. The opposite of a kinase.

Phosphorescence: The emission of radiation in a similar manner to fluorescence but on a longer timescale so that emission continues after excitation ceases.

Photon: A quantum of light energy.

Photosynthesis: The process by which carbohydrates are synthesized from atmospheric CO_2 and water using light as the source of energy.

Photosystem: A functional unit of the light-dependent reactions of photosynthesis. Each membrane-embedded photosystem contains a reaction center, which forms the core of the photosystem, and a pool of light-absorbing antenna pigments.

Pitch: The distance required to complete one turn of a helical structure.

pI: See isoelectric point.

PKa: A logarithmic value that indicates the strength of an acid. pKa is defined as the negative logarithm of the acid dissociation constant, Ka. The pH at which a titratable group is half ionized.

Plasmid: A small circular DNA that replicates independently of the genome. It can be used for cloning and expression by carrying the gene of interest. It generally contains a gene coding for resistance to some antibiotics to provide a selectable marker to prevent the plasmid from getting lost from the bacterial host.

Plate number (N): An index that indicates column chromatography efficiency to determine column efficiency based on calculation in which the larger the theoretical plate number, the sharper the peaks.

Polar: Having uneven distribution of charge. A molecule or functional group is polar if its center of negative charge does not coincide with its center of positive charge.

Polarization of fluorescence: Also known as anisotropy. A measurement of the changing orientation of a molecule in space with respect to the time between the absorption and emission events.

Polymerase chain reaction (PCR): A technique for increasing the amount of DNA, or amplifying it. In the PCR reaction, oligonucleotides complementary to the ends of the desired DNA sequence are used as primers for multiple rounds of DNA synthesis.

Polyacrylamide gel electrophoresis (PAGE): A technique used to separate molecules of different charge and/or size based on their migration through a highly cross-linked gel matrix in an electric field.

Polypeptide: A peptide containing many (usually more than twenty) amino acid residues.

Posttranslational modification (PTM): Covalent modification of a protein that occurs after elongation of the polypeptide is complete.

Preferential hydration: If there is a deficiency of ligand binding, i.e., a greater affinity for water, K'ex <1, the measured binding is negative and there is preferential exclusion of the ligand and an excess of water in the protein domain.

Promiscuous: An enzyme that catalyzes two different reactions.

Prosthetic group: A coenzyme that is tightly bound to an enzyme. A prosthetic group, unlike a co-substrate, remains bound to a specific site of the enzyme throughout the catalytic cycle of the enzyme.

***Protease*:** An enzyme that catalyzes hydrolysis of peptide bonds. The physiological substrate of proteases are proteins.

Protein: A biopolymer (macromolecule) consisting of one or more polypeptide chains.

***Protein glycosylation**:* The covalent addition of carbohydrates to proteins. In N-glycosylation, the carbohydrate is attached to the amide group of the side chain of an asparagine residue. In 0-glycosylation, the carbohydrate is attached to the hydroxyl group of the side chain of a serine or threonine residue.

***Protein glycation**:* The spontaneous, nonenzymatic reaction of proteins with simple reducing sugars and related metabolites.

Protein kinase: An enzyme that catalyzes the phosphorylation of protein substrates.

Protein phosphatase: An enzyme that catalyzes the hydrolytic removal of a phosphoryl group from a phosphoprotein.

Proximity effect: The increase in the rate of nonenzymatic or enzymatic reactions attributable to high effective concentrations of reactants, which results in more frequent formation of transition states.

Pseudo-first-order reaction: A multi-reactant reaction carried out under conditions where the rate depends on the concentration of only one reactant.

Pseudogene: A gene that does not contain introns.

Ψ ***(psi):*** The angle of rotation around the bond between the α-carbon and the carbonyl carbon of a peptide group.

PTM: See posttranslational modification.

Quantum yield: Defined as the ratio of the number of photons emitted to the number adsorbed. Anything with a quantum yield greater than 0.1 could be considered a reasonable strong fluorescent molecule.

Quaternary structure: The organization of two or more polypeptide chains within a multi-subunit protein.

R group: A part of the molecule not explicitly shown in a chemical structure, such as an alkyl group or the side chain of an amino acid.

R state: The more active conformation of an allosteric protein: opposite of T state.

Ramachandran plot: A diagram plotting the protein backbone dihedral angles ϕ and Ψ for amino acids.

Racemization: The change of a chiral molecule to make a 1:1 mixture of D and L isomers. For the amino acid, this can be done by removing the proton or the α-carbon and replacing it randomly on either face.

Radius of gyration (r_g): Defined as the root mean square distance of the atoms from the center of gravity; it effectively is the radius of a uniform sphere that has the same mass and density as the protein and therefore is the average radius. It is often abbreviated r_g.

Random coil: A peptide in random coil conformation has its residues populating the ϕ and Ψ map randomly according to their relative free energies.

Rate acceleration: The ratio of the rate constant for a reaction in the presence of enzyme (k_{cat}) divided by the rate constant for that reaction in the absence of enzyme (k_n). The rate acceleration value is a measure of the efficiency of an enzyme.

Rate-determining step: The slowest step in a chemical reaction. The rate- determining step has the highest activation energy among the steps leading to the formation of a product from the substrate.

Rayleigh scattering: elastic scattering of light, or other electromagnetic radiation, by particles with a size much smaller than the wavelength of the radiation. See static light scattering (SLS).

Reaction mechanism: The step-by-step atomic or molecular events that occur during chemical reactions.

Receptor: A cell membrane protein that binds a specific ligand, leading to some cellular response.

Red shift: The position shift of a peak or signal to a longer wavelength (lower energy).

Reducing agent: A substance that loses electrons in an oxidation-reduction reaction and thereby becomes oxidized.

Reduction: The gain of electrons by a substance through transfer from another substance (the reducing agent). Reductions can take several forms, including the loss of oxygen from a compound, the addition of hydrogen to a double bond of a compound, or a decrease in the valence of a metal.

Refractive index: A parameter in optics that determines the speed by which light travels through a medium other than a vacuum.

Relative molecular mass (M_r): The mass of a molecule relative to 1/12th the mass of ^{12}C.

Renaturation: The restoration of the native conformation of biological macromolecules, usually resulting in restoration of biological activity.

Residue: A single component within a polymer. The chemical formula of a residue is that of the corresponding monomer minus the elements of water. E.g., a lysine in a protein is called a lysine residue.

***Restriction enzyme*:** A nuclease that cuts DNA at a specific short sequence.

Reverse transcriptase: A type of DNA polymerase that catalyzes the synthesis of a strand of DNA from an RNA template.

Ribonucleic acid (RNA): A polymer consisting of ribonucleotide residues joined by 3'-5' phosphodiester bonds. The sugar moiety in RNA is ribose. Genetic information contained in DNA is transcribed in the synthesis of RNA, some of which (mRNA) is translated in the synthesis of proteins.

Ribosome: The large molecular machine composed of ribosomal RNA plus protein that translates messenger RNA into protein.

RNA: See ribonucleic acid.

Ribozyme: An RNA molecule with enzymatic activity.

Rise: The distance between one residue and the next along the axis of a helical macromolecule.

Rossmann fold: A protein fold originally described by Michael Rossmann, containing a characteristic GXGXXG sequence motif, which functions to bind to nucleic acids.

S: See entropy and Svedberg unit.

Salt bridge: An electrostatic attraction between two charged groups in a protein, between Asp, Glu, or the C terminus of Lys, Arg, His, or the N terminus.

Sanger reagent: Fluorodinitrobenzene (FDNB).

Scaffold protein: A protein that acts as a bridge between two proteins, bringing them into proximity so that they bind or react together.

Schiff base: A complex formed by the reversible condensation of a primary amine with an aldehyde (an aldimine) or a ketone (ketimine).

Schlieren optics: An optical technique that allows us to see small changes in the refractivity of air and other transparent media.

Screening: The solvation of a charge by water or ions leads to an effective decrease in the charge, described by the word screening.

SDS-PAGE (sodium dodecyl sulfate polyacrylamide gel electrophoresis): A gel is made from polyacrylamide into which is incorporated sodium dodecyl sulfate (SDS). When a voltage is placed across the gel, proteins move down the gel, by a distance depending on their size. The method separates proteins by size.

Secondary structure: The regularities in local conformations within macromolecules. In proteins, secondary structure is maintained by hydrogen bonds between carbonyl and amide groups of the backbone. α-helix and β-strand are examples of secondary structures.

Second order reaction: A reaction whose rate depends on the concentration of two reactants.

Sequential reaction: An enzymatic reaction in which all the substrates must be bound to the enzyme before any product is released.

Serine protease: A protease with an active-site serine that acts as a nucleophile during catalysis.

Site-directed mutagenesis: An in vitro procedure by which one nucleotide residue in a gene is replaced by another, resulting in the production of an altered protein sequence.

SLS: see static light scattering.

Solvation: A state in which solvent molecules surround a molecule or ion.

Solvation sphere: The shell of solvent molecules that surround an ion or solute.

Specific heat: The amount of heat required to raise the temperature of 1 gram of a substance by 1°C.

Specific heat capacity change (ΔCp): The amount of heat that must be added to one unit of mass of the substance (protein) to cause an increase of one unit in temperature), normalized to the number of residues in the protein.

Specificity constant: The ratio kcat/Km. This determines how specific an enzyme is for a particular substrate and is also the apparent second-order rate constant for a bimolecular reaction (for example, an enzyme plus a single substrate).

SPR: See surface plasmon resonance.

Standard free energy change ($\Delta G°$): The free energy change for a reaction under biochemical standard state conditions.

Standard state: In chemistry, a standard state is defined so that all chemical reactions can be compared under a defined set of reference conditions. The standard state for liquids and solids is the most stable form at 298 K under 1 atm pressure.

Static light scattering (Rayleigh scattering) is a technique for measuring the intensity of scattered light and obtaining information about the average molecular weight and size of macromolecules in solution. It involves passing a laser beam through the solution and analyzing the scattered photons at various angles.

Steady state: A state in which the rate of synthesis of a compound is equal to its rate of utilization or degradation.

Stereoisomers: Compounds with the same molecular formula but different spatial arrangements of their atoms.

Stereospecificity: The ability of an enzyme to recognize and act upon a single stereoisomer of a substrate.

Stokes radius: See hydrodynamic radius.

Substrate: A reactant in a chemical reaction. In enzymatic reactions, substrates are specifically acted upon by enzymes, which catalyze the conversion of substrates to products.

Supersaturated solution: A solution in which the solute's concentration is greater than its solubility. In these circumstances, the thermodynamic preference is for the solute to either precipitate or crystallize; kinetic factors may mean that either or both processes may be slow.

Super secondary structure: A grouping of elements of regular secondary structure. These include antiparallel helix, coiled-coil, β- hairpin.

Surfactant: See detergent.

Surface plasmon resonance (SPR): An optical technique used to measure molecular interactions in real time. SPR signal is directly dependent on the refractive index of the medium on the sensor chip. The binding of biomolecules results in changes in the refractive index on the sensor surface.

Svedberg unit (S): A unit of 10^{-13} seconds used for expressing the sedimentation coefficient, a measure of the rate at which a large molecule or particle sediments in an ultracentrifuge. Large S values usually indicate large masses.

System: A part of the universe chosen for study. It will have spatial boundaries, but it may be open or closed with respect to the transfer of matter. It may or may not be thermally insulated from its surroundings. If insulated, it is an adiabatic system.

T state: The less active conformation of an allosteric enzyme; opposite to R state.

Termination codon: Also known as a stop codon. One of the three mRNA codons that code not for an amino acid but for the end of the polypeptide chain. These are UAG (amber), UAA (ochre), and UGA (umber), the last of which can also code for selenocysteine.

Tertiary structure: The compacting of polymeric chains into one or more globular units within a macromolecule. In proteins, tertiary structure is stabilized by hydrophobic interactions between side chains.

Thermodynamics: The branch of physical science that studies transformations of heat and energy.

Thin-layer chromatography: A chromatographic technique used to separate components of a mixture on a thin layer of porous material.

Time-of-flight (TOF): A mass analyzer that utilizes an electric field to accelerate generated ions through the same electrical potential and then measures the time each ion takes to reach the detector.

Tm: See melting temperature.

TOF: See time-of-flight.

Topology: The arrangement of membrane-spanning segments and connecting loops in an integral membrane protein.

Transaminase: An enzyme that catalyzes the transfer of an amino group from an α-amino acid to an α-keto acid. Transaminases require the coenzyme pyridoxal phosphate.

Transition state: An unstable high-energy arrangement of atoms in which chemical bonds are being formed or broken. Transition states have structures between those of the substrates and the products of the reaction.

Transition-state analog: A compound that resembles a transition state. Transition-state analogs bind extremely tightly to the active sites of appropriate enzymes and thus act as potent inhibitors.

Transition-state stabilization: The increased binding of transition states to enzymes relative to the binding of substrates or products. Transition-state stabilization lowers the activation energy and thus contributes to catalysis.

Tyndall effect: The scattering of a light beam by a medium containing microscopic suspended particles.

Turnover number: See catalytic constant.

Uncompetitive inhibition: The inhibition of an enzyme-catalyzed reaction by an inhibitor that binds only to the enzyme-substrate complex, not to the free enzyme.

v: See initial velocity.

v2⁻: See partial specific volume.

Van der Waals force: A weak intermolecular force produced between neutral atoms by transient electrostatic interactions. Van der Waals attraction is stronger when the atoms are separated by the sum of their Van der Waals radii; strong Van der Waals repulsion precludes a closer approach.

Van der Waals radius: The effective size of an atom. The distance between the nuclei of two nonbonded atoms at the point of maximal attraction is the sum of their Van der Waals radii.

Vmax: See maximum velocity.

Vant' Hoff equation: Relates the change in temperature (T) to the change in the equilibrium constant (K) given the enthalpy of the reaction.

X-ray crystallography: A technique used to determine secondary, tertiary, and quaternary structure of biological macromolecules. In X-ray crystallography, a crystal of the macromolecule is bombarded with X-rays, which are diffracted and then detected electronically or on a film. The atomic structure is deduced by mathematical analysis of the diffraction pattern.

Zero-order reaction: A reaction whose rate is independent of reactant concentration. The rate does not change when the concentration of the reactants increases or decreases.

Zwitterion: A compound carrying both a positive and negative charge.

Zymogen: A catalytically inactive enzyme precursor that must be modified by limited proteolysis to become enzymatically active. Also known as proenzyme.

ENDNOTES

1. Molecular mass and molecular weight are not the same, although they are used interchangeably. As a synonym of molar mass, molecular weight is defined as the mass of a sample of that compound divided by the amount of substance in that sample, measured in moles (g/mol).
2. In the early 1980s, Sidney Altman and Thomas Cech independently found that RNA molecules can function as catalysts for chemical reactions. Before then, only proteins were thought to function in this way. The catalytic RNAs were called ribozymes. Altman and Cech were awarded the 1989 Nobel Prize in chemistry for their discovery.
3. The term daltons (abbreviated Da) is in honor of John Dalton (1766–1844), who developed the atomic theory of matter.
4. Schlieren optic is one of the methods used to analyze the distribution of molecules in the analytical ultracentrifuge (Model E). The Schlieren system depends on the fact that light passing through the regions of uniform concentration is not deviated, but when passing through the regions of varying concentration it is deviated because of the change in index of refraction (which varies with concentration).
5. All fluorescence of a protein is due to tryptophan, tyrosine, and phenylalanine, unless the protein is known to contain another fluorescent component.
6. The absorption of polarized light by a chromophore is maximal when the plane of polarization is parallel to a particular axis of the chromophore, called the electric dipole moment.
7. Energy transfer from tryptophan to phosphopyridoxylated residues can be carried out using excitation at 295 nm and monitoring the fluorescence emission at 396 nm.

REFERENCES

Abbatiello SE., and Porter TJ. (1997). Anion-mediated precipitation of recombinant human morphogenetic protein-2 (rhBMP-2) is dependent upon the heparin binding N-terminal region. *Prot. Sci.* 6., 99.

Abeles RH. in *Enzyme-activated irreversible inhibitors* (N.Seiler, M.J. Jung, and J. Kosk-Weser, Eds.) Elsevier North-Holland.1978.

Abeles RH. (1981). Suicide enzyme inactivators and others. *Pure and Applied Chemistry,* 53 (1), 149–160. https://doi.org/10.1351/pac198153010149

Ackers GK., and Halvorson HR. (1974). The linkage between oxygenation and subunit dissociation in human hemoglobin. Proc Natl Acad Sci U S A. Nov;71(11):4312–6. doi: 10.1073/pnas.71.11.4312. PMID: 4530985; PMCID: PMC433872.

Ackers GK. (1979). Linked functions in allosteric proteins: an exact theory for the effect of organic phosphates on oxygen affinity of hemoglobin. *Biochemistry.* Jul 24;18(15):3372–80. doi: 10.1021/bi00582a025. PMID: 465477.

Ackers GK., and Smith FR. (1987). The hemoglobin tetramer: a three-state molecular switch for control of ligand affinity. *Annu Rev Biophys Biophys Chem.* 16:583–609. doi: 10.1146/annurev.bb.16.060187.003055. PMID: 3593506.

Adnan A., Acharya S., Alenazy LA, de Las Vecillas L., Giavina Bianchi P., Picard M, Calbache-Gil L, Romero-Pinedo S., Abadí A-Molina AC., Kerr W., Pedicone C., Nagai J., Hollers E., Dwyer D., and Castells M. (2003). Multistep IgE Mast Cell Desensitization Is a Dose- and Time-Dependent Process Partially Regulated by SHIP-1 (2003) *J Immunol.* 15;210(6):709–720. doi: 10.4049/jimmunol.2100485. PMID: 36881903; PMCID: PMC9986054.

Aeschbach R., Amadò R., and Neukom H. (1976). Formation of dityrosine cross-links in proteins by oxidation of tyrosine residues. *Biochim Biophys Acta.* Aug 9;439(2):292–301. doi: 10.1016/0005-2795(76)90064-7. PMID: 952964.

Alberty RA., and Di Cera E., and Wyman J (1901–1995) a biographical memory. In Biographical memories, volume 83. The National Academies Press Washington, DC, 2003.

Andreu JM. and Timasheff SN. (1981). The Ligand- and Microtubule assembly- induced GTPase Activity of Purified Calf Brain Tubulin. *Arch Biochem. Biophys.* 211 (1):151–157.

Andreu JM., and Timasheff SN. (1982). Conformational states of tubulin liganded to Colchicine, Tropolone Methyl Ether, and Podophyllotoxin. *Biochemistry*: 24: 6465–6476.

Andreu JM., Gorbunoff MJ, Lee JC., and Timasheff SN. (1984). Interaction of tubulin with bifunctional colchicine analogues: an equilibrium study. *Biochemistry.* Apr 10;23(8):1742–52. doi: 10.1021/bi00303a025. PMID: 6722122.

Andreu JM., Gorbunoff, MJ., Medrano, FJ., Rossi, M., and Timasheff SN. (1991). Mechanism of Colchicine Binding to Tubulin. Tolerance of Substituents in Ring C' of Biphenyl Analogues. *Biochemistry* 30:3777–3786.

Anfinsen CB. (1973). Principles that govern the folding of protein chains. *Science.* Jul 20;181(4096):223–30. doi: 10.1126/science.181.4096.223. PMID: 4124164.

Anholt R., Lindstrom, J., and Montal M. (1980). Functional equivalence of monomeric and dimeric forms of purified acetylcholine receptors from Torpedo californica in reconstituted lipid vesicles. *European Journal of Biochemistry,* 109(2): 481–487.

Antollini SS., and Barrantes FJ. (2007). Laurdan studies of membrane lipid-nicotinic acetylcholine receptor protein interactions. *Methods Mol Biol.* 400: 531–42. doi: 10.1007/978-1-59745-519-0_36. PMID: 17951758.

Arakawa T., and Timasheff SN. (1982). Stabilization of protein structure by sugars. *Biochemistry.* Dec 7;21(25):6536–44. doi: 10.1021/bi00268a033. PMID: 7150574.

Arakawa T., and Timasheff SN. (1985). Mechanism of Poly (ethylene glycol) interaction with proteins. *Biochemistry.* 24:6756–6761).

Arakawa T., Bhat R., and Timasheff SN.(1990). Preferential interactions determine protein solubility in three-component solutions: The $MgCl_2$ system. *Biochemistry* 29: 1914–1923.

Arnautova YA., Vorobjev YN., Vila JA., and Scheraga HA. (2009). Identifying native-like protein structures with scoring functions based on all-atom ECEPP force fields, implicit solvent models, and structure relaxation. *Proteins.* Oct;77(1):38–51. doi: 10.1002/prot.22414. PMID: 19384995; PMCID: PMC4502597.

Aronstam RS. In Progress in Cholinergic Biology: Model Cholinergic Synapses (Hanin, T., and Goldberg, A.M. eds) pp45–77, Raven Press, New York. 1982.

Ashraf GM., Greig NH., Khan TA., Hassan I., Tabrez S., Shakil S., Sheikh IA., Zaidi SK., Akram M, Jabir NR, Firoz CK, Naeem A, Alhazza IM., Damanhouri GA., and Kamal MA. (2014). Protein misfolding and aggregation in Alzheimer's disease and type 2 diabetes mellitus. *CNS Neurol Disord Drug Targets.*13(7):1280–93. doi: 10.2174/1871527313666140917095514. PMID: 25230234; PMCID: PMC5193501.

REFERENCES

Aune K., Lee J., Prakash V., Bhat R., Andreu JM., Monasterio O., Pérez-Ramírez B., Shearwin K., Arakawa T., Carpenter J., Crowe J., Crowe L., Somero G., Gagnon P., and Charles MT. (2021). A tribute to Dr. Serge N. Timasheff, our mentor. *Biophys Rev.* Jul 6;13(4):459–484. doi: 10.1007/s12551-021-00814-9. PMID: 34471434; PMCID: PMC8355303.

Bae W., Yoon TY., and Jeong C. (2021) Direct evaluation of self-quenching behavior of fluorophores at high concentrations using an evanescent field. *Plos one.* 16(2):e0247326. DOI: 10.1371/journal.pone.0247326. PMID: 33606817; PMCID: PMC7895399.

Bah A., Vernon RM., Siddiqui Z., Krzeminski M., Muhandiram R., Zhao C., Sonenberg N., Kay LE., and Forman-Kay JD. (2015). Folding of an intrinsically disordered protein by phosphorylation as a regulatory switch. *Nature.* Mar 5;519(7541):106–9. doi: 10.1038/nature13999. Epub 2014 Dec 22. PMID: 25533957.

Baldwin RL., and Eisenberg D. (In: Protein Engineering (Oxender DE and Fox CFEds) Liss, New York p127–148. 1987.

Baldwin RL. (1996). How Hofmeister ion interactions affect protein stability. *Biophys J.* 71(4):2056–63. doi: 10.1016/S0006-3495(96)79404-3. PMID: 8889180; PMCID: PMC1233672.

Baldwin., RL. (2002). John Schellman, and his scientific work. Biophysical Chemistry 101–102: 9–13.

Barrantes FJ. (1983). Recent developments in the structure and function of the acetylcholine receptor. *Int Rev Neurobiol.* 24:259–341. doi: 10.1016/s0074-7742(08)60224-x. PMID: 6317598.

Barrantes FJ., 1988. Muscle endplate cholinoreceptors. *Pharmacology & therapeutics*, 38(3): 331–385.

Bascos NA., and Landry SJ. (2019). A History of Molecular Chaperone Structures in the Protein Data Bank. *International Journal of Molecular Sciences.* 20 (24): 6195.

Bedbrook JR., and Bogorad L. (1976) Endonuclease recognition

sites mapped on Zea mays chloroplast DNA. *Proc Natl. Acad Sci USA* 73: 4309–4313

Bedbrook R, Kolodner R., and Bogorad L. (1977) Zea mays chloroplast ribosomal RNA genes are part of a 22,000 base pair inverted repeat. *Cell* 11: 739–749.

Benjin X., and Ling L. (2020). Developments, applications, and prospects of cryo-electron microscopy. *Protein Sci.* Apr;29(4):872–882. doi: 10.1002/pro.3805. Epub 2019 Dec 26. PMID: 31854478; PMCID: PMC7096719.

Bertz M., and Rief M. (2008). Mechanical unfoldons as building blocks of maltose-binding protein. *J Mol Biol.* Apr 25;378(2):447–58. doi: 10.1016/j.jmb.2008.02.025. Epub 2008 Feb 21. PMID: 18355837.

Bhat R., and Timasheff SN. (1992). Steric exclusion is the principal source of the preferential hydration of proteins in the presence of polyethylene glycols. *Protein Sci.* Sep;1(9):1133–43. doi: 10.1002/pro.5560010907. PMID: 1304392; PMCID: PMC2142180.

Bellmaine S., Schnellbaecher A., and Zimmer A.(2020). Reactivity and degradation products of tryptophan in solution and proteins. *Free Radical Biology and Medicine.* 160, 20 November 2020, Pages 696–718.

Bernal J., Fankuchen I. and Perutz, M. (1938). An X-Ray Study of Chymotrypsin and Hemoglobin. *Nature* 141: 523–524.

Bichsel VE., Liotta LA., and Petricoin EF 3rd. (2001). Cancer proteomics: from biomarker discovery to signal pathway profiling. *Cancer J.*7: 69–78.

Bilbrough T., Piemontese E., and Seitz O. (2002) Dissecting the role of protein phosphorylation: a chemical biology toolbox. *Chem Soc Rev.* 4;51(13):5691–5730. doi: 10.1039/d1cs00991e. PMID: 35726784.

Blackstock WP.,and Weir MP. (1999). Proteomics: quantitative and physical mapping of cellular proteins. *Trends Biotechnol.*17:121–127.

Blanco MA., Perevozchikova T., Martorana V., Manno M., and Roberts, CJ. (2014). Protein–protein interactions in dilute to concentrated solutions: α-Chymotrypsinogen in acidic conditions." *The Journal of physical chemistry B* 118 (22): 5817–5831.

Bobrosk, K., Hug GL., Pogocki D., Marciniak B., and Schoeneich, CH. (2007). Stabilization of disulfide cations through complexation with the peptide bond: Mechanisms relevant to oxidation of proteins containing multiple methionine residues. *J. Phy. Chem B* 111: 9608–9620.

Bode W., Schwager P., and Huber R. (1978). Transition of bovine trypsinogen to a trypsin-like state upon strong ligand binding. The refined crystal structures of the bovine trypsinogen-pancreatic trypsin inhibitor complex and of its ternary complex with Ile-Val at 1.9 Å resolution. *J. Mol. Biol.* 118: 99–112.

Bogorad L. (2003) Photosynthesis research: advances through molecular biology – the beginnings, 1975–1980's and on. *Photosynth Res* 76: 13–33. (http://www.life.uiuc.edu/ govindjee/Part2/02_Bogorad.pdf).

REFERENCES

Bolen DW., and Santoro MM. (1988). Unfolding free energy changes determined by the linear extrapolation method. 2. Incorporation of delta G. degree. N-U values in a thermodynamic cycle. *Biochemistry,* 27(21): 8069–8074.

Borgia A., Borgia MB., Bugge K., Kissling VM., Heidarsson PO., Fernandes CB., Sottini A., Soranno A., Buholzer KJ., Nettels D., Kragelund BB., Best RB., and Schuler B. (2018). Extreme disorder in an ultrahigh-affinity protein complex. *Nature.* Mar 1;555(7694):61–66. doi: 10.1038/nature25762. Epub 2018 Feb 21. PMID: 29466338; PMCID: PMC6264893.

Borowitzka LJ. ,and Brown AD. (1974). The salt relations of marine and halophilic species of the unicellular green alga, Dunaliella . *Arch. Microbiol.* 96: 37–52 https://doi.org/10.1007/BF00590161

Borowitzka LJ. Glycerol and other carbo- hydrate osmotic effectors. In Transport Processes - Iono- and Osmoregulation, pp, 437–453. Edited by R. Gilles and M. Gilles-Baillien. Berlin: Springer-Verlag. 1985.

Borroni V., and Barrantes FJ. (2011). Cholesterol modulates the rate and mechanism of acetylcholine receptor internalization. *J Biol Chem.* May 13;286(19):17122–32. doi: 10.1074/jbc.M110.211870. Epub 2011 Feb 28. PMID: 21357688; PMCID: PMC3089556.

Boström M, Tavares FW., Finet S., Skouri-Panet F., Tardieu A., and Ninham BW. (2005). Why forces between proteins follow different Hofmeister series for pH above and below pI. *Biophys Chem.* 117(3):217–24.

Brackes JP., Berg DK., and Hall ZW. (1975). *Cold Spring Harbor Symp. Quant. Biol.* 40: 253–262.

Bradshaw, JM. and Waksman G. (1998). Calorimetric investigation of proton linkage by monitoring both the enthalpy and association constant of binding: application to the interaction of the Src SH2 domain with a high-affinity tyrosyl phosphopeptide. *Biochemistry,* 37(44):15400–15407.

Britannica, The Editors of Encyclopedia. Optical activity. Encyclopedia Britannica, 24 Jan. 2018, https://www.britannica.com/science/optical-activity. Accessed 14 August 2023.

Browne WJ., North AC., Phillips DC., Brew K., Vanaman TC., Hill RL. (1969). A possible three-dimensional structure of bovine alpha-lactalbumin based on that of hen's egg-white lysozyme. *J Mol Biol.* May 28;42(1):65–86. doi: 10.1016/0022-2836(69)90487-2. PMID: 5817651.

Bryngelson JD., and Wolynes PG. (1987). Spin glasses and the statistical mechanics of protein folding. *Proc Natl Acad Sci U S A*. Nov;84(21):7524–8. doi: 10.1073/pnas.84.21.7524.

Bryngelson JD., Onuchic JN., Socci ND., and Wolynes, PG. (1995). Funnels, pathways, and the energy landscape of protein folding: A synthesis. *Proteins* 21: 167–195.

Buckler DR., Haas E., and Scheraga HA. (1993). C-terminal labeling of ribonuclease A with an extrinsic fluorescent probe by carboxypeptidase Y-catalyzed transpeptidation in the presence of urea. *Anal Biochem*. Feb 15;209(1):20–31. doi: 10.1006/abio.1993.1078. PMID: 8465956.

Burt DW., and Law AS. (1994). Evolution of the transforming growth factor-β superfamily. *Prog. Growth Factor Res.* 5: 99–118.

Burzio LO., Riquelme PT., and Koide SS. (1979). ADP ribosylation of rat liver nucleosomal core histones. *J Biol Chem.* 1979 254(8):3029–37. PMID: 218926.

Cabantchik IZ., Balshin M., Breuer W., and Rothstein, A. (1975) Pyridoxal phosphate. An anionic probe for protein amino groups exposed on the outer and inner surfaces of intact human red blood cells. *Journal of Biological Chemistry*, 250: (13), 5130–5136.

Callaway E. (2015). The revolution will not be crystallized: a new method sweeps through structural biology. *Nature.* Sep 10;525(7568):172–4. doi: 10.1038/525172a. PMID: 26354465.

Callaway E. (2020). Revolutionary cryo-EM is taking over structural biology. *Nature.* Feb;578(7794):201. doi: 10.1038/d41586-020-00341-9. PMID: 32047310.

Cann JR., and Hinman ND. (1976). Hummel-Dreyer gel chromatography procedure as applied to ligand-mediated association. *Biochemistry* 15: 4614–4622.

Carlier MF., Didry D., Simon C., and Pantaloni D. (1989). Mechanism of GTP hydrolysis in tubulin polymerization: characterization of the kinetic intermediate microtubule-GDP-Pi using phosphate analogues. *Biochemistry.* Feb 21;28(4):1783–91. doi: 10.1021/bi00430a054. PMID: 2719934.

Carlin G., and Bang NU. (1980). Enhancement of plasminogen activation and hydrolysis of purified fibrinogen and fibrin by dextran 70. *Thromb Res.*;19(4–5):535–46. doi: 10.1016/0049-3848(80)90026-2. PMID: 6160642.

REFERENCES

Carmona A., Schwager S., Juliano M., Juliano L., and Sturrok, ED.(2006). A continuous fluorescence resonance energy transfer angiotensin I-converting enzyme assay. *Nat Protoc* 1: 1971–1976. https://doi.org/10.1038/nprot.2006.306

Casassa EF. and Eisenberg H., (1960). On the Definition of Components in Solutions Containing Charged Macromolecular Species1. *The Journal of Physical Chemistry,* 64(6):753–756.

Casassa EF. and Eisenberg H., (1961). Partial Specific Volumes and Refractive Index Increments in Multicomponent Systems1. *The Journal of Physical Chemistry,* 65(3): pp.427–433.

Casassa EF. and Eisenberg H., (1964). Thermodynamic analysis of multicomponent solutions. *Advances in protein chemistry,* 19:287–395.

Cech TR. (2002). Ribozymes, the first 20 years. *Biochem Soc Trans.* Nov;30(Pt 6):1162–6. doi: 10.1042/bst0301162. PMID: 12440996.

Chait BT., and Kent SB. (1992). Weighing naked proteins: practical, high-accuracy mass measurement of peptides and proteins. *Science.* 257:1885–1894.

Changeux J. (2012). The Nicotinic Acetylcholine Receptor: The Founding Father of the Pentameric Ligand-gated Ion Channel Superfamily. *J.Biol.Chem..* 287(48):40207–40215. http://dx.doi.org/10.1074/jbc.r112.407668

Chari R., Jerath K., Badkar AV., and Kalonia DS. (2009). Long- and short-range electrostatic interactions affect the rheology of highly concentrated antibody solutions. *Pharm Res.* Dec;26(12):2607–18. doi: 10.1007/s11095-009-9975-2. Epub 2009 Oct 1. PMID: 19795191.

Chelius D., Jing K., Lueras A., Rehder DS., Dillon TM., Vizel A., Rajan R., Li T., Treuheit MJ. and Bondarenko PV. (2006). Formation of pyroglutamic acid from N-terminal glutamic acid and in immunoglobulin gamma antibodies. *Anal Chem.* 78: 2370–2336.

Chen H., Gu F., and Huang Z. (2006). Improved Chou-Fasman method for protein secondary structure prediction. *BMC Bioinformatics* 7 (Suppl 4), S14. https://doi.org/10.1186/1471-2105-7-S4-S14

Chiou SH., and Wu CY. (2011). Clinical proteomics: current status, challenges, and future perspectives. *Kaohsiung J Med Sci.* 27(1):1–14.

Chiu W., Knapek E., and Dietrick I.(1981). Electron radiation damage of a thin protein crystal at 4K. *Ultramicroscopy.* 6:291–295.

Chou PY., and Fasman GD. (1974). Prediction of protein conformation. *Biochemistry.* 13 (2): 222–245. doi:10.1021/bi00699a002. PMID 4358940.

Chou PY., and Fasman GD. (1978). Empirical Predictions of Protein Conformation. *Annual Review of Biochemistry.* 47:251–276. https://doi.org/10.1146/annurev.bi.47.070178.001343

Clarke JH., and Martinez-Carrion M. (1986). Labeling of functionally sensitive sulfhydryl-containing domains of acetylcholine receptors from Torpedo californica membranes. *J Biol Chem.* 261(22):10063–72. PMID: 3733702.

Clarke J., Garcia-Borron JC., and Martinez-Carrion M. (1987). (1-Pyrene)sulfonyl azide: a fluorescent probe for measuring the transmembrane topology of acetylcholine receptor subunits *Arch. Biochem. Biophys.* 256: 101–109

Claudio T., Ballivet M., Patrick J. and Heinemann, S., (1983). Nucleotide and deduced amino acid sequences of Torpedo californica acetylcholine receptor gamma subunit. *Proceedings of the National Academy of Sciences, USA* 80(4): 1111–1115.

Conti-Tronconi BM., Hunkapiller MW., Lindstrom J., and Raftery, MA. (1982) Subunit structure of the acetylcholine receptor from Electrophorus electricus. *Proc Natl Acad Sci U S A.* 79:(21):6489-6493.

Correia IR. (2010). Stability of IgG isotypes in serum. *MAbs.* May-Jun;2(3):221–32. doi: 10.4161/mabs.2.3.11788. Epub 2010 May 16. PMID: 20404539; PMCID: PMC2881250.

Criado M., Eibl H., and Barrantes, FJ. (1982). Effects of lipids on acetylcholine receptor. Essential need of cholesterol for maintenance of agonist-induced state transitions in lipid vesicles. *Biochemistry* 21:15, 3622–3629.

Coen DM., Bedbrook JR., Bogorad L. and Rich A. (1977) Maize chloroplast DNA fragment encoding the large subunit of ribulose bisphosphate carboxylase. *Proc Natl Acad Sci USA.* 74: 5487–5491

Cohen G. and Eisenberg H., (1968). Deoxyribonucleate solutions: Sedimentation in a density gradient, partial specific volumes, density and refractive index increments, and preferential interactions. *Biopolymers: Original Research on Biomolecules,* 6(8):1077–1100.

Cooper A., Nutley MA., and Wadood A. Differential Scanning Calorimetry, Chapter 11, In: Protein-Ligand Interactions. Hydrodynamics and calorimetry. Edited by Harding, S.E., and Chowdhry, B.Z. Oxford University press, New York, p287–318. 2020.

Cournoyer JJ., Lin C., and O'Connor, PB. (2006) Detecting deamidation products by electron capture dissociation. *Anal. Chem.* 78: 1264–1271.

Cournoyer JJ., Lin C., Bowman MJ., and O'Connor P. B. (2007). Quantitating the relative abundance of isoaspartyl residues in deamidated proteins by electron capture dissociation. *J. Am. Mass Spectrom*, 18: 48-56.

Cox WM., and Wolfenden JH., (1934). The viscosity of strong electrolytes measured by a differential method. *Proceedings of the Royal Society of London. Series A.*, 145(855):475–488.

Crabtree MD., Mendonça CATF., Bubb QR., and Clarke J., (2018). Folding and binding pathways of BH3-only proteins are encoded within their intrinsically disordered sequence, not templated by partner proteins. *J. Biol. Chem.* 293, 9718–9723.

Daniel E., and Weber G. 1966. Cooperative effects in binding by bovine serum albumin. I. The binding of 1-anilino-8-naphthalenesulfonate. Fluorimetric titrations. *Biochemistry.* 5:1893–1900.

Das RK., and Pappu RV. (2013). Conformations of intrinsically disordered proteins are influenced by linear sequence distributions of oppositely charged residues. *Proc. Natl. Acad. Sci. U.S.A.* 110: 13392–13397.

Debye P., and Hückel E. (1923). Zur Theorie der Elektrolyte. I. Gefrierpunktserniedrigung und verwandte Erscheinungen. Physikalische Zeitschrift. 24: 185–206.

Drescher DG, Ramakrishnan NA, Drescher MJ. (2009). Surface plasmon resonance (SPR) analysis of binding interactions of proteins in inner-ear sensory epithelia. *Methods Mol Biol.*493:323–43. doi: 10.1007/978-1-59745-523-7_20. PMID: 18839357; PMCID: PMC2864718.

De Chadarevian S. (2018). John Kendrew and myoglobin: Protein structure determination in the 1950s. *Protein Sci.* 27: 1136–1143.

Demeler B., and van Holde KE. (2004). Sedimentation velocity analysis of highly heterogeneous systems. *Analytical Biochemistry*, 335(2), 279–288

Dickerson RE. (2008). Max Perutz and the secret of life, by Georgina Ferry. *Protein Sci.* (2):377–9. doi: 10.1110/ps.073363908. PMCID: PMC2222719.

Dill KA., and Chan HS. (1997) From Levinthal to pathways to funnels. *Nat Struct Biol.* 4:10–19.

Dinman JD. (2012). Control of gene expression by translational recording. *Advances in Protein Chemistry and Structural Biology.* 86: 129–149.

Drobnak I., De Jonge N., Haesaerts S., Vesnaver G., Loris R., and Lah J (2013). Energetic basis of uncoupling folding from binding for an intrinsically disordered protein. *J. Am. Chem. Soc.* 135, 1288–1294.

Dubochet J., Adrian M., and Schultz P. (1988). Cryo-electron microscopy of vitrified specimens. *Q. Rev. Biophys.* 21:129–228.

Durup J. (1998) On Levinthal paradox and the theory of protein folding. *Journal of Molecular Structure:* THEOCHEM. 424, Issues 1–2, 9 February, 157–169.

Dwyer BP. (1988). Evidence for the extramembranous location of the putative amphipathic helix of acetylcholine receptor. *Biochemistry,* 27(15): 5586–5592.

Eberharter A., and Becker PB. (2002). Histone acetylation: a switch between repressive and permissive chromatin. Second in review series on chromatin dynamics. *EMBO Rep.* 3(3):224–9. doi: 10.1093/embo-reports/kvf053. PMID: 11882541; PMCID: PMC1084017.

Ecker DM., Jones SD. and Levine HL. (2015). The therapeutic monoclonal antibody market. MAbs 7 (1): 9–14.

Edison A. (2001). Linus Pauling and the planar peptide bond. *Nat Struct Mol Biol* 8, 201–202. https://doi.org/10.1038/84921

Edsall JT. (1935). Apparent Molal Heat Capacities of Amino Acids and Other Organic Compounds *J. Am. Chem. Soc.* 57(8):1506–1507

Edsall JT., and Sockmayer W. (1980). George Scatchard 1892–1973. A Biographical Memoir. *National Academy of Sciences*, Washington DC p 333–378

Eftink MR and Ghiron CA. (1976a). Fluorescence quenching of indole and model micelle systems *The Journal of Physical Chemistry* 80 (5), 486–493. DOI: 10.1021/j100546a014

Eftink MR, Ghiron CA.(1976b). Exposure of tryptophanyl residues in proteins. Quantitative determination by fluorescence quenching studies. *Biochemistry.* Feb 10;15(3):672–80. doi: 10.1021/bi00648a035. PMID: 1252418.

Eger R., and Rifkin DB., (1977). The preparation and use of pyridoxal [32P] phosphate as a labeling reagent for proteins on the outer surface of membranes. *Biochimica et Biophysica Acta (BBA)-Biomembranes*, 470(1):70–83.

Eisenberg H., Cohen G. (1968). An interpretation of the low-angle X-ray scattering of DNA solutions. *J. Mol Biol.* Oct 28;37(2):355–62. doi: 10.1016/0022-2836(68)90274-x. PMID: 5760499.

Eichler J. (2019). Protein glycosylation. Curr Biol. Apr 1;29(7): R229–R231. doi: 10.1016/j.cub.2019.01.003. PMID: 30939300.

Ekins S., Mestres J., and Testa B. (2007). In silico pharmacology for drug discovery: applications to targets and beyond. *Br J Pharmacol.* Sep;152(1):21–37. doi: 10.1038/sj.bjp.0707306. Epub 2007 Jun 4. PMID: 17549046; PMCID: PMC1978280.

REFERENCES

Ellis RJ. (1979). The most abundant protein in the world. *Trends Biochem. Sci.* 4: 241–244.

Engelborghs Y. (1998). General features of the recognition by tubulin of colchicine and related compounds. *Eur Biophys J.* 27(5):437–45. doi: 10.1007/s002490050154. PMID: 9760725.

Englander SW., Downer, NW., and Teitelbaum H.(1972). Hydrogen exchange. *Annual Review of Biochemistry*, 41(1), pp.903–924.

Farach Jr HA., Mattingly Jr, JR. and Martinez-Carrion, M., (1983). Specific labeling of the active site of cytosolic aspartate aminotransferase through the use of a cofactor analog N-(bromoacetyl) pyridoxamine. *Biochemistry*, 22(5):1034–1039.

Feng H., Zhou, Z., and Bai, Y. (2005). A protein folding pathway with multiple folding intermediates at atomic resolution. *Proceedings of the National Academy of Sciences*, 102(14), 5026–5031. https://doi.org/10.1073/pnas.0501372102

Fersht AR. (1989) Structure and Mechanism in Protein Science. W. H. Freeman and Company, New York.

Ferry G. Max Perutz and the Secret of Life. Woodbury, New York: Cold Spring Harbor Laboratory Press, ix + 368 pp, 1st edition (October 30, 2007)

Fine JL., and Pearlman RE. (2023). On the origin of life: an RNA-focused synthesis and narrative. *RNA.* Aug;29(8):1085–1098. doi: 10.1261/rna.079598.123. Epub 2023 May 4. PMID: 37142437; PMCID: PMC10351881.

Finer-Moore J., and Stroud RM. (1984). Amphipathic analysis and possible formation of the ion channel in an acetylcholine receptor. *Proc Natl Acad Sci U S A.* Jan;81(1):155–159. doi: 10.1073/pnas.81.1.155. PMID: 6320162; PMCID: PMC344629.

Főrster T. (1959). 10th Spiers Memorial Lecture. Transfer mechanisms of electronic excitation. *Disc. Faraday Soc.* 27: 7–17.

Fox SW., and Harada, K. (1958). The thermal copolymerization of amino acids to a product resembling protein. *Science* 128: 1214.

Fox SW. (1964). Thermal copolymerization of amino acids and production of formed microparticles on lava. *Nature* 201: 336–337.

Fox SW., and Harada, K. (1960). The thermal copolymerization of amino acids common to protein. *Journal of American Chemical Society 82:* 3745–3751.

Fox SW., and Windsor CR. (1970). Synthesis of amino acids by the heating of formaldehyde and ammonia. *Science* 170: 984–986.

Franson J., Florin-Robertsson E., Axelson K., and Nyhlen C. (1996). Oxidation of human insulin-like growth factor I in formulation studies: Kinetics of methionine oxidation in aqueous solution and in solid state. *Pharm Res.*13: 1252–1257.

Franson J., and Hegeman, A. (1996). Oxidation of insulin-like growth factor I in formulation studies. II Effects of oxygen, visible light, and phosphate on methionine oxidation in aqueous solution and evaluation of possible mechanisms. *Pharm Res.*13: 1476–1481.

Frequently Asked Questions on Patents and Exclusivity. *Food and Drug Administration.* February 5, 2020. https://www.fda.gov/drugs/development-approval-process-drugs/frequently-asked-questions-patents-and-exclusivity#What_is_the_difference_between_patents_a

Friend SH., and Gurd FR. (1979a) Electrostatic stabilization in myoglobin. Interactive free energies between individual sites. *Biochemistry.* Oct 16;18(21):4620–30. doi: 10.1021/bi00588a024. PMID: 40593.

Friend SH., and Gurd FR. (1979b). Electrostatic stabilization in myoglobin. pH dependence of summed electrostatic contributions. *Biochemistry.* Oct 16;18(21):4612–9. doi: 10.1021/bi00588a023. PMID: 40592.

Fuxreiter M. (2018). Fold or not to fold upon binding-Does it really matters? *Curr. Opin. Struct. Biol.*, 54: 19–25.

Garidel P., Blume A., and Wagner M. (2015). Prediction of colloidal stability of high concentration protein formulations. *Pharm Dev Technol.* May;20(3):367–74. doi: 10.3109/10837450.2013.871032. Epub 2014 Jan 6. PMID: 24392929.

Garcia-Borron JC., Bieber AL., and Martinez-Carrion M. (1987). Reductive methylation as a tool for the identification of the amino groups in alpha-bungarotoxin interacting with nicotinic acetylcholine receptor. *Biochemistry.* 26(14):4295–303. doi: 10.1021/bi00388a017. PMID: 3663591.

Garcia-Mira MM, Sadqi M., Fischer N., Sanchez-Ruiz JM., and Muñoz V. (2002). Experimental identification of downhill protein folding. *Science.* Dec 13;298(5601):2191–5. doi: 10.1126/science.1077809. PMID: 12481137.

Garnier J., Osguthorpe DJ., and Robson B. (1978). Analysis of the accuracy and implications of simple methods for predicting the secondary structure of globular proteins. *J Mol Biol.* Mar 25;120(1):97–120. doi: 10.1016/0022-2836(78)90297-8. PMID: 642007.

Garnier J. (2017). Where the Name "GOR" Originates: A Story. *Methods Mol Biol.* 1484:1–5. doi: 10.1007/978-1-4939-6406-2_1. PMID: 27787815.

Gehlen MH. (2020). The centenary of the Stern-Volmer equation of fluorescence quenching: From the single line plot to the SV quenching map. Journal of Photochemistry and Photobiology C: *Photochemistry Reviews* 42: 100338

Gekko K., and Timasheff SN. (1981). Mechanism of protein stabilization by glycerol: preferential hydration in glycerol-water mixtures. *Biochemistry.* Aug 4;20(16):4667–l76. doi: 10.1021/bi00519a023. PMID: 7295639.

Geiger T., and Clarke S. (1987). Deamidation, isomerization, and racemization at asparaginyl and aspartyl residues in peptides. Succinimide-linked reactions that contribute to protein degradation. *Journal of Biological Chemistry,* 262(2): 785–794. https://doi.org/10.1016/S0021-9258(19)75855-4

Gill SJ., Dec SF., Olofsson G., and Wadsoe I. (1985). Anomalous heat capacity of hydrophobic solvation. *J. Phys. Chem.* 89, (17): 3758–3761.

Glaeser RM.(2008). Retrospective: radiation damage and its associated "information limitations" *J. Struct. Biol.* 163:271–276.

Goetz R., Koster R., Winkler C., Raulf F., Lottspeich F., Schartl M., and Thoenen H. (1994). Neurotrophin-6 is a new member of the nerve growth factor family. *Nature* 372: 266–269.

Goh GK., Dunker AK., and Uversky VN. (2009). Protein intrinsic disorder and influenza virulence: the 1918 H1N1 and H5N1 viruses. *Virol J.* Jun 3; 6:69. doi: 10.1186/1743-422X-6-69. PMID: 19493338; PMCID: PMC2701943.

Goldbeter A. (2003). Ilya Prigogine (1917–2003) *Journal of Biosciences,* 28 (6): 657–59.

Gonzalez-Ros JM., Paraschos A., and Martinez-Carrion, M. (1980). Characterization of acetylcholine receptor isolated from Torpedo californica electroplax through the use of an easily removable detergent, beta-D-octylglucopyranoside. *Proc. Natl. Acad. Sci. USA.* 77: 1796–1800

Goodson HV., and Jonasson EM. (2018). Microtubules and Microtubule-Associated Proteins. *Cold Spring Harb Perspect Biol.* Jun 1;10(6):a022608. doi: 10.1101/cshperspect. a022608. PMID: 29858272; PMCID: PMC5983186.

Goswami S., Wang W., Arakawa T., and Ohtake S. (2013). Developments and Challenges for mAb-Based Therapeutics. *Antibodies.* 2(3):452–500. https://doi.org/10.3390/antib2030452

Gräff J., and Tsai, LH. (2013). Histone acetylation: molecular mnemonics on the chromatin. *Nat Rev Neurosci* 14: 97–111. https://doi.org/10.1038/nrn3427

Grandy Jr, W. Entropy and the Time Evolution of Macroscopic Systems (International Series of Monographs on Physics). Oxford University Press; Reprint edition (July 22, 2012) 224 pages. 2012.

Greene R F., and Pace C N. (1974). Urea and guanidine hydrochloride denaturation of ribonuclease, lysozyme, α-chymotrypsin, and b-lactoglobulin. *Journal of Biological Chemistry*, 249(17): 5388–5393.

Green R., Krause J., Ptak S., Briggs AW., Ronan MT., Simons J.F., Du L., Egholm M., Rothberg J.M., Paunovic M., and Pääbo S. (2006). Analysis of one million base pairs of Neanderthal DNA. *Nature* 444, 330–336. https://doi.org/10.1038/nature05336

Gurney RW. Ionic Processes in Solution McGraw-Hill, New York. 1953.

Gutfreund H. Enzymes: Physical principles. John Wiley and sons, Bristol 1972.

Guy HR. (1984). A structural model of the acetylcholine receptor channel based on partition energy and helix packing calculations. *Biophys J.* Jan;45(1):249–61. doi: 10.1016/S0006-3495(84)84152-1. PMID: 6324907; PMCID: PMC1435233.

Guziewicz NG., Trierweiler G., Johnson K., Simler R., and Pérez-Ramírez B. Probing the solution behavior of therapeutic proteins by differential scanning calorimetry: Applications to pre-formulation and formulation development. In: Current Trends in Microcalorimetry. Boston MA, Microcal, LLC. 2007.

Guziewicz N., Best A., Pérez-Ramírez B., and Kaplan DL. (2011). Lyophilized silk fibroin hydrogels for the sustained local delivery of monoclonal antibodies. *Biomaterials.* 32: 2642–2650.

Guziewicz NA., Massetti AJ., Pérez-Ramírez BJ., and Kaplan DL. (2013). Mechanisms of monoclonal antibody stabilization and release from silk biomaterials. *Biomaterials.* Oct;34(31):7766–75. doi: 10.1016/j.biomaterials.2013.06.039. Epub 2013 Jul 13. PMID: 23859659; PMCID: PMC3756937.

Habermann SM., and Murphy KP. (1996). Energetics of hydrogen bonding in proteins: a model compound study. *Protein Sci.* 5: 1229–39.

Hadži S., Loris R., and Lah J. (2021). The sequence-ensemble relationship in fuzzy protein complexes. *Proc Natl Acad Sci U S A.* Sep 14;118(37):e2020562118. doi: 10.1073/pnas.2020562118. PMID: 34504009; PMCID: PMC8449401.

Hanukoglu I., (2015). Proteopedia: Rossmann Fold: A Beta-Alpha-Beta Fold at Dinucleotide Binding Sites. *Biochemistry and Molecular Biology Education.*43: 206–209.

Hao Z., Jin DY., Stafford DW., and Tie JK. (2020). Vitamin K-dependent carboxylation of coagulation factors: insights from a cell-based functional study. *Haematologica.* Aug;105(8):2164–2173. doi: 10.3324/haematol.2019.229047. Epub 2019 Oct 17. PMID: 31624106; PMCID: PMC7395276.

Han Y., Chaudhary AG., Chordia MD., Sackett DL., Pérez-Ramírez B., Kingston DG., and Bane S. (1996). Interaction of a fluorescent derivative of paclitaxel (Taxol) with microtubules and tubulin-colchicine. *Biochemistry.* Nov 12;35(45):14173–83. doi: 10.1021/bi960774l. PMID: 8916903.

Han X., Jin M., Breuker K., and McLafferty FW.(2006). Extending top-down mass spectrometry to proteins with masses greater than 200 kilodaltons. *Science.* 314:109–112.

Harding SE. and Rowe AJ. (2010). Insight into protein-protein interactions from analytical ultracentrifugation. *Biochemical Society Transactions*, 38(4), 901–907

Harkness DR. (1976). Cyanate and sickle-cell anemia. *Trends in Biochemical Sciences,* 1(2):73–76.

Harms MJ., Castañeda CA., Schlessman JL., Sue GR., Isom DG., Cannon BR., and García-Moreno E B. (2009), The pK(a) values of acidic and basic residues buried at the same internal location in a protein are governed by different factors. *J Mol Biol.* May 29;389(1):34-47. doi: 10.1016/j.jmb.2009.03.039. Epub 2009 Mar 24. PMID: 19324049; PMCID: PMC3373015.

Hartley H.(1951). "Origin of the Word 'Protein.'". *Nature.* 168 (4267): 244.

Hawe A., Sutter M., and Jiskoot W. (2008). Extrinsic Fluorescent Dyes as Tools for Protein Characterization. Pharm Res 25, 1487–1499. https://doi.org/10.1007/s11095-007-9516-9

Hawkes R., Grutter M.G., and Shellman J. (1984). Thermodynamic stability and point mutations of bacteriophage T4 Lysozyme. *J. Mol. Biol.* 175: 195-212.

Hayward SB., and Glaeser RM. (1979). Radiation damage of purple membrane at low temperature. *Ultramicroscopy.* 4:201–210.

He B., Wang K., Liu Y., Xue B., Uversky VN., and Dunker AK. (2009). Predicting intrinsic disorder in proteins: an overview. *Cell Res.* Aug;19(8):929–49. doi: 10.1038/cr.2009.87. PMID: 19597536.

Herrera R., and Rosen OM. (1986). Autophosphorylation of the insulin receptor in vitro. Designation of phosphorylation sites and correlation with receptor kinase activation. *J. Biol Chem.* Sep 15;261(26):11980–5. PMID: 3745176.

Hedstrom L., Moorman AR., Dobbs J., and Abeles RH. (1984). Suicide inactivation of chymotrypsin by benzoxazinones *Biochemistry* 23 (8), 1753–1759. DOI: 10.1021/bi00303a026

Hennessey JP., Johnson WC., Bahler C., and Wood HG. (1982). Subunit interactions of transcarboxylase as studied by circular dichroism. *Biochemistry 21*: 642–646.

Henzler-Widman K., and Kern D. (2007). Dynamic personalities of proteins. *Nature* 450: 964–972.

Hilser VJ., and Thompson EB., (2007). Intrinsic disorder as a mechanism to optimize allosteric coupling in proteins. *Proc. Natl. Acad. Sci. U.S.A.* 104, 8311–8315.

Hines DJ., and Kaplan, DL. (2011). Mechanism of controlled release from silk fibroin films. *Biomacromolecules.* 12: 804–812.

Hlasiwetz H., and Habermann J. (1873). On the protein substances: second treatise. *Annalen der Chemie und Pharmacie* 169, 150 [as translated and excerpted in Mikulás Teich. (1992) *A Documentary History of Biochemistry,* 1770–1940 (Rutherford, NJ: Fairleigh Dickinson University Press.]

Ho JG., Middelberg AP., Ramage P., and Kocher HP. (2003). The likelihood of aggregation during protein denaturation can be assessed using the second virial coefficient. *Protein Sci.*Apr;12(4):708–16. doi: 10.1110/ps.0233703. PMID: 12649429; PMCID: PMC2323847.

Hofer M., Winter G., and Myschik J. (2012). Recombinant spike silk particles for controlled release of protein drugs. *Biomaterials.* 33:1554–1562.

Hofmeister F. (1888a). Zur Lehre von der Wirkung der Salze. *Arch. Exp. Pathol. Pharmakol. (Leipzig) 24:* 247–260

Hofmeister F. (1888b). Zur Lehre von der Wirkung der Salze. *Arch. Exp. Pathol. Pharmakol. (Leipzig) 25*: 1–30.

Hofmann M., Winzer M., Weber C., and Gieseler H. (2016). Prediction of protein aggregation in high concentration protein solutions utilizing protein-protein interactions determined by low volume static light scattering. *Journal of Pharmaceutical Sciences* 105 (6): 1819–1828.

Honig B., and Yang A.-S. (1995). Free energy balance in protein folding. *Adv. Protein Chem.*, 46: 27–58.

Hoppe H. (2000). Cerezyme-recombinant protein treatment for Gaucher's disease. *J Biotechnol.* Jan 21;76(2–3):259–61. doi: 10.1016/s0168-1656(99)00165-0. PMID: 10656341.

Horan RL., Antle K., Collette AL., Wang Y., Huang J., Moreau JE., Volloch V., Kaplan DL., and Altman GH. (2005). In vitro degradation of silk fibroin. *Biomaterials.* Jun;26(17):3385–93. doi: 10.1016/j.biomaterials.2004.09.020. PMID: 15621227.

Horan RL, Antle K., Collette AL., Wang Y., Huang J., Moreau JE., Volloch V., Kaplan DL., and Altman GH. (2005). In vitro degradation of silk fibroin. Biomaterials. Jun;26(17):3385–3393. doi: 10.1016/j.biomaterials.2004.09.020. PMID: 15621227.

Hribar B., Southall NT., Vlachy V. and Dill KA. (2002). How ions affect the structure of water. *J. Am. Chem. Soc.* 124: 12302–12311.

Huzil JT., Chen K., Kurgan L., and Tuszynski JA. The roles of beta-tubulin mutations and isotype expression in acquired drug resistance. Cancer Inform. 2007 Apr 27; 3:159–81. PMID: 19455242; PMCID: PMC2675838.

Idakieva K., Parvanova K., and Todinova S. (2005)

Differential scanning calorimetry of the irreversible denaturation of Rapana thomasiana (marine snail, Gastropod) hemocyanin. *Biochimica et Biophysica Acta.*1548:50–56.

Inoue S., Tanaka K., Arisaka F., Kimura S., Ohtomo K., and Misuno,S. (2000). Silk fibroin of bombyx mori is secreted, assembling a high molecular mass elementary unit consisting of h-chain, l-chain and p25, with a 6:6:1 molar ratio. *J. Biol. Chem.* 275: 40517–40528.

Jackson SE., elMasry N., and Fersht AR. (1993). Structure of the hydrophobic core in the transition state for folding of chymotrypsin inhibitor 2: A critical test of the protein engineering method of analysis. *Biochemistry* 32 (42): 11270–11278. DOI: 10.1021/bi00093a002

Jameson DM., and Ross JA. (2010) Fluorescence Polarization/Anisotropy in Diagnostics and Imaging. *Chemical Reviews* 110 (5), 2685–2708

DOI: 10.1021/cr900267p

Jan A., Hartley DM., and Lashuel HA. (2010) Preparation and characterization of toxic aβ aggregates for structural and functional studies in Alzheimer's disease research. *Nat Protoc* 5(6):1186–1209.

Jencks WP. Catalysis in Chemistry and enzymology. Dover. New York. 1987.

Jentoft N., and Dearborn DG. (1979). Labeling of proteins by reductive methylation using sodium cyanoborohydride. *J. Biol. Chem.* 254, 4359–4365.

Jentoft N., and Dearborn DG. (1983). Protein labeling by reductive alkylation. *Methods Enzymol* 91:570–579. doi: 10.1016/s0076-6879(83)91052-2. PMID: 6855602.

Johnson CS., and Gabriel DA. Light Scattering. Dover Publications, New York. 1981

Kabsch W., and Sander C. (1983). "How good are predictions of protein secondary structure?". *FEBS Lett.* 155 (2): 179–82. doi:10.1016/0014-5793(82)80597-8. PMID 6852232. S2CID 41477827.

Kanaori K., Tajiri Y., Tsuneshige A., Ishigami I., Ogura T., Tajima K., Neya S., and Yonetani T. (2011). T-quaternary structure of oxy human adult hemoglobin in the presence of two allosteric effectors, L35 and IHP. *Biochim Biophys Acta.* Oct;1807(10):1253–61. doi: 10.1016/j.bbabio.2011.06.004. Epub 2011 Jun 15. PMID: 21703224.

Karan R., Capes MD., and Dassarma S. (2012). Function and biotechnology of extremophilic enzymes in low water activity. *Aquat Biosyst.* Feb 2;8(1):4. doi: 10.1186/2046-9063-8-4. PMID: 22480329; PMCID: PMC3310334.

Karplus M. (2011). Behind the folding funnel diagram. *Nat. Chem. Biol.* 7, 401–404).

Kauffman G.B., and Kauffman L.M. (1996). An Interview with Linus Pauling. *Journal of Chemical Education.* 73 (1), 29.DOI: 10.1021/ed073p29

Kauzmann W. (1959). Some factors in the interpretation of protein denaturation. *Adv. Prot. Chem.* 14: 1–63.

Kendrew JC., Bodo G., Dintzis HM., Parrish RG, Wyckoff H., and Phillips DC. (1958) A three-dimensional model of the myoglobin molecule obtained by X-ray analysis. *Nature* 181:662–666.

Kendrew JC., Dickerson RE., Strandberg BE., Hart RG., Davies DR., Phillips DC., and Shore VC. (1960). Structure of myoglobin: a three-dimensional Fourier synthesis at 2 Å resolution. *Nature* 185:422–427.

Kendrew JC. Myoglobin and the structure of proteins. Nobel lecture, December 11, 1962 in: Nobel lectures, including presentation speeches and laureates' biographies. Chemistry, 1942–1962. Amsterdam: Elsevier, pp. 676–698, on p. 651, 696. 1964.

Kendrew JC. (1961). The three-dimensional structure of a protein molecule. *Sci Am* 205:96–110.

Kendrick BS., Chang BS., Arakawa T., Peterson B., Randolph TW., Manning MC., and Carpenter JF. (1997). Preferential exclusion of sucrose from recombinant interleukin-1 receptor antagonist: role in restricted conformational mobility and compaction of native state. *Proc Natl Acad Sci U S A.* Oct 28;94(22):11917–22. doi: 10.1073/pnas.94.22.11917. PMID: 9342337; PMCID: PMC23655.

Kielley WW., and Harrington, WF. (1960). A model for the myosin molecule. *Biochim Biophys Acta.* Jul 15; 41:401–421. doi: 10.1016/0006-3002(60)90037-8. PMID: 14408979.

King J. (2022). Using T4 Genetics and Laemmli's Development of High Resolution SDS Gel Electrophoresis to Reveal Structural Protein Interactions Controlling Protein Folding and Phage Self-Assembly. *J. Biol. Chem.* 298(10): 102463. 1–10.

Kingsbury JS., and Laue TM. (2011). Fluorescence-detected sedimentation in dilute and highly concentrated solutions. *Methods Enzymol.* 492:283–304. doi: 10.1016/B978-0-12-381268-1.00021-5. PMID: 21333796.

Kingsbury JS., Saini A., Auclair SM., Fu L., Lantz MM., Halloran KT., Calero-Rubio C., Schwenger W., Airiau CY., Zhang J., and Gokarn YR. (2020). A single molecular descriptor to predict solution behavior of therapeutic antibodies. *Sci Adv.* Aug 5;6(32):eabb0372. doi: 10.1126/sciadv. abb0372. PMID: 32923611; PMCID: PMC7457339.

Kirkwood JG. and Goldberg RJ., (1950). Light scattering arising from composition fluctuations in multi-component systems. *The Journal of Chemical Physics,* 18(1): pp.54–57.

Kirkwood JG. and Buff FP. (1951) The Statistical Mechanical Theory of Solutions. I *J. Chem. Phys.* 19, 774–777.

Kirkwood JG., and Shumaker JB. (1952a). The Influence of Dipole Moment Fluctuations on the Dielectric Increment of Proteins in Solution. *Proc. Natl. Acad. Science. USA* 38: 855–862.

Kirkwood JG., and Shumaker JB. (1952b). Forces between Protein Molecules in Solution Arising from Fluctuations in Proton Charge and Configuration. *Proc. Natl. Acad. Science. USA* 38: 863–871.

Kleppe K., Ohtsuka E., Kleppe R., Molineux I., and Khorana HG. (1971). "Studies on polynucleotides *1, *2XCVI. Repair replication of short synthetic DNAs as catalyzed by DNA polymerases". *Journal of Molecular Biology.* 56 (2): 341–361.

Kobiilka BK., and Gether U. (2002) Use of fluorescence spectroscopy to study conformational changes in the β2-adrenoceptor. *Methods in Enzymology.* 343:170–182.

Kojic N., Pritchard EM., Tao H., Brenckle MA., Mondia JP., Panilaitis B., Omenetto F., and Kaplan DL. (2012). Focal Infection Treatment using Laser-Mediated Heating of Injectable Silk Hydrogels with Gold Nanoparticles. *Adv Funct Mater.* Sep 25;22(18):3793–3798. doi: 10.1002/adfm.201200382. PMID: 24015118; PMCID: PMC3760432.

Korsak M. and Kozyreva T., 2015. Beta amyloid hallmarks: from intrinsically disordered proteins to Alzheimer's disease. *Intrinsically Disordered Proteins Studied by NMR Spectroscopy*, pp.401–421.

Kresge N., Simoni RD., and Hil RL. (2008). A New Method for Estimating Protein Stability: The Work of Nick Pace. *J Biol Chem.* Sep 19;283(38): e15. PMCID: PMC3258848.

Krimm S., and Bandekar J. (1986). Vibrational spectroscopy and conformation of peptides, polypeptides, and proteins. *Adv Protein Chem.*38:181–364. doi: 10.1016/s0065-3233(08)60528-8. PMID: 3541539.

Krishnankutty R., Bhat A., Azmi A., Souchelnytskyi S., and Uddin S. (2016) An Overview of Proteomics Techniques and its Application as a Tool in Biomarker and Drug Discovery. *J Proteomics Enzymol.*2:2.

Kumar S., Tsai CJ., and Nussinov R. (2003). Temperature range of thermodynamic stability for the native state of reversible two-state proteins. *Biochemistry*, 42, (17) 4864–4873.

Ladbury JE., Klebe G., and Freire E. (2010). Adding calorimetric data to decision making in lead discovery: a hot tip. *Nat Rev Drug Discov.*;9:23–27.

Laemmli UK. (1970). Cleavage of structural proteins during the assembly of the head of bacteriophage T4. *Nature.* 227(5259):680–5.

Lagmuir I. (1938a). Protein monolayers. *Cold Spring Harbor Symp. Quant. Biol.* 6:171–189.

Lagmuir I. (1938b). The properties and structure of proteins films. Proc. Roy. *Inst.Gt Britain* 30: 483–496.

Lahm HW., and Langen H. (2000). Mass spectrometry: a tool for the identification of proteins separated by gels. *Electrophoresis.* 21:2105–2114.

Lakowicz JR. (1999). 2nd. Plenum; New York: 1999. Principles of Fluorescence Spectroscopy.

Langer RS. (1980). Polymeric delivery systems for controlled drug release. *Chemical Engineering Communications* 6: 1–48.

Laskey RA., Honda BM., Mills AD., and Finch JT. (1978). "Nucleosomes are assembled by an acidic protein which binds histones and transfers them to DNA". *Nature.* 275 (5679): 416–20.

Laue TM. Sedimentation equilibrium as thermodynamic tool In Johnson M. L. & Ackers G. K. (Eds.), *Methods in Enzymol.* 259: 427–452. Cambridge, MA: Academic Press. 1995.

La Verde V., Dominici P., and Astegno A. (2017). Determination of Hydrodynamic Radius of Proteins by Size Exclusion Chromatography. *Bio Protoc.* Apr 20;7(8): e2230. doi: 10.21769/BioProtoc.2230. PMID: 34541230; PMCID: PMC8410290.

Lazar KL., Patapoff TW. and Sharma V. (2010). Cold denaturation of monoclonal antibodies. *mAbs* 2 (1): 45–52.

Lecanda F., Avioli V., and Chen, S. (1997). Regulation of bone matrix protein expresión and induction of differentiation of human osteoblasts and human marrow stromal cells by bone morphogenetic protein-2. *J. Cell Biochem.* 67: 386–398.

Lee JC. and Timasheff SN. (1974). Partial specific volumes and interactions with solvent components of proteins in guanidine hydrochloride *Biochemistry* 13: 257–265.

Lee JC., Harrison D., and Timashef, SN. (1975) Interaction of Vinblastine with Calf Brain Microtubule Protein. *J. Biol. Chem.* 250 (24): 9276–9282.

Lee JC., Gekko K., and Timasheff SN. (1979). Measurements of preferential solvent interactions by densimetric techniques. *Methods Enzymol* 61:26–49.

Lee JC., and Timasheff SN. (1981). The stabilization of proteins by sucrose. *J. Biol. Chem.* 256, 7193–7201.

Lee LLY and Lee JC. (1987). Thermal stability of proteins in the presence of poly (ethylene glycols). *Biochemistry.* 26: 7813–7819.

Lee J., Scheraga HA. and Rackovsky S. (1998). New optimization method for conformational energy calculations on polypeptides: Conformational space annealing. *Journal of Computational Chemistry* 18(9):1222–1232.

Lee N., Bessho Y., Wei K., Szostak JW., and Suga H. (2000) Ribozyme-catalyzed tRNA aminoacylation. Nat Struct Biol. Jan;7(1):28–33. doi: 10.1038/71225. PMID: 10625423.

Lee YC., Chiang CC., Huang PY., Chung CY, Huang TD., Wang CC., Cheng CL, Chang RS., Liao CH., and Reiz RR. (2017). Evidence of preserved collagen in an Early Jurassic sauropodomorph dinosaur revealed by synchrotron FTIR microspectroscopy. *Nat Commun* **8**, 14220. https://doi.org/10.1038/ncomms14220

Leeuwenburgh C., Rasmussen JE., Hsu FF., Mueller DM., Pennathur S., and Heinecke JW. (1997). Mass spectrometric quantification of markers for protein oxidation by tyrosyl radical, copper, and hydroxyl radical in low density lipoprotein isolated from human atherosclerotic plaques. *J Biol Chem.* Feb 7;272(6):3520–6. doi: 10.1074/jbc.272.6.3520. PMID: 9013599.

le Maire M., Aggerbeck LP., Monteilhet C., Andersen JP. and Møller JV.(1986). The use of high-performance liquid chromatography for the determination of size and molecular weight of proteins: a caution and a list of membrane proteins suitable as standards. *Anal. Biochem* 154(2): 525–535.

Lermyte F. (2020). Roles, Characteristics, and Analysis of Intrinsically Disordered Proteins: A Minireview. *Life (Basel).* Nov 30;10(12):320. doi: 10.3390/life10120320. PMID: 33266184; PMCID: PMC7761095.

Lentz TL., and Wilson PT. (1988a). Neurotoxin-binding site on the acetylcholine receptor. *Int Rev Neurobiol.*;29:117–60. doi: 10.1016/s0074-7742(08)60085-9. PMID: 3042662.

Lentz TL., and Wilson PT. (1988b). Neurotoxin-binding site on the acetylcholine receptor. *Int Rev Neurobiol.* 29:117–60. doi: 10.1016/s0074-7742(08)60085-9. PMID: 3042662.

Lindstrom J., Shelton G., and Fujii Y. (1988) Myasthenia gravis. *Adv. Immunol.* 42: 233–284.

Levinthal C. (1969) How to fold graciously. Mössbauer Spectroscopy. *Biological Systems Proceedings* 67: 22–24.

Li J., and Hilser VJ. (2018). Assessing allostery in intrinsically disordered proteins with ensemble allosteric model. *Methods Enzymol.* 611, 531–557.

Lin TY., and Timasheff SN. (1994). Why do some organisms use a urea-methylamine mixture as osmolyte? Thermodynamic compensation of urea and trimethylamine N-oxide interactions with protein. *Biochemistry.* Oct 25;33(42):12695–701. doi: 10.1021/bi00208a021. PMID: 7918496.

Lindstrom J., Tzartos S., Gullick W., Hochschwender S., Swanson L., Sargent P., Jacob M., and Montal M. (1983). Use of monoclonal antibodies to study acetylcholine receptors from electric organs, muscle, and brain and the autoimmune response to receptor in myasthenia gravis. *Cold Spring Harb Symp Quant Biol.* 48 Pt 1:89–99. doi: 10.1101/sqb.1983.048.01.012. PMID: 6586364.

Lipfert J., Doniach S., Das R., and Herschlag D. (2014). Understanding nucleic acid-ion interactions. *Annu Rev Biochem.*83:813–41. doi: 10.1146/annurev-biochem-060409-092720. Epub 2014 Mar 5. PMID: 24606136; PMCID: PMC4384882.

Litman Y., Chiang KY., Seki T., Nagata, Y., and Bonn M. (2024). Surface stratification determines the interfacial water structure of simple electrolyte solutions. *Nat. Chem.* https://doi.org/10.1038/s41557-023-01416-6

Liu JK. (2014). The history of monoclonal antibody development–progress, remaining challenges, and future innovations. *Annals of medicine and surgery*, 3(4): 113–116.

Lowe JB. and Marth JD. (2003). A genetic approach to mammalian glycan function. *Annu. Rev. Biochem.* 72, 643–691.

Lowry OH., Rosebrough NJ., Farr AL., and Randall, RJ. (1951). Protein Measurement with the Folin Phenol Reagent. *J. Biol. Chem.193, 265–275.*

Lu H, Wang Y, Wu Y, Yang P, Li L., and Li S. (2008). Hydrogen-bond network and local structure of liquid water: An atoms-in-molecules perspective. *J Chem Phys.* Sep 28;129(12):124512. doi: 10.1063/1.2985605. PMID: 19045041.

Lu X., Nobrega RP., Lynaugha,H., Jainb T., Barlow K., and Boland, T., Sivasubramanianb, A., Vásquez, M., and Xu, Y. (2019). Deamidation and isomerization liability analysis of 131 clinical-stage antibodies. *MABS*, 11(1): 45–57.

Lu RM., Hwang YC., Liu IJ., Lee CC., Tsai HZ., Li HJ., and Wu HC. (2020). Development of therapeutic antibodies for the treatment of diseases. *J Biomed Sci.* Jan 2;27(1):1. doi: 10.1186/s12929-019-0592-z. PMID: 31894001; PMCID: PMC6939334.

Lu, A., Lumeng L., and Li TK. (1986). Preparation of [^{14}C] pyridoxal 5'-phosphate. *Methods Enzymol.*122:97–102. doi: 10.1016/0076-6879(86)22154-0. PMID: 3702706.

Lundblad RL., and Noyes CM. Chemical Reagents for Protein Modification. Volume I, II. CRC Press, Inc. Boca Raton. 1984.

Lundblad JR., Laurance M., and Goodman RH., (1996). Fluorescence polarization analysis of protein-DNA and protein-protein interactions. *Molecular Endocrinology*, 10(6), pp.607–612.

Madan B., and Sharp, K. (1996). Heat Capacity Changes Accompanying Hydrophobic and Ionic Solvation: A Monte Carlo and Random Network Model Study *J. Phys. Chem.* 100, (18): 7313–7713.

Makarov M., Sanchez Rocha AC., Krystufek, R., Cherepashuk I., Dzmitruk V., Charnavets T., Faustino AM., Lebl M., Fujishma K., Fried SD., and Hlouchova K. (2023). Early Selection of the Amino Acid Alphabet Was Adaptively Shaped by Biophysical Constraints of Foldability. *Journal of the American Chemical Society* 145 (9): 5320–5329.

DOI: 10.1021/jacs.2c12987

Makowski L., Rodi D.J., Mandava S., Minh DDL., Gore DB., and Fischetti RF.(2008). Molecular Crowding Inhibits Intramolecular Breathing Motions in Proteins. *J. Mol. Biol.* 375 (2): 529–546.

Malhotra P, and Udgaonkar JB. (2016). How cooperative are protein folding and unfolding transitions? *Protein Sci.* Nov;25(11):1924–1941. doi: 10.1002/pro.3015. Epub 2016 Sep 13. PMID: 27522064; PMCID: PMC5079258.

Manning MC., Patel,K., and Borchard RT. (1989). Stability of protein pharmaceuticals, *Pharm. Res.* 6: 903–918.

Manning MC., Chou DK., Murphy BM., Payne RW., and Katayama DS. (2010). Stability of protein pharmaceuticals: an update. *Pharm Res.* Apr;27(4):544–75. doi: 10.1007/s11095-009-0045-6. Epub 2010 Feb 9. PMID: 20143256.

Manning MC., Liu J., Li T, and Holcomb RE. (2018). Rational Design of Liquid Formulations of Proteins. *Adv Protein Chem Struct Biol.* 112:1–59.

doi: 10.1016/bs.apcsb.2018.01.005. Epub 2018 Mar 5. PMID: 29680234.

Mao AH., Crick SL., Vitalis A., Chicoine CL., and Pappu RV. (2010). Net charge per residue modulates conformational ensembles of intrinsically disordered proteins. *Proc. Natl Acad. Sci.USA.* 2010; 107:8183–8188. https://doi.org/10.1073/pnas.0911107107.

Mark WY., Liao JC., Lu Y., Ayed A., Laister R., Szymczyna B., Chakrabartty A., and Arrowsmith CH. (2005). Characterization of segments from the central region of BRCA1: An intrinsically disordered scaffold for multiple protein–protein and protein–DNA interactions? *J. Mol. Biol.*345:275–287. doi: 10.1016/j.jmb.2004.10.045.

Marcus Y. (2009). Effect of ions on the structure of water: structure making and breaking. *Chem Rev.* Mar 11;109(3):1346–70. doi: 10.1021/cr8003828. PMID: 19236019.

Marquez J., Iriarte A., and Martinez-Carrion M., (1989). Covalent Modification of a Critical Sulfhydryl Group in the Acetylcholine Receptor: Cysteine-222 of the a-Subunit1 *Biochemistry*, 28: 7433–7439.

Martin W., Baross J., Kelley D., and Russell, M.J. (2008). Hydrothermal vents and the origin of life, *Microbiology* 6: 805–814.

Martinez-Carrion M., and Raftery MA. (1973). Use of a fluorescent probe for the study of ligand binding by the isolated cholinergic receptor of Torpedo californica. *Biochem Biophys Res Commun.* Dec 19;55(4):1156–64. doi: 10.1016/s0006-291x(73)80016-6. PMID: 4771989.

Martinez-Carrion M. (1975). 31P nuclear-magnetic-resonance studies of pyridoxal and pyridoxamine phosphates. Interaction with cytoplasmic aspartate transaminase. *Eur J Biochem.* May;54(1):39–43. doi: 10.1111/j.1432-1033. 1975.tb04111.x. PMID: 238848.

Martinez-Carrion M., and Boettcher B. (1975). Fluorescence energy transfer between active sites in aspartate amino transaminase. *Bioch. Biophys. Res. Comm.* 67(4):1480–1487.

Martinez-Carrion M., Sator V., and Raftery MA. (1975). The molecular weight of an acetylcholine receptor isolated from Torpedo californica. *Biochem Biophys Res Commun.* Jul 8;65(1):129–37. doi: 10.1016/s0006-291x (75)80070-2. PMID: 1147976.

Massova I., and Kollman PA. (1999). Computational alanine scanning to probe protein-protein interactions: A novel approach to evaluate binding free energies. *J. Am. Chem. Soc.* 121: 8133–8143.

Mátrai J., Verheyden G., Krüger P., and Engelborghs Y. (2004). Simulation of the activation of alpha-chymotrypsin: analysis of the pathway and role of the propeptide. *Protein Sci.* 4 Dec;13(12):3139–50. doi: 10.1110/ps.04825004. PMID: 15557259; PMCID: PMC2287318.

Mathews JB., Friend SH., and Gurd FRN. (1981). Electrostatic effects in hemoglobin: electrostatic energy associated with allosteric transition and effector binding *Biochemistry* 20: 571–580.

Matthews JB., and Gurd FRN. (1986). Stabilization and destabilization of protein structure by charge interactions. *Methods Enzymol.* 130: 437–453.

Matthews BW., Nicholson H., and Becktel WJ., (1987). Enhanced protein thermostability from site-directed mutations that decrease the entropy of unfolding. *Proceedings of the National Academy of Sciences, USA.* 84(19): 6663–6667.

Matthews JM., and Fersht AR., (1995). Exploring the energy surface of protein folding by structure-reactivity relationships and engineered proteins: observation of Hammond behavior for the gross structure of the transition state and anti-Hammond behavior for structural elements for unfolding/folding of barnase. *Biochemistry*, 34(20), pp.6805–6814.

McCarthy MP., Earnest JP., Young EF., Choe S., and Stroud RM. (1986). The molecular neurobiology of the acetylcholine receptor. *Annu Rev Neurosci.*9:383–413. doi: 10.1146/annurev.ne.09.030186.002123. PMID: 2423008.

McCrea PD., Popot JL., and Engelman DM. (1987). Transmembrane topography of the nicotinic acetylcholine receptor delta subunit. *EMBO J.* Dec 1;6(12):3619–26. doi: 10.1002/j.1460-2075. 1987.tb02693. x. PMID: 3428268; PMCID: PMC553829.

McLoughlin EC., and O'Boyle NM. (2020). Colchicine-Binding Site Inhibitors from Chemistry to Clinic: A Review. *Pharmaceuticals (Basel).* Jan 3;13(1):8. doi: 10.3390/ph13010008. Erratum in: Pharmaceuticals (Basel). 2020 Apr 20;13(4): PMID: 31947889; PMCID: PMC7168938.

Means GE., and Feeney RE. Chemical Modification of Proteins, Holden-Day Inc., San Francisco, pp. 132–134. 1971.

Medrano FJ., Andreu JM., Gorbunoff JM., and Timasheff SN. (1989). Roles of Colchicine Ring B and C in the Binding Process to Tubulin. *Biochemistry,* 28: 5589–5599.

Medrano FJ., Andreu JM., Gorbunoff JM., and Timasheff SN. (1991). Roles of Ring C Oxygens in the Binding of Colchicine to Tubulin. *Biochemistry* 30:3770–3777.

Meech SR. (2009). Excited state reactions in fluorescent proteins. *Chem. Soc. Rev.*; 38:2922–2934.

Menendez M., Laynez J., Medrano FJ., and Andreu JM. (1989.) A Thermodynamic Study of the Interaction of Tubulin with Colchicine Site Ligands. *J. Biol. Chem.* 264, (28), 16367–16371.

Meunier JC., Olsen R., Menez A., Morgat JL., Fromageot P., Ronseray AM., Boquet P., and Changeux JP. (1971). Quelques propriétés physique de la protéine réceptrice de l'acétylcholine étudiées à l'aide d'une neurotoxine radioactive [Some physical properties of the acetylcholine protein receptor studied with a radioactive neurotoxin]. *C R Acad Hebd Seances Acad Sci D.* Aug 2;273(5):595–8. French. PMID: 5001139.

Michalet X., Weiss S., and Jäger M. (2006). Single-molecule fluorescence studies of protein folding and conformational dynamics. *Chem. Rev.* 106:1785–1813.

Miklos AC., Li C., Sharaf NG., and Pielak JG., (2010). Volume Exclusion and Soft Interaction Effects on Protein Stability under Crowded Conditions. *Biochemistry* 49 (33): 6984–6991.

Mille, SL., and Urey HC. (1959). Organic Compound Synthesis on the Primitive Earth. *Science.* 130 (3370): 245–51.

Miller SL. (1953). A production of amino acids under possible primitive Earth conditions. *Science,* 117: 528–529.

Miller SL. (1955). Production of some organic compounds under possible primitive Earth conditions. *Journal of American Chemical Society* 77: 2351–2361.

Miller SL., and Bada JL. (1988). Submarine hot springs and the origin of life. *Nature* 334: 609–611.

Miller SL., and Lazcano A. (1995). The origin of life—did it occur at high temperatures? *Journal of Molecular Evolution* 41: 689–692.

Mills FC., and Ackers GK. (1979). Quaternary enhancement in binding of oxygen by human hemoglobin. *Proc Natl Acad Sci U S A.* Jan;76(1):273–7. doi: 10.1073/pnas.76.1.273. PMID: 284341; PMCID: PMC382921.

Minezaki Y., Homma K., and Nishikawa K. (2005). Genome-wide survey of transcription factors in prokaryotes reveals many bacteria-specific families not found in archaea. *DNA Res.* 12: 269–280

Minton AP. (2006). Macromolecular crowding. *Curr Biol.* 16:R269–271.

Monod J., Wyman J., and Changeux J-P. (1965). On the Nature of Allosteric Transitions — a Plausible Model. *J. Mol. Biol.* 12 (1): 88–118. doi:10.1016/S0022-2836(65)80285-6. PMID 14343300.

Moore S., Spackman DH., and Stein WH. (1958) Chromatography of Amino Acids on Sulfonated Polystyrene Resins, *Analytical Biochemistry,* Vol. 30 (7): 1185–1190.

Morales-Pérez C., Noviello C. and Hibbs R. (2016). X-ray structure of the human α4β2 nicotinic receptor. *Nature* 538: 411–415. https://doi.org/10.1038/nature19785

Morin R., Kaplan DL, and Pérez-Ramírez B. (2006). Bone morphogenetic protein-2 binds as multilayers to a collagen delivery matrix. An equilibrium thermodynamic analysis. *Biomacromolecules.* 7: 131–138.

Morrison KL. and Weiss GA. (2001). Combinatorial alanine-scanning. *Curr. Opin. Chem. Biol.* 5: 302–307.

Mulder GJ. (1838). "Ueber die Zusammensetzung einiger thierischen Substanzen". *Journal für Praktische Chemie* (in German). 16: 129–152.

Müller MM., Allison JR., Hongdilokkul N., Gaillon L., Kast P., Van Gunsteren WF., Marlière P., and Hilver D.. (2013). Directed Evolution of a Model Primordial Enzyme Provides Insights into the Development of the Genetic Code. *PLoS Genet* 9(1): e1003187doi: 10.1371/journal.pgen.1003187.

Muñoz V., and Sánchez-Ruiz JM. (2004). Exploring protein-folding ensembles: a variable-barrier model for the analysis of equilibrium unfolding experiments. *Proc Natl Acad Sci U S A.* Dec 21;101(51):17646–51. doi: 10.1073/pnas.0405829101. Epub 2004 Dec 9. PMID: 15591110; PMCID: PMC539728.

Murphy LR., Matubayasi N., Payne VA., and Levy RM (1998). Protein hydration and unfolding – insights from experimental partial specific volumes and unfolded protein models. *Folding and Design* 3(2):105–118.

Murphy KP., and Gill SJ. (1990). Group additivity thermodynamics for dissolution of solid cyclic dipeptides into water. *Thermochimica Acta*, 172:11–20.

Murphy KP., and Gill SJ., (1991). Solid model compounds and the thermodynamics of protein unfolding. *Journal of molecular biology*, 222(3):699–709.

Murphy KP. and Freire E. (1992). Thermodynamics of structural stability and cooperative folding behavior in proteins. *Advances in protein chemistry,* 43:313–361.

Murphy KP., Bhakuni V., Xie D. and Freire E. (1992). Molecular basis of co-operativity in protein folding: III. Structural identification of cooperative folding units and folding intermediates. *Journal of molecular biology,* 227(1):293–306.

Murphy RM. (1997). Static and dynamic light scattering of biological macromolecules: what can we learn? *Current Opinion in Biotechnology*, 8(1): 25–30.

Myers JK., Pace CN., and Scholtz JM. (1995). Denaturant m values and heat capacity changes: relation to changes in accessible surface areas of protein unfolding. *Protein Sci.* Oct;4(10):2138–48. doi: 10.1002/pro.5560041020. Erratum in: *Protein Sci* 1996 May;5(5):981. PMID: 8535251; PMCID: PMC2142997.

Na GC., and Timasheff SN. (1981). Interaction of Calf brain tubulin with glycerol. *J. Mol. Biol.* 151:165–178.

Na GC., and Timasheff SN. (1985). Measurement and analysis of ligand binding isotherms linked to protein self-associations. *Methods Enzymol.* 117: 496–519

Na GC, and Timasheff SN. (1986). Interaction of Vinblastine with Calf Brain Tubulin: Multiple Equilibria. *Biochemistry* 25: 6214–6222.

Naganathan AN., Sanchez-Ruiz JM., and Muñoz V. (2005). Direct measurement of barrier heights in protein folding. *J Am Chem Soc.* Dec 28;127(51):17970–1. doi: 10.1021/ja055996y. PMID: 16366525.

Neira JL., Bintz J., Arruebo M., Rizzuti B., Bonacci T., Vega S., Lanas A., Velázquez-Campoy A., Iovanna J.L., and Abián O. (2017). Identification of a Drug Targeting an Intrinsically Disordered Protein Involved in Pancreatic Adenocarcinoma. *Sci. Rep.* 7, 39732.

Nevskaya NA., and Chirgadze YN. (1976). Infrared spectra and resonance interactions of amide-I and II vibration of alpha-helix. *Biopolymers.* Apr;15(4):637–48. doi: 10.1002/bip.1976.360150404. PMID: 1252599.

Nguyen LT., Wiencek JM., and Kirsch LE. (2003). Characterization methods for the physical stability of biopharmaceuticals. *PDA J Pharm Sci Technol.* Nov-Dec;57(6):429–45. PMID: 14765559.

Nichol LW., and Winzor DJ. (1976) Ligand-induced polymerization. *Biochemistry* 15: 3015–3019.

Nicholson WI. (2009). Ancient micronauts: Interplanetary transport of microbes by cosmic impacts. *Trends in Microbiology 17*: 243–250.

Nicoli, G., and Prigogine I. Self-Organization in Nonequilibrium Systems: From Dissipative Structures to Order through Fluctuations 1st Edition. 1977. Wiley; 1st edition (May 13, 1977). New York. p441–442. 1977.

Nishikawa K. (1983). Assessment of secondary-structure prediction of proteins comparison of computerized chou-fasman method with others. *Biochimica et Biophysica Acta (BBA) - Protein Structure and Molecular Enzymology.* 748 (2): 285–299.

Noda M., Takahashi H., Tanabe T., Toyosato M., Kikyotami S., Hirose T., Asai M., Takashima H., Inayama S., Miyata T., and Numa, S. (1983). Primary structures of β- and δ-subunit precursors of *Torpedo californica* acetylcholine receptor deduced from cDNA sequences *Nature (London)* 301: 251–255

Nogales E., Wolf SG., and Downing KH.(1998). Structure of the alpha beta tubulin dimer by electron crystallography. Nature. 1998 Jan 8;391(6663):199–203. doi: 10.1038/34465. Erratum in: *Nature* May 14;393(6681):191. PMID: 9428769.

Nozaka M., Kuwajima K., Nitta K.,and Sugai S. (1978). Detection and characterization of the intermediate on the folding pathway of human alpha-lactalbumin. *Biochemistry.* Sep 5;17(18):3753–8. doi: 10.1021/bi00611a013. PMID: 698195.

Ohkawara B., Lemura S., Dijke P., and Ueno N. (2002). Action range of rhBMP-2 is defined by its N-terminal basic amino acid core. *Curr. Biol.* 12: 205–209.

Oldfield CJ., and Dunker AK. (2014). Intrinsically disordered proteins, and intrinsically disordered protein regions. *Ann. Rev. Biochem.* 83, 553–584.

Onuchic JN., Luthey-Schulten Z., and Wolynes PG. (1997). Theory of protein folding: the energy landscape perspective. *Annu Rev Phys Chem.* 48:545–600. doi: 10.1146/annurev.physchem.48.1.545. PMID: 9348663.

Oparin AI. The Origin of Life on Earth. New York: Academic Press Inc.; Reprinted in 1959 edition (January 1, 1957). 1957.

Ortea I., O'Connor G., and Maquet, A. (2016). Review on proteomics for food authentication. *J Proteomics.* 147:212–225.

Ovádi J., Mohamed Osman IR., and Batke J. (1983). Interaction of the dissociable glycerol-3-phosphate dehydrogenase and fructose-1,6-bisphosphate aldolase. Quantitative analysis by an extrinsic fluorescence probe. *Eur J Biochem.* Jun 15;133(2):433–7. doi: 10.1111/j.1432-1033.1983.tb07482.x. PMID: 6406231.

Pace CN., and Vanderburg KE. (1979). Determining globular protein stability: guanidine hydrochloride denaturation of myoglobin. *Biochemistry 18* (2): 288–292

DOI: 10.1021/bi00569a008

Pace CN., Grimsley GR., and Scholtz JM. (2009). Protein ionizable groups: pK values and their contribution to protein stability and solubility. *J Biol Chem.* May 15;284(20):13285–9. doi: 10.1074/jbc. R800080200. Epub 2009 Jan 21. PMID: 19164280; PMCID: PMC2679426.

Pandey A., and Mann M. (2000). Proteomics to study genes and genomes. *Nature.* 405:837–846.

Panilaitis B., Altman GH., Chen J., Jin HJ., Karageorgiu V., and Kaplan DL. (2003). Macrophage responses to silk. *Biomaterials.* 24: 3079–3085.

Parker ET., and Lollar P. (2021). Measurement of the Translational Diffusion Coefficient and Hydrodynamic Radius of Proteins by Dynamic Light Scattering. *Bio Protoc.* Oct 20;11(20): e4195. doi: 10.21769/BioProtoc.4195. PMID: 34761067; PMCID: PMC8554817.

Pauling L., and Niemann C. (1939). The structure of proteins. *Journal of the American Chemical Society* 61 (7): 1860–1867. DOI: 10.1021/ja01876a065

Pauling L. 1939. The Nature of the Chemical Bond and the Structure of Molecules and Crystals: An Introduction to Mode. Cornell University Press; Third edition (February 15, 1960).

Pauling L., (2015a). The discovery of the alpha helix. Culture of Chemistry: The Best Articles on the Human Side of 20th-Century Chemistry from the *Archives of the Chemical Intelligencer,* pp.161–167.

Pauling L. The Discovery of the Alpha Helix. In: Hargittai, B., Hargittai, I. (eds) Culture of Chemistry. Springer, Boston, MA. (2015b). https://doi.org/10.1007/978-1-4899-7565-2_35.

Pauling L., Corey RB., and Branson HR. (1951). The structures of proteins: Two hydrogen-bonded helical configurations of the polypeptide chain. *Proc. Natl. Acad. Sci. USA.* 37: 205–210.

Pauling L., and Corey RB. (1951). The Pleated sheet, a new layer configuration of polypeptide chains. *Proc. Natl. Acad. Sci. USA.* 37: 251–256

Pérez de Buerres Ramirez B., (2020). *COVID-19: El Gran Terrorista*. Boston, Massachusetts. Novo Codex Editions.

Pérez-Ramírez B., and Martinez-Carrion M. (1987) A novel method to study the topography of transmembrane proteins using pyridoxal 5'-phosphate. in Biochemistry of Vitamin B6 (Korpela, T. and Christen, P. eds.) Birkhauser, Basel Switzerland. pp 375–378.

Pérez-Ramírez B., and Martinez-Carrion, M. (1989) Pyridoxal phosphate as a probe of the cytoplasmic domains of transmembrane proteins: application to the nicotinic acetylcholine receptor. *Biochemistry* 28: 5034–5040.

Pérez-Ramírez B. (1990). PhD dissertation. University of Missouri.

Pérez-Ramírez B., and Castells M. (1991). In vitro biosynthesis of rat sperm outer dense fiber components. *Life Sciences.*49 (21) 1549–1554.

Pérez-Ramírez B. (1994). Thermal stability of Torpedo californica acetylcholine receptor in a cholesterol lipid environment. *Mol Cell Biochem.* 132: 91–99. https://doi.org/10.1007/BF00926917

Pérez-Ramírez B., Shearwin KE., and Timasheff SN. (1994). The colchicine-induced GTPase activity of tubulin: state of the product. Activation by microtubule-promoting cosolvents. *Biochemistry.* May 24;33(20):6253–61. doi: 10.1021/bi00186a027. PMID: 8193140.

Pérez-Ramírez B., and Timasheff SN. (1994). Cosolvent modulation of the tubulin-colchicine GTPase-activating conformational change: strength of the enzymatic activity. *Biochemistry.* May 24;33(20):6262–7. doi: 10.1021/bi00186a028. PMID: 8193141.

Pérez-Ramírez B., Iriarte A., and Martinez-Carrion M. (1994). Residues 377-389 from the delta subunit of Torpedo californica acetylcholine receptor are located in the cytoplasmic surface. *J. Protein Chem.* Jan;13(1):67–76. doi: 10.1007/BF01891994. PMID: 8011073.

Pérez-Ramírez B., and Timasheff S.N. (1995). Stoichiometric and substoichiometric inhibition of microtubules by colchicine analogs: A linked equilibrium model *Protein Sci.* 4. Suppl. 2, 410S.

Pérez-Ramírez B., Andreu JM., Gorbunoff MJ., and Timasheff, SN. (1996). Stoichiometric and substoichiometric inhibition of tubulin self-assembly by colchicine analogues. *Biochemistry*, 35(10):3277–3285.

Pérez-Ramírez B., and Stecker J. Probing reversible self-association of therapeutic proteins by sedimentation velocity in the analytical ultracentrifuge In: Therapeutic Proteins: Methods and Protocols, 308, chapter 24, pp301–318 (C.Mark Smales & David C. James, Eds.). Humana Press, Inc. 2005.

Pérez-Ramírez B., Guziewicz N., and Simler R. Pre-formulation Research: Assessing Protein solution behavior early during therapeutic development, in: “Formulation and Process Strategies for Manufacturing of a Biopharmaceutical”. Edited by Feroz Jameel and Susan Hershenson. John Wiley & Sons, New York, pp 119–146. 2010.

Pérez-Ramírez B., Guziewicz N., Simler, R., and Sreedhara A. Approaches for early developability assessment of proteins to guide quality by design of liquid formulations, in: Quality by Design for Biopharmaceutical Drug Product Development. Edited by Feroz Jameel, Susan Hershenson, Mansoor A. Khan and Sheryl Martin-Moe. Springer, New York,. pp 87–114. 2015.

Pérez-Ramírez B., and Simler RB. (2019). Thermodynamic probes of cold instability: Application to therapeutic proteins. *Eur. Phar. Rev.* 24 (05): 26–29.

Pérez-Ramírez B. Enabling Biologics Combination products: Device ability in protein therapeutics. In *Development of Biopharmaceutical Drug-Device Products* (AAPS Advances in the Pharmaceutical Sciences Series, 35). Edited by Feroz Jameel, John W. Skoug, and Robert R. Nesbitt. Springer, New York. Page 49–69. 2020.

Permyakov EA. (1993) Luminescent Spectroscopy of Proteins. CRC Press, Inc., Boca Raton, 38.

Perry E., Morris C., Court J., Cheng A., Fairbairn A., McKeith I., Irving D., Brown A., and Perry R. (1995) Alteration in nicotine binding sites in Parkinson’s disease, Lewy Body Dementia and Alzheimer’s disease, possible index of early neuropathology. Neuroscience 64, 385–395.

Petrassi HM., Johnson SM., Purkey HE., Chiang KP., Walkup T., Jiang X., Powers ET., and Kelly JW. (2005). Potent and selective structure-based dibenzofuran inhibitors of transthyretin amyloidogenesis: kinetic stabilization of the native state. *J Am Chem Soc.* 127(18):6662–71. doi: 10.1021/ja044351f. PMID: 15869287.

Pillai AS., Chandler SA., Liu Y., Signore AV., Cortez-Romero CR., Benesch JLP., Laganowsky A., Storz JF., Hochberg GKA., and Thornton, JW.(2020). Origin of complexity in haemoglobin evolution. Nature 581, 480–485. https://doi.org/10.1038/s41586-020-2292-y

Plaza del Pino IM., Ibarra-Molero, B., and Sánchez-Ruiz JM. (2000). Lower kinetic limit to protein thermal stability a proposal regarding protein stability in vivo and its relation with misfolded diseases. *Proteins,* 40: 58–70.

Prakash V., and Timasheff SN. (1992). Aging of Tubulin at neutral pH: The destabilizing effect of vinca alkaloids. *Arch. Bichem. Biophys.* 295 (1), 137–145.

Price NC., and Stevens L. (1999) Fundamentals of enzymology. The Cell and molecular biology of catalytic proteins (3rd edition). Oxford University Press, Oxford. 1999.

Pritchard EM., and Kaplan DL. (2011). Silk fibroin materials for controlled release drug delivery. *Expert Opinion in Drug Delivery.* 8: 797–811.

Pritchard J.F., Jurima-Romet M., Reimer M.L.J., Mortimer E., Rolfe B., and Cayen M.N. (2003). Making better drugs: Decision gates in non-clinical drug development. *Nature Reviews Drug Discovery* 2: 542–553

Privalov P., and Khechinashvili N. (1974). A thermodynamic approach to the problem of stabilization of globular protein structure: A calorimetric study. *Journal of Molecular Biology,* 86(3): 665–684. https://doi.org/10.1016/0022-2836(74)90188-0

Privalov PL. (1979). Stability of proteins: small globular proteins. *Adv Protein Chem.* 33:167–241. doi: 10.1016/s0065-3233(08)60460-x. PMID: 44431.

Privalov PL., Kafiani KA., Monaselididze DR. (1964). ISSLEDOVANIE TEPLOVO I DENATURATSII DNK S POMOSHCH'IU ADIABATNOGO MIKROKALORIMETRA [STUDY OF HEAT DENATURATION OF DNA BY USE OF THE ADIABATIC MICROCALORIMETER]. Dokl Akad Nauk SSSR. Jun 1; 156:951–3. Russian. PMID: 14191312.

Privalov PL., and Gill SJ. (1988). Stability of protein structure and hydrophobic interaction. *Adv Protein Chem.*39:191–234. doi: 10.1016/s0065-3233(08)60377-0. PMID: 3072868.

Privalov PL. (1990). Cold denaturation of proteins. *Crit Rev Biochem Mol Biol.* 25(4):281–305. doi: 10.3109/10409239009090612. PMID: 2225910.

Privalov PL.,and Makhatadze GI. (1990). Heat capacity of proteins. II. Partial molar heat capacity of the unfolded polypeptide chain of proteins: protein unfolding effects. *J Mol Biol.* May 20;213(2):385–91. doi: 10.1016/S0022-2836(05)80198-6. PMID: 2160545.

Quesada-Soriano I., García-Maroto F., and García-Fuentes L. (2006).Kinetic study on the irreversible thermal denaturation of Schistosoma japonicum glutathione s-transferase. *Biochim. Biophys. Acta* 1764: Issue 5, 979–984.

Quigley A., and Williams DR. (2015). The second virial coefficient as a predictor of protein aggregation propensity: A self-interaction chromatography study. *Eur J Pharm Biopharm.* Oct; 96:282–90. doi: 10.1016/j.ejpb.2015.07.025. Epub 2015 Aug 7. PMID: 26259782; PMCID: PMC4644993.

Raftery MA., Schmidt J., Clark DG. (1972). Specificity of -bungarotoxin binding to Torpedo californica electroplax. *Arch Biochem Biophys.* Oct;152(2):882–886. doi: 10.1016/0003-9861(72)90285-8. PMID: 4635795.

Raftery MA., Hunkapiller MW., Strader CD., and Hood LE. (1980). Acetylcholine receptor: complex of homologous subunits. Science. Jun 27;208(4451):1454–6. doi: 10.1126/science.7384786. PMID: 7384786.

Rahman MS., Akhtar N., Jamil HM., Banik RS., and Asaduzzaman SM. (2015). TGF-β/BMP signaling and other molecular events: regulation of osteoblastogenesis and bone formation. *Bone Res.* Apr 14; 3:15005. doi: 10.1038/boneres.2015.5. PMID: 26273537; PMCID: PMC4472151.

Ramani K., Purohit V., Middaugh RC., and Balasubramanian SV. (2005). Aggregation Kinetics of Recombinant Human FVIII (rFVIII). *Journal of Pharmaceutical Sciences,* 94: 2023–2029.

Rao ST., and Rossmann MG. (1973) Comparison of super-secondary structures in proteins. *J Mol Biol* 76: 241–56

Rasmussen BF., Stock AM., Ringe D. and Petsko GA., (1992). Crystalline ribonuclease A loses function below the dynamical transition at 220 K. *Nature,* 357(6377): 423–424.

Ray WJ. Jr., and Kosland DE. Jr. (1961). A method for characterizing the type and numbers of groups involved in enzyme action. *J. Biol. Chem.* 236: 1973–1979.

Rappu P., Salo AM., Myllyharju J., and Heino J. (2019). Role of prolyl hydroxylation in the molecular interactions of collagens. *Essays Biochem.* Sep 13;63(3):325–335. doi: 10.1042/EBC20180053. PMID: 31350381; PMCID: PMC6744578.

Reddi AH. (1997) Bone morphogenetic proteins: an unconventional approach to isolation of firsts mammalian morphogens. *Cytokine Growth Factor Res.* 8: 11–20.

Reddi AH. (1998). Role of morphogenetic proteins in skeletal tissue engineering and regeneration. *Nat. Biotechnol.* 16: 247–252.

Reissner KJ., and Aswald DW. (2003) Deamidation and isoaspartic formation in proteins. Unwanted alterations or surreptitious signals? *Cell Mol. Life Sci.* 60: 1281–1295.

Reynolds JA. and Tanford C., (1970). Binding of dodecyl sulfate to proteins at high binding ratios. Possible implications for the state of proteins in biological membranes. *Proceedings of the National Academy of Sciences*, 66(3):1002–1007.

Reysenbach AI., and Shock E. (2002). Merging genomes with geochemistry in hydrothermal ecosystems. *Science* 296: 1077–1082.

Rhie A., Nurk S., Cechova M. et al. (2023) The complete sequence of a human Y chromosome. *Nature.* https://doi.org/10.1038/s41586-023-06457-y

Rhogani M., Mansukhani A., Dell'Era P., Bellosta P., Basilico C., Rifkin DB., and Moscatelli D. (1994). Heparin increases the affinity of basic fibroblast growth factor for its receptor but is not required for binding. *J. Biol. Chem.* 269: 3976–3984.

Riquelme PT., Burzio LO and Koide SS. (1979). ADP Ribosylation of Rat Liver Lysine-rich Histone in Vitro *J. Biol. Chem.* 254: 3028–3078.

Rivas G., Fernández JA., and Milton AP. (1999). Direct observation of dilute proteins in the presence of inert macromolecules at high concentration via a tracer sedimentation equilibrium: Theory, experiment, and biological significance. *Biochemistry* 38: 9379–9388.

Ripamonti U., and Reddi AH. (1997). Tissue engineering morphogenesis and regeneration of the periodontal tissues by bone morphogenetic proteins. *Crit. Rev. Oral Biol Med.* 8: 154–163.

Ripoll DR., Vorobjev YN., Liwo A., Vila JA., and Scheraga HA. (1996). Coupling between folding and ionization equilibria: effects of pH on the conformational preferences of polypeptides. *J Mol Biol.* Dec 13;264(4):770–83. doi: 10.1006/jmbi.1996.0676. PMID: 8980685.

Robertson AD., and Murphy KP. (1997) Protein Structure and the Energetics of Protein Stability. *Chem Rev.* Aug 5;97(5):1251–1268. doi: 10.1021/cr960383c. PMID: 11851450.

Robertson AD. (2002). Intramolecular interactions at protein surfaces and their impact on protein function. *Trends in Biochemical Sciences,* 27(10): 521–526. https://doi.org/10.1016/S0968-0004(02)02184-9

Robinson DR., and Jencks WP. (1965). The Effect of Concentrated Salt Solutions on the Activity Coefficient of Acetyltetraglycine Ethyl Ester *Journal of the American Chemical Society* 87 (11), 2470–2479 DOI: 10.1021/ja01089a029

Robinson RA., and Stokes RH. Electrolyte Solutions. (Second Revised Edition). Dover Publications. Inc. Mineola, New York, p27. 2002.

Roberts WJ., Hubert E., Iriarte A., and Martinez-Carrion M. (1988). Site-specific Methylation of a Strategic Lysyl Residue in Aspartate Aminotransferase *J. Biol. Chem.* 263: 7196–7208.

Rodriguez-Larrea D., Minning S., Borchert TV., and Sanchez-Ruiz JM. (2006). Role of solvation barriers in protein kinetic stability. *J Mol Biol.* Jul 14;360(3):715–24. doi: 10.1016/j.jmb.2006.05.009. Epub 2006 May 19. PMID: 16784752.

Rose GD., and Wolfenden R. (1993) Hydrogen bonding, hydrophobicity, packing, and protein folding. *Annual Review of Biophysics and Biomolecular Structure* 22 (1): 381–415.

Rosen OM., Herrera R., Olowe Y., Petruzzelli LM., and Cobb MH. (1983). Phosphorylation activates the insulin receptor tyrosine protein kinase. *Proc Natl Acad Sci U S A.* Jun;80(11):3237–40. doi: 10.1073/pnas.80.11.3237. PMID: 6574482; PMCID: PMC394015

Rossi M., Gorbunoff MJ., Caruso F., Wing B., Pérez-Ramírez B., and Timasheff SN.(1996). Structural analysis of the substoichiometric and stoichiometric microtubule-inhibiting biphenyl analogues of colchicine. *Biochemistry.* Mar 12;35(10):3286–9. doi: 10.1021/bi950524p. PMID: 8605165.

Rubio Gomez MA., and Ibba M. (2020). Aminoacyl-tRNA synthetases. *RNA.* Aug;26(8):910–936. doi: 10.1261/rna.071720.119. Epub 2020 Apr 17. PMID: 32303649; PMCID: PMC7373986.

Ruff KM., Pappu RV., and Holehouse A S. (2019). Conformational preferences, and phase behavior of intrinsically disordered low complexity sequences: Insights from multiscale simulations. *Curr. Opin. Struct. Biol.* 56, 1–10.

Ruppert R., Hoffman E., and Sebald W. (1996). Human bone morphogenetic protein-2 contains a heparin-binding site which modifies its biological activity. *Eur. J. Biochem.* 237: 295–302.

Salgia R., Jolly MK., Dorff T., Lau,C., Wenniger K., Orban J., and Kulkarni P. (2018). Prostate-associated gene 4 (PAGE4): Levering the conformational dynamics of a dancing protein cloud as a therapeutic target. *J. Clin. Med.* Jun 17;7(6):156. doi: 10.3390/jcm7060156.

Saluja A., Badkar AV., Zeng DL., Nema S. and Kalonia DS., (2007). Ultrasonic storage modulus as a novel parameter for analyzing protein-protein interactions in high protein concentration solutions: correlation with static and dynamic light scattering measurements. *Biophysical journal,* 92(1): 234–244.

Saluja A., Yang M., and Pérez-Ramírez B. (2018). Enabling manufacture of robust biologic drug products. *European Pharm. Rev.* 02: 18–20.

Sanchez-Ruiz JM., and Martinez-Carrion M. (1986). Ionization state of the coenzyme 5'-phosphate ester in cytosolic aspartate aminotransferase. A Fourier transform infrared spectroscopic study. *Biochemistry.* May 20;25(10):2915–20. doi: 10.1021/bi00358a027. PMID: 3718929.

Sanchez-Ruiz JM., and Martinez-Carrion M. (1988). A Fourier-transform infrared spectroscopic study of the phosphoserine residues in hen egg phosvitin and ovalbumin. Biochemistry. May 3;27(9):3338–42. doi: 10.1021/bi00409a033. PMID: 3134047.

Sanchez-Ruiz JM. (2010). Protein kinetic stability. *Biophys Chem.* (2010). May;148(1–3):1–15. doi: 10.1016/j.bpc.2010.02.004. Epub Feb 11. PMID: 20199841.

Sandow J., Landgraf W., Becker R, and Seipke G. (2015). Equivalent Recombinant Human Insulin Preparations, and their Pla.ce in Therapy. *Eur Endocrinol.* Apr;11(1):10–16. doi: 10.17925/EE.2015.11.01.10. Epub 2015 Apr 11. PMID: 29632560; PMCID: PMC5819055.

Sanger F. (1945). The free amino groups of insulin. *Biochem J.* 39(5):507–15. doi: 10.1042/bj0390507. PMID: 16747948; PMCID: PMC1258275.

Sange, F., and Thompson EO. (1953a). The amino-acid sequence in the glycyl chain of insulin. I. The identification of lower peptides from partial hydrolysates. *Biochem J.* Feb;53(3):353–66. doi: 10.1042/bj0530353. PMID: 13032078; PMCID: PMC1198157.

Sanger F., and Thompson EO. (1953b). The amino-acid sequence in the glycyl chain of insulin. II. The investigation of peptides from enzymic hydrolysates. *Biochem J.* Feb;53(3):366–74. doi: 10.1042/bj0530366. PMID: 13032079; PMCID: PMC1198158.

Santoro MM., and Bolen DW. (1988). Unfolding free energy changes determined by the linear extrapolation method. 1. Unfolding of phenylmethanesulfonyl. alpha.-chymotrypsin using different denaturants. *Biochemistry,* 27(21): 8063–8068.

Scatchard G. (1946). Physical chemistry of protein solutions; derivation of the equations for the osmotic pressure. *J. am. Chem. Soc.* 68,(11): 2315–9.

Schachman HK. (1957). Ultracentrifugation, diffusion, and viscometry. *Methods in Enzymology,* Vol 4, p32. Edited by S. Colowick and N.O Kaplan. New York, Academic Press.

Schellman JA., and Kauzmann W. (1951). The dielectric polarization of ice, *Phys. Rev.*82: 315.

Schellman JA. (1955). The stability of hydrogen-bonded peptide structures in aqueous solution, *C.R. Trav. Lab Carlsberg Ser.Chim.*29: 230–259.

Schellman JA. (1978). Solvent denaturation, *Biopolymers* 17 1305–1322.

Schellman GC., and Jaenicke R. Protein folding." The alphaL conformation at the ends of helices. Jaenicke R, editor. Amsterdam: Elsevier pp. 53–61. 1980.

Schellman JA., Hawkes RB., Jaenicke R. (Ed.), Protein Folding, Elsevier-North Holland, Amsterdam pp. 331–341. 1980.

Schellman JA. (1987). The thermodynamic stability of proteins. *Annu Rev Biophys Biophys Chem.* 16:115–137. doi: 10.1146/annurev.bb.16.060187.000555. PMID: 3297085.

Schellman JA., (1990). A simple model for salvation in mixed solvents: applications to the stabilization and destabilization of macromolecular structures. *Biophysical chemistry*, 37(1–3), pp.121–140.

Schellman JA. (1994). The thermodynamics of solvent exchange. *Biopolymers.* Aug;34(8):1015–26. doi: 10.1002/bip.360340805. PMID: 8075384.

Schellman JA., and Somero GN. (1996) SergeTimasheff: the man with a genius for solution in biology. *Biophys J.* 71 (4):1985–1993.

Schellman JA. (2002). Fifty years of solvent denaturation. *Biophys Chem.* May 2;96(2–3):91–101. doi: 10.1016/s0301-4622(02)00009-1. PMID: 12034431.

Schellman JA. (2005). Destabilization and stabilization of proteins. *Quarterly reviews of Biophysics,* 38(4): pp.351–361.

Schilling S., Wasternack C., and Demuth H. (2008). Glutamyl cyclases from animals and plants: a case of functionally convergent protein evolution. *Biol Chem* 389: 983–991.

Schoellmann G., and Shaw, E. (1962). A new method for labelling the active center of chymotrypsin. *Biochemical and Biophysical Research Communications,* 7 (1), 36–40.

Schoenich C., Hgeman MJ., and Borchard RT. (1997). Stability of peptides and proteins, in Controlled Drug Delivery Challenges and Strategies, Park, K., American Chemical Society, Washington DC, pp 205–228.

Schjoldager KT., Narimatsu Y., Joshi HJ., and Clausen H. (2020). Global view of human protein glycosylation pathways and functions. *Nat Rev Mol Cell Biol.* Dec;21(12):729–749. doi: 10.1038/s41580-020-00294-x. Epub 2020 Oct 21. PMID: 33087899.

Schoellmann G., and Shaw E. (1963). Direct evidence for the presence of histidine in the active center of chymotrypsin. *Biochemistry* 2: 252–255

Schuck P. (1998). Sedimentation analysis of noninteracting and self-associating solutes using numerical solutions to the Lamm equation. *Biophysical Journal,* 75(3), 1503–1512.

Schuck P., and Rossmanith P. (2000). Determination of the sedimentation coefficient distribution by least-squares boundary modeling. *Biopolymers.* Oct 15;54(5):328–41. doi: 10.1002/1097-0282(20001015)54:5<328:AID-BIP40>3.0.CO;2-P. PMID: 10935973.

Seelig J., and Seelig A. (2021). Molecular understanding of calorimetric protein unfolding experiments. *Biophys Rep.* Dec 6;2(1):100037. doi: 10.1016/j.bpr.2021.100037. PMID: 36425081; PMCID: PMC9680786.

Segel IH. in Enzyme Kinetics. Behavior and analysis of rapid equilibrium and steady state enzyme systems. John Wiley & sons. New York. 1975.

Shire SJ., McKay P., Leung DW., Cachianes GJ., Jackson E., and Wood WI (1990). Preparation and properties of recombinant DNA derived tobacco mosaic virus coat protein. *Biochemistry* 29: 5119–5126.

Shirley BA., Stanssens P., Hahn U., and Pace CN. (1992). Contribution of hydrogen bonding to the conformational stability of ribonuclease T1. *Biochemistry* 31 (3): 725–732. DOI: 10.1021/bi00118a013

Sillen A., Verheyden S., Delfosse L., Braem T., Robben J., Volckaert G., and Engelborghs Y. (2003). Mechanism of fluorescence and conformational changes of the sarcoplasmic calcium binding protein of the sand worm Nereis diversicolor upon Ca2+ or Mg2+ binding. *Biophys J.* 85:1882–1893.

Silverman RB., and Abeles RH. (1976). Inactivation of pyridoxal phosphate dependent enzymes by mono- and polyhaloalanines. *Biochemistry* 1976: 15, 21, 4718–4723

Slavik J. (1982). Anilinonaphtalene sulfonate as a probe of membrane composition and function. *Biochim. Biophys. Acta.* 694:1–25.

Slebe JC., and Martinez-Carrion M. (1976). Carbamylation of aspartate transaminase and the pK value of the active site lysyl residue. *J Biol Chem.* Sep 25;251(18):5663–5669. PMID: 965383.

Smith FR., and Ackers GK. (1985). Experimental resolution of cooperative free energies for the ten ligation states of human hemoglobin. *Proceedings of the National Academy of Sciences USA* 82(16): 5347–5351.

Smith L., Lind MJ., Welham KJ., and Cawkwell L. (2006). Cancer Biology Proteomics Group. Cancer proteomics and its application to discovery of therapy response markers in human cancer. *Cancer.* 107:232–241.

Spackman DH., Stein WH., and Moore S. (1958). Automatic Recording Apparatus for Use in the Chromatography of Amino Acids. *Analytical Chemistry,* 30: 1190–1205.

Somero GN. (1986) Protons, osmolytes, and fitness of internal milieu for protein function. *American Journal of Physiology,* 251: R197–213.

Sousa FL., Thiergart T., Landan G., Nelson-Sathi S., Pereira LAC., Allen JF., Lane N., and Martin WF., (2013). Early bioenergetic evolution. *Philosophical Transactions of the Royal Society* B 368.

Spiro RG. (2002). Protein glycosylation: nature, distribution, enzymatic formation, and disease implications of glycopeptide bonds. *Glycobiology* 12, 43R–56R.

Spivak-Kroizman T., Lemmon MA., Dikie L., Ladbury JE., Pinchasi D., Huang J., Jaye M., Crumley G., Schlessinger J., and Lax, I. (1994). Heparin induced oligomerization of FGF molecule responsible for FGF receptor dimerization, activation, and cell proliferation. *Cell* 79: 1015–1024.

Srinivas PR., Srivastava S., Hanash S., and Wright GL Jr. (2001). Proteomics in early detection of cancer. *Clin Chem.*47:1901–1911.

Stabenfeldt SE., Gossett JJ. and Barker TH. (2010). Building better fibrin knob mimics: an investigation of synthetic fibrin knob peptide structures in solution and their dynamic binding with fibrinogen/fibrin holes. *Blood,* 116 (8): 1352–1359.

Stanley P. (2016). What have we learned from glycosyltransferase knockouts in mice? *J. Mol. Biol.* 428: 3166–3182.

Staudinger H. (1920). "Über Polymerisation". *Ber. Dtsch. Chem. Ges.* 53 (6): 1073–1085. doi:10.1002/cber.1920053062.

Steinlein O., Mullev J., Propping P., Wallace R., Phillips H., Sutherland G., Scheffer I., and Berkovic S. (1995) A missense mutation in the neuronal nicotinic acetylcholine receptor (t4 subunit is associated with autosomal dominant nocturnal frontal lobe epilepsy. *Nature Genet.* 11:201–203.

Stefani M., and Rigacci S. (2013). Protein folding and aggregation into amyloid: the interference by natural phenolic compounds. *Int. J. Mol. Sci.* Jun 13;14(6):12411–57. doi: 10.3390/ijms140612411. PMID: 23765219; PMCID: PMC3709793.

Stenflo J., Fernlund P., Egan W., and Roepstorff P. (1974). Vitamin K dependent modifications of glutamic acid residues in prothrombin. *Proc Natl Acad Sci US A.*;71(7):2730–2733.

Stillinger FH. (1980). Water revisited. *Science.* Jul 25;209(4455):451–7. doi: 10.1126/science.209.4455.451. PMID: 17831355.

Stockmayer WH., (1950). Light scattering in multi-component systems. *The Journal of Chemical Physics*, 18(1):58–61.

Stryer L. (1965). The interaction of a naphthalene dye with apomyoglobin and apohemoglobin. A fluorescent probe of nonpolar binding sites. *J. Mol. Biol.* 13:482–495.

Sumner JB. (1926). The isolation and crystallization of the enzyme urease: Preliminary paper. *J. Biol. Chem.* 69: 435–441.

Svedberg T., and Pedersen KO. The Ultracentrifuge. Oxford, Clarendon Press, (Johnson Reprint Corporation, New York). 1940.

Sydow JF., Lipsmeier F., Larraillet V., Hilger M., Mautz B., Mølhøj M., Kuentzer J., Klostermann S., Schoch J., Voelger HR., Regula JT., Cramer P., Papadimitriou A., and Kettenberger H. (2014). Structure-based prediction of asparagine and aspartate degradation sites in antibody variable regions. *PLoS One.* Jun 24;9(6): e100736. doi: 10.1371/journal.pone.0100736. PMID: 24959685; PMCID: PMC4069079.

Takai K., Nakamura K., Susuki K., Inogaki F., Nealson KH., and Kumagal H. (2006). Ultramafics-hydrothermalism-hydrogenesis-hipersaline (UltaH3) linkage: a key insight into early microbial ecosystem in the Archean deep-sea hydrothermal systems. *Paleontological Research* 10: 269–282.

Takakusa H., Kikuchi K., Urano Y., Sakamoto S., Yamaguchi K., and Nagano T.(2002). Design and Synthesis of an Enzyme-Cleavable Sensor Molecule for Phosphodiesterase Activity Based on Fluorescence Resonance Energy Transfer *Journal of the American Chemical Society.* 124 (8): 1653–1657 DOI: 10.1021/ja011251q

Tams JW., and Welinder KG. (1998). Glycosylation and thermodynamic versus kinetic stability of horseradish peroxidase. *FEBS Lett.* Jan 16;421(3):234–6. doi: 10.1016/s0014-5793(97)01573-1. PMID: 9468313.

Tan CS., Jørgensen C., and Linding R. (2010). Roles of "junk phosphorylation" in modulating biomolecular association of phosphorylated proteins? *Cell Cycle.* Apr 1;9(7):1276–80. doi: 10.4161/cc.9.7.11066. PMID: 20234177.

Tanford C., Hauenstein J.D. and Rands DG., (1955). Phenolic Hydroxyl Ionization in Proteins. II. Ribonuclease1a. *Journal of the American Chemical Society,* 77(24):6409–6413.

Tanford C. in Physical Chemistry of Macromolecules. Wiley, New York, p130. 1961.

Tanford C. (1964). Isothermal Unfolding of Globular Proteins in Aqueous Urea Solutions. *Journal of the American Chemical Society* 86 (10): 2050–2059. DOI: 10.1021/ja01064a028

Tanford C. (1969). Extension of the theory of linked functions to incorporate the effects of protein hydration. *Journal of molecular biology,* 39(3): 539–544.

Tanford C. The Hydrophobic Effect: Formation of Micelles and Biological Membranes. Wiley; 2nd edition (January 18, 1980).

Tanford C., and Reynolds J. (2001). Nature's Robots: a history of proteins. Oxford University Press. ISBN 9780198606949.

Tao T., and Cho J. (1979). Fluorescence lifetime quenching studies on the accessibilities of actin sulfhydryl sites. *Biochemistry* 18 (13): 2759–2765

DOI: 10.1021/bi00580a011

Taylor LS., York P., Williams AC., Edwards HG., Mehta V., Jackson GS., Badcoe IG., and Clarke AR. (1995). Sucrose reduces the efficiency of protein denaturation by a chaotropic agent. *Biochim Biophys Acta.* Nov 15;1253(1):39–46. doi: 10.1016/0167-4838(95)00142-h. PMID: 7492597.

Teich MA. Documentary History of Biochemistry, 1770–1940 (Rutherford, NJ: Fairleigh Dickinson University Press, 579 pages. 1992.

Thomadakis G., Ramoshebi LN., Crooks J., Rueger DC., and Ripamonti U. (1999). Immunolocalization of bone morphogenetic protein-2 and 3 and osteogenesis protein-1 during murine tooth morphogenesis and in other craniofacial structures. *Eur. J. Oral Sci.* 107:368–377.

Thompson RW Jr., Latypov RF., Wang Y., Lomakin A., Meyer JA., Vunnum S., and Benedek GB. (2016). Evaluation of effects of pH and ionic strength on colloidal stability of IgG solutions by PEG-induced liquid-liquid phase separation. *J Chem Phys.* Nov 14;145(18):185101. doi: 10.1063/1.4966708. PMID: 27846698.

Tiktopulo EI., and Privalov PL. (1978). Papain denaturation is not a two-state transition *FEBs Lett.* 91: 57–58.

Timasheff SN., Dintzis HM., Kirkwood JG., and Coleman BD. (1957). Light Scattering Investigation of Charge Fluctuations in Isoionic Serum Albumin Solutions1. *Journal of the American Chemical Society,* 79(4): 782–791.

Timasheff SN., and Kronman MJ. (1959). The extrapolation of light-scattering data to zero concentration. *Arch Biochem Biophys.* Jul;83(1):60–75. doi: 10.1016/0003-9861(59)90010-4. PMID: 13661992.

Timasheff SN, In electromagnetic Scattering (Kerker, M.ed.), Pergamon, New York, pp337–355. 1963.

Timasheff SN., (1966). Specific interactions in proteins due to proton fluctuations. *Biopolymers: Original Research on Biomolecules*, 4(1): 107–120.

Timasheff SN. Some Physical Probes of Enzyme Structure in Solution. Volume 2. The Enzymes Editor(s): Paul D. Boyer, Academic Press, New York, pp 371–443. 1970.

Timasheff SN. (1992). Water as ligand: Preferential binding and exclusion of denaturants in protein unfolding. *Biochemistry* 31:9857–9864.

Timasheff SN., (1993). The control of protein stability and association by weak interactions with water: how do solvents affect these processes? *Annual review of biophysics and biomolecular structure,* 22(1), pp.67–97.

Timasheff SN. In Protein Solvent Interactions (Gregory, R. ed.). Marcel Dekker, New York pp445–482. 1994.

Timasheff SN., and Xie G. (2003). Preferential interactions of urea with lysozyme and their linkage to protein denaturation. *Biophysical Chemistry*, 105, Issues 2–3, 421–448.

Timasheff SN. (2004). The tribulations of a stateless European child in his discovery of America and his thorny path to protein thermodynamics. *Comprehensive Biochemistry* 43, 393–492

Tirard S. (2017) J. B. S. Haldane and the origin of life. *J Genet.* 96(5):735–739.

Tompa P. (2002). Intrinsically unstructured proteins. *Trends Biochem, Sci,* 2002, 27, 527–533.

Tompa P. (2005). The interplay between structure and function in intrinsically unstructured proteins. *FEBS Lett* 579: 3346–3354.

Toto A., Malagrinò F., Visconti L., Troilo F., Pagano L., Brunori M., Jemth P., and Gianni S. (2020). Templated folding of intrinsically disordered proteins. *J Biol Chem.* May 8;295(19):6586–6593. doi: 10.1074/jbc. REV120.012413. Epub 2020 Apr 6. PMID: 32253236; PMCID: PMC7212662.

Toyoshima C., and Unwin N. (1990). Three-dimensional structure of the acetylcholine receptor by cryoelectron microscopy and helical image reconstruction. *J Cell Biol.* Dec;111(6 Pt 1):2623–35. doi: 10.1083/jcb.111.6.2623. PMID: 2277076; PMCID: PMC2116367.

Traube J., (2002). The attraction pressure. *The Journal of Physical Chemistry*, 14(5), pp.452–470.

Tran PT., Walker EA.,. and Salmon ED. (1997). A Metastable Intermediate State of Microtubule Dynamic Instability That Differs Significantly between Plus and Minus Ends. *J Cell Bio*l 138 (1): 105–117

Tsioris K., Raja WK., Pritchard EM., Panilaitis B., Kaplan DL. and Omenetto F.G., (2012). Fabrication of silk microneedles for controlled-release drug delivery. *Advanced Functional Materials*, 22(2): pp.330–335.

Tsytlonok M., Sanabria H., Wang Y., Felekyan S., Hemmen K., Phillips AH., Yun MK., Waddell MB., Park CG., Vaithiyalingam S., Iconaru .L, White SW., Tompa P., Seidel CAM., and Kriwacki R. (2019). Dynamic anticipation by Cdk2/Cyclin A-bound p27 mediates signal integration in cell cycle regulation. *Nat Commun.* Apr 11;10(1):1676. doi: 10.1038/s41467-019-09446-w. PMID: 30976006; PMCID: PMC6459857.

Tuena de Gómez-Puyou M., and Gómez-Puyou A. (1998) Enzymes in Low Water Systems, *Critical Reviews in Biochemistry and Molecular Biology*, 33:1, 53–89, DOI: 10.1080/10409239891204170.

Uhrich KE., Cannizzaro SM., Langer RS., and Shakesheff KM. (1999). Polymeric systems for controlled drug release. Chem Rev. Nov 10;99(11):3181–98. doi: 10.1021/cr940351u. PMID: 11749514.

Uludag H., Friess W., Williams D., Porter TJ., Timony G., D'Augusta D., Blake C., Palmer R., Biron, R., and Wozney JM. (1999). Collagen sponges as osteoinductive devices: Effect ofin vitro sponge characteristics and protein pl on in vitro rhBMP pharmacokinetic. *Ann. NY Acad. Sci.* 875:369–378.

Unwin N. (1995). Acetylcholine receptor channel imaged in the open state. *Nature* 373, 37–43. https://doi.org/10.1038/373037a0

Unwin N. (2003). Structure and action of the nicotinic acetylcholine receptor explored by electron microscopy. *FEBS Letters* 555 91–95.

Ureta T. Estrategias moleculares de la evolución: Origen y evolución de proteínas y enzimas. Editorial Universitaria, Santiago, Chile pp408. 2010.

Uversky VN., Gillespie JR. and Fink AL., (2000). Why are "natively unfolded" proteins unstructured under physiologic conditions? *Proteins: structure, function, and bioinformatics,* 41(3): pp.415–427.

Uversky VN. (2002). Native Unfolded proteins: A point where biology waits for physics. *Protein Sci.* 11, 739–756.

Uversky VN. (2008). Amyloidogenesis of natively unfolded proteins. *Curr Alzheimer Res.* Jun;5(3):260–87. doi: 10.2174/156720508784533312. PMID: 18537543; PMCID: PMC2719783.

Uversky VN. (2011). Multitude of binding modes attainable by intrinsically disordered proteins: a portrait gallery of disorder-based complexes. *Chem Soc Rev.* Mar;40(3):1623–34. doi: 10.1039/c0cs00057d. Epub 2010 Nov 3. PMID: 21049125.

Uversky VN. (2013). A decade and a half of protein intrinsic disorder: Biology still waits for physics. *Protein Sci* 22: 693–724.

Uversky VN. (2014). Wrecked regulation of intrinsically disordered proteins in diseases: pathogenicity of deregulated regulators. *Front Mol Biosci.* Jul 25; 1:6. doi: 10.3389/fmolb.2014.00006. PMID: 25988147; PMCID: PMC4428494.

Uversky VN. (2016). p53 proteoforms and intrinsic disorder: An illustration of the protein structure-function continuum concept. *Int. J. Mol. Sci.* 17: 1874.

Van Holde KE. and Baldwin RL. (1958). Rapid Attainment of sedimentation equilibrium. *J. Phys. Chem.* 62:734–743.

Vandlen RL., Wu WC., Eisenach JC., and Raftery MA. (1979). Studies of the composition of purified Torpedo californica acetylcholine receptor and of its subunits. *Biochemistry.* May 15;18(10):1845–54. doi: 10.1021/bi00577a001. PMID: 435447.

Varadi M., Anyango S., Deshpande M., Nair S., Natassia C., Yordanova G., Yuan D., Stroe O., Wood G., Laydon A., Žídek A., Green T., Tunyasuvunakool K., Petersen S., Jumper J., Clancy E., Green R., Vora A., Lutfi M., Figurnov M., Cowie A., Hobbs N., Kohli P., Kleywegt G., Birney E., Hassabis D., and Velankar S. (2022). AlphaFold Protein Structure Database: massively expanding the structural coverage of protein-sequence space with high-accuracy models. *Nucleic Acids Res.* Jan 7;50(D1):D439–D444. doi: 10.1093/nar/gkab1061. PMID: 34791371; PMCID: PMC8728224.

Vauquelin LN., and Robiquet PJ. (1806). "La découverte d'un nouveau principe végétal dans le suc des asperges". *Annales de Chimie* (in French). 57: 88–93.

Vlasak J., Bussat M.C., Wang S., Wagner-Rousset E., Schaefer M., Klinguer-Hamour C., Kirchmeier M., Corvaïa N., Ionescu R. and Beck A., (2009). Identification and characterization of asparagine deamidation in the light chain CDR1 of a humanized IgG1 antibody. *Analytical biochemistry,* 392(2): 145–154.

Velázquez Campoy A., and Freire E. (2005). ITC in the post-genomic era...? Priceless. *Biophys Chem.* Apr 1;115(2–3):115–24. doi: 10.1016/j.bpc.2004.12.015. Epub 2004 Dec 25. PMID: 15752592.

Veronese FM., Piszkiewicz D. and Smith EL., (1972). Inactivation of bovine glutamate dehydrogenase by carbamyl phosphate and cyanate. *Journal of Biological Chemistry,* 247(3):754–759.

Wahba MEK., El-Enany N., and Belal F. (2015). Application of the Stern–Volmer equation for studying the spectrofluorimetric quenching reaction of eosin with clindamycin hydrochloride in its pure form and pharmaceutical preparations. *Anal. Methods.*;7:10445–10451.

Wakankar AA., and Borchardt RT. (2006). Formulation considerations for proteins susceptible to asparagine deamidation and aspartate isomerization. J Pharm Sci. 95(11): 2321–2336.

Wang EA., Rosen V., D'Alessandro JS., Bauduy M., Cordes P., Harada T., Israel DI., Hewick RM., Kerns KM., LaPan P., Luxemburg DP., McQuaid D., Moutsatsos LK., Nove J., and Wozney JM. (1990). Recombinant human bone morphogenetic protein induces bone formation. *Proc. Natl. Acad. Sci. USA.* 87: 2220–2224.

Wang Y., Rudym DD., Walsh A., Abrahamsen L., Kim HJ., Kim HS., Kirker-Head C., and Kaplan DL. (2008). In vivo degradation of three-dimensional silk fibroin scaffolds. *Biomaterials.* Aug-Sep;29(24–25):3415–28. doi: 10.1016/j.biomaterials.2008.05.002. Epub 2008 May 27. PMID: 18502501; PMCID: PMC3206261.

Wang Y., Lomakin A., Latypov R F., and Benedek GB. (2011). Phase separation in solutions of monoclonal antibodies and the effect of human serum albumin. *Proceedings of the National Academy of Sciences USA,* 108(40), 16606–16611.

Wang Y., Latypov RF., Lomakin A., Meyer JA., Kerwin BA., Vunnum S., and Benedek, GB. (2014). Quantitative evaluation of colloidal stability of antibody solutions using PEG-induced liquid–liquid phase separation. *Molecular pharmaceutics,* 11(5): 1391–1402.

Wang Y., Lomakin A., Kanai S., Alex R., and Benedek GB. (2017). Liquid–Liquid Phase Separation in Oligomeric Peptide Solutions. *Langmuir* 33 (31): 7715–7721.

Wang Y., and Latypov, RF. (2019). Quantitative Evaluation of Protein Solubility in Aqueous Solutions by PEG-Induced Liquid–Liquid Phase Separation. In: McManus, J. (eds) *Protein Self-Assembly. Methods in Molecular Biology,* vol 2039. Humana, New York, NY. https://doi.org/10.1007/978-1-4939-9678-0_3

Wang B., Zhang L., Dai T., Qin Z., Lu H., Zhang L., and Zhou F. (2021). Liquid-liquid phase separation in human health and diseases. *Signal Transduct Target Ther.* Aug 2;6(1):290. doi: 10.1038/s41392-021-00678-1. PMID: 34334791; PMCID: PMC8326283.

Wang YN., Xia GY., Wang L.Y, Ge GB, Zhang HW., Zhang JF., Wu YZ., and Lin S. (2018). Purpurolide A, 5/5/5 Spirocyclic Sesqu.iterpene Lactone in Nature from the Endophytic Fungus Penicillium purpurogenum. *Org Lett.* 2018 Nov 16;20(22):7341–7344. doi: 10.1021/acs.orglett.8b03323. Epub 2018 Nov 5. PMID: 30394758.

Wang W., and Wang D. (2019). Extreme Fuzziness: Direct Interactions between Two IDPs. *Biomolecules.* Feb 26;9(3):81. doi: 10.3390/biom9030081. PMID: 30813629; PMCID: PMC6468500.

Wang T., Gao X., Huai Z., Gong Z., Mao T., Fa, X., Wu X., Duan Z., Wang X., Du J. and Yao M., (2023). Revolutionizing antibody discovery industry with highly efficient and accurate AI-based epitope-specific antibody de novo design workflow. *Antibody Therapeutics*, *6*(Suppl 1). Jul 24;6(Suppl 1):tbad014.024. doi: 10.1093/abt/tbad014.024. PMCID: PMC10370440.

Ward JJ., Sodhi JS., McGuffin LJ., Buxton BF., and Jones DT. (2004). Prediction and functional analysis of native disorder in proteins from the three kingdoms of life. *J. Mol. Biol.* 337: 635–645

Watson JD., and Crick FH. (1974). Molecular structure of nucleic acids: a structure for deoxyribose nucleic acid. J.D. Watson and F.H.C. Crick. Published in Nature, number 4356 April 25, 1953. *Nature.* Apr 26;248(5451):765. doi: 10.1038/248765a0. PMID: 4599080.

Weber G. (1960). Fluorescence-polarization spectrum and electronic-energy transfer in proteins. *Biochemical Journal*, 75(2): 345–352.

Weber G. (1975). Energetics of Ligand binding to proteins. *Adv. Prot. Chem.* 29: 1–83.

Webster T., Tsai H., Kula M., Mackie GA., and Schimmel P. (1984). Specific sequence homology and three-dimensional structure of an aminoacyl transfer RNA synthetase. *Science.* Dec 14;226(4680):1315–7. doi: 10.1126/science.6390679. PMID: 6390679.

Weisel JW.,and Litvinov RI. (2017). Fibrin Formation, Structure and Properties. *Subcell Biochem.*;82:405–456. doi: 10.1007/978-3-319-49674-0_13. PMID: 28101869; PMCID: PMC5536120.

Westheimer FH., and Schmidt Jr DE. (1971). pK of the lysine amino group at the active site of acetoacetate decarboxylase. *Biochemistry,* 10(7):1249–1253.

Whitehouse P., Matino A., Marcus K., Zweig R., Singer H., Price D., and Kellar K. (1988) Reduction in acetylcholine and nicotine binding in several degenerative diseases. *Arch. Neurol.* 45: 722–724.

Wills PR., Jacobsen MP.,and Winzor DJ. (2000). Analysis of sedimentation equilibrium distributions reflecting nonideal macromolecular associations. *Biophys J.* Oct;79(4):2178–87. doi: 10.1016/S0006-3495(00)76466-6. PMID: 11023922; PMCID: PMC1301108.

Winzor DJ., Dinu V., Scott DJ., and Harding SE. (2023). Retrospective rationalization of disparities between the concentration dependence of diffusion coefficients obtained by boundary spreading and dynamic light scattering. *Eur Biophys* J. Jul;52(4-5):333-342. doi: 10.1007/s00249-023-01664-x. Epub 2023 Jul 7. PMID: 37414903; PMCID: PMC10444695.

Wright PE., and Dyson HJ. (1999). Intrinsically unstructured proteins: re-assessing the protein-structure function paradigm. *J. Mol. Biol.* 293: 321–331.

Wright MR. (2007). An Introduction to Aqueous Electrolyte Solutions. John Wiley & Sons, New York.

Wright PE., and Dyson HJ. (2009). Linking folding and binding. *Curr Opin Struct Biol.* 19: 31–38.

Wu, S., Wang D., Liu, J., Feng,Y,; Weng,J., Li,Y., Gao,X., Liu,J., and Wang W. (2017). The dynamic multisite interactions between two intrinsically disordered proteins. *Angew. Chem. Int. Ed. Engl.*56:7515–7519.

Wyatt PJ. (1993). Light scattering and the absolute characterization of macromolecules. *Analytica Chimica Acta.* 272 (1): 1–40.

Wyman, J. 1948. Heme Proteins. *Adv. Prot. Chem* 4:407–531

Wyman Jr J. (1964). Linked functions and reciprocal effects in hemoglobin: a second look. *Adv. Prot. Chem.* 19:223–286.

Wyman J., and Gill SJ. (1990). Binding and linkage, University Science Books, Mill Valley, CA.

Xiao S., Patsalo V., Shan B., and Raleigh DP. (2013). Rational modification of protein stability by targeting surface sites leads to complicated results. *J. Am. Chem. Soc.* 110: 11337–11342.

Xie G., and Timasheff SN. (1997). The thermodynamic mechanism of protein stabilization by trehalose. *Biophysical Chemistry*, 64, Issues 1–3, 25–43.

Xue B., Blocque, D., Habchi J., Uversky AV., Kurgan L., Uversky VN., and Longhi S. (2014). Structural Disorder in Viral Proteins. *Chemical Reviews.*114 (13): 6880–6911.DOI: 10.1021/cr4005692

Yancey PH., Clark ME., Hand SC., Bowlus RD., and Somero GN. (1982). Living with water stress: evolution of osmolyte systems. *Science.* Sep 24;217(4566):1214–22. doi: 10.1126/science.7112124. PMID: 7112124.

Yancey PH. (2005). Organic osmolytes as compatible, metabolic, and counteracting cytoprotectants in high osmolarity and other stresses. *J Exp Biol.* Aug;208(Pt 15):2819–30. doi: 10.1242/jeb.01730. PMID: 16043587.

Yang D., Kroe-Barrett R., Singh S., and Laue T. (2019). IgG Charge: Practical and Biological Implications. *Antibodies* Mar 14;8(1):24. doi: 10.3390/antib8010024. PMID: 31544830; PMCID: PMC6640702.

Yamauchi M., and Sricholpech M. (2012). Lysine post-translational modifications of collagen. *Essays Biochem.*;52:113–33. doi: 10.1042/bse0520113. PMID: 22708567; PMCID: PMC3499978.

Yphantis DA. (1964). Equilibrium ultracentrifugation of dilute solutions. *Biochemistry.* Mar; 3:297–317. doi: 10.1021/bi00891a003. PMID: 14155091.

Yucel T., Cebe P., and Kaplan DL. (2009). Vortex-induced injectable silk fibroin hydrogels. *Biophys J.* 97: 2044–2050.

Zhao H., and Zastrow ML. (2022). Transition metals induce quenching of monomeric near-infrared fluorescent proteins. *Biochemistry.* 61:494–504.

Zosel F., Mercadante D., Nettels D., and Schuler B. (2018). A proline switch explains kinetic heterogeneity in a coupled folding and binding reaction. *Nat. Commun.* 9, 3332.

Zwanzig R., Szabo A. and Bagchi B. (1992). Levinthal's paradox. *Proc. Natl. Acad. Sci. USA* 89: 20–22.

APPENDIX

Conversion Factors

- Energy: 1 Joule = 10^7 ergs = 0.239 cal.

  1 cal. = 4.184 Joule

- Length: 1 nm = 10 Å = 1 x 10^{-7} cm = 1×10^{-9} m

- Pressure: 1 Atm = 760 torr = 14.696 psi

- Temperature K = °C + 273

Fundamental Constants

Avogadro's number	No = 6.0230 x 10^{23}
Boltzmann constant	k_B = 1.3844 x 10^{-16} erg/deg
Electronic charge	e = 4.80286 x10^{-10} esu
Electron mass	m = 9.1083 x 10^{-28} gm
Faraday constant	f = 96,490 coulomb/equiv
Gas constant	R = 1.9872 cal deg^{-1} $mole^{-1}$ 8.3144 j deg^{-1} $mole^{-1}$
Planck's constant	h = 6.6252 x 10^{-27} erg sec
Protein partial specific volume	v2⁻ = 0.708 mL/g (average)
Standard acceleration of gravity	g = 980.665 x cm/sec^2
Velocity of light	c = 2.997930 x 10^{10} cm/sec
Water ionization constant	Kw = 10^{-14}

Symbols for the amino acids			pK_1	pK_2	pK_R
Alanine	Ala	A	2.35	9.69	-
Arginine	Arg	R	2.17	9.04	12.48
Asparagine	Asn	N	2.02	8.8	-
Aspartic acid	Asp	D	2.09	9.82	3.86
Cysteine	Cys	C	1.71	10.78	8.33
Glutamic acid	Glu	E	2.19	9.67	4.25
Glutamine	Gln	Q	2.17	9.13	-
Glycine	Gly	G	2.34	9.6	-
Histidine	His	H	1.82	9.17	6.0
Isoleucine	Ile	I	2.36	9.68	-
Leucine	Leu	L	2.36	9.60	-
Lysine	Lys	K	2.18	8.95	10.53
Methionine	Met	M	2.38	9.21	-
Phenylalanine	Phe	F	1.83	9.13	-
Proline	Pro	P	1.99	10.6	-
Serine	Ser	S	2.21	9.15	-
Threonine	Thr	T	2.63	10.43	-
Tryptophan	Trp	W	2.38	9.39	-
Tyrosine	Tyr	Y	2.20	9.11	10.07
Valine	Val	V	2.32	9.62	-

pK_1 (carboxyl group), pK_2 (amino group), pK_R (side chain).

ABOUT THE AUTHOR

Bernardo Pérez-Ramírez is a scientist, professor, writer, and philanthropist. He holds a PhD in biochemistry and postdoctoral studies in physical biochemistry of proteins, with more than thirty years of experience in the structure and function of proteins in solution and development of formulations for therapeutic proteins. He is an adjunct professor at the Department of Biomedical Engineering, Tufts University in Boston, and has received an Honorary Professorship from the University of Nottingham in the United Kingdom. He is the author of more than forty scientific articles and several books on diverse subjects covering protein science, history, heraldry, poetry, and essays. He is a member of the American Society for Biochemistry and Molecular Biology (ASBMB), The New York Academy of Sciences, The Spanish Society of Biochemistry and Molecular Biology (SEBBM), Fellow, Royal Society of Chemistry (FRSC), and Academician of the Institute of Historical Studies Bances and Valdés of Spain. The author resides in Boston, Massachusetts.

INDEX

A

D

E

F

G

H

I

N

O

P

W

X

Z